INTRODUCTION TO AGRICULTURAL ECONOMICS

INTRODUCTION TO AGRICULTURAL ECONOMICS

SEVENTH EDITION

John B. Penson, Jr.
Texas A&M University

Oral Capps, Jr.
Texas A&M University

C. Parr Rosson III
Texas A&M University

Richard T. Woodward
Texas A&M University

330 Hudson Street, NY, NY 10013

Vice President, Portfolio Management: Andrew Gilfillan
Portfolio Manager: Pamela Chirls
Editorial Assistant: Lara Dimmick
Development Editor: Melissa Mashburn
Senior Vice President, Marketing: David Gesell
Field Marketing Manager: Thomas Hayward
Marketing Coordinator: Elizabeth MacKenzie-Lamb
Director, Digital Studio and Content Production: Brian Hyland
Managing Producer: Cynthia Zonneveld
Managing Producer: Jennifer Sargunar
Content Producer: Rinki Kaur

Content Producer: Ruchi Sachdev
Manager, Rights Management: Johanna Burke
Operations Specialist: Deidra Smith
Full-Service Management and Composition: iEnergizer Aptara®, Ltd.
Full-Service Project Manager: Rakhshinda Chishty
Cover Design: Studio Montage
Cover Photo: James Pintar/Fotolia
Printer/Binder: LSC Communications
Cover Printer: Lehigh-Phoenix
Text Font: Garamond 3 LT Pro, 11/13

Library of Congress Cataloging-in-Publication Data

Names: Penson, John B., Jr., author. | Capps, Oral, author. | Rosson, C.
    Parr, author. | Woodward, Richard T., author.
Title: Introduction to agricultural economics/John B. Penson, Jr., Oral
    Capps, Jr., C. Parr Rosson III, Richard T. Woodward.
Description: Seventh edition. | Hoboken: Pearson, 2017. | Includes
    bibliographical references and index.
Identifiers: LCCN 2017002085 | ISBN 9780134602820 | ISBN 013460282X
Subjects: LCSH: Agriculture—Economic aspects.
Classification: LCC HD1415 .P374 2017 | DDC 338.1—dc23 LC record available at https://lccn.loc.gov/2017002085

1 17

ISBN 13: 978-0-13-460282-0
ISBN 10:   0-13-460282-X

We thank our families for their patience and support,
and dedicate this book to them:

My wife Donna; children Matt, John, and Laura;
and my mother Mary Elizabeth for her interest in literature JBP

My wife Debbie, sons Kevin and Eric, and my mother Shirley and grandmother
May Manuel—my most ardent supporters. I am forever grateful to them for inspiring
me to do my best and to always finish strong! OCJ

My wife Helen and sons CP, Henry, and Jonathan CPR

My wife Rosie and children Christopher and Sophia RTW

contents

5
Measurement and Interpretation of Elasticities 74

part three
Business Behavior and Market Equilibrium

6
Introduction to Production and Resource Use 90

7
Economics of Input and Product Substitution 109

part five
Macroeconomics of Agriculture

 12

Product Markets and National Output 226

 13

Macroeconomic Policy Fundamentals 248

17

Why Nations Trade 342

18

Agricultural Trade Policy and Preferential Trading Arrangements 354

Preface

The purpose of this book is to provide beginning students in agriculture with a systematic introduction to basic economic concepts and issues as they relate to a major segment of the U.S. economy—the food and fiber industry. This process requires an understanding of the microeconomic and macroeconomic forces influencing the decisions of producers and consumers of food and fiber products, including (1) farmers and ranchers, (2) the agribusinesses that supply them with production inputs and credit, (3) the agribusinesses that process food products and manufacture fiber products, and (4) the agribusinesses that provide marketing and related services at the wholesale and retail levels to both domestic consumers and overseas markets.

The integration of micro-macro-trade linkages provided in our book helps students understand these linkages which are so important in today's economy. The coverage of demand and supply concepts with examples helps students understand later concepts covered in macroeconomics and trade. References back to microeconomic concepts when discussing macroeconomic and trade concepts help cement these interrelationships for students.

NEW TO THIS EDITION

The 7th edition has been thoroughly revised to provide students with the most up-to-date coverage of this dynamic sector within the global economy. Key updates include the following:

- A new four-color design to better capture student interest and make the economic concepts expressed within the art program easier to grasp;
- Expanded TEQ (Testing Your Economic Quotient) questions at the end of each chapter to aid student self-study efforts;
- Major restructuring of Chapter 11 dealing with government's role in agriculture, including updated coverage of current farm commodity policy;
- Extensive revision of Chapters 12–15 dealing with key macroeconomic topics and how macroeconomic events affect agriculture;
- All tables and figures have been updated to reflect the most currently available data.

CONCEPTUAL OVERVIEW

This book goes beyond the farm gate to address the entire food and fiber industry, which represents a notable percentage of the U.S. national output. This book places a strong emphasis on the macroeconomics of agriculture, the role of government in agriculture, and international agricultural trade. Experience over the last several decades certainly has shown that farmers and ranchers, agribusinesses, financial institutions, and consumers of food and fiber products are significantly affected by macroeconomic policies and trade agreements.

Conceptually, the book begins with defining the field of agricultural economics with examples and the structure of the nation's food and fiber industry. The book then builds from microeconomics to macroeconomics to international economics. The student is initially introduced to concepts and examples of consumer behavior and behavior of the firm, concluding with derivation of the market demand and supply under perfect competition and then imperfect competition. The book offers extensive coverage of the important concept of elasticity and its relevance to understanding revenue and economic welfare implications for both consumers and producers. Microeconomic coverage also includes addressing concepts in natural resources and government programs, important facets of agricultural economics. The extensive coverage of microeconomic concepts including imperfect competition helps students understand the coverage of macroeconomics at the firm, market, and economy level. Finally, the coverage of international trade helps students understand the events in today's global economy.

TEXT STRUCTURE AND CHAPTER PEDAGOGY

Our book is divided into six parts and includes 18 chapters. Part 1 consists of two chapters and serves as an introduction to the material. We begin the book by answering the question raised in Chapter 1, "What is agricultural economics?" We define the field of economics and then develop our definition of agricultural economics based on the role agricultural economists play at the micro and macro levels. Chapter 2 provides a historical background by discussing the changing structure of agriculture during the post–World War II period and of the sectors that supply farmers and ranchers with inputs, process their output, sell value-added products to domestic consumers, and trade food and fiber products in the global marketplace.

Part 2 helps students understand the economic decisions made by consumers of food and fiber products. Topics include the forces influencing consumer behavior (Chapter 3); the concept of market demand for a particular product (Chapter 4); and the elasticity of demand (Chapter 5). The specification of key elasticity measures is supplemented by empirical examples and their relevance to decision-making in the food and fiber industry, including the potential magnitude of consumer response and its implication on producer revenue.

Part 3 covers the supply side of the market. Chapter 6 focuses on issues related to resource use and production responses by businesses in the short run. Chapter 7 discusses the economic forces underlying the firm's input use, the expansion of the firm, and the choice of commodities. An introduction to the market supply curve and determination of market clearing prices and quantities under perfect competition (Chapter 8) and imperfect competition (Chapter 9) completes this part. This section of the book includes empirical examples that illustrate the magnitude and applicability of the relationships covered in these chapters.

Part 4 addresses the role of government in the food and fiber industry. Natural resources, the environment, and agriculture are covered in Chapter 10. This chapter includes the role of government regulation, which reflects the increasing recognition that natural resources and the environment are scarce resources and require careful management. The government's role in providing subsidies to agriculture, curbing market power, and providing for a secure and safe food supply is addressed in Chapter 11.

Part 5 focuses on the macroeconomics of agriculture. Chapter 12 outlines the general linkages between product markets and national output. Chapter 13 documents the importance of monetary and fiscal policy to the performance of the economy. The consequences of business fluctuations in the economy are covered in Chapter 14. Chapter 15 covers the relationship between macroeconomic policy and its effects on the economic performance of agriculture.

Part 6 focuses on international agricultural trade issues. Chapter 16 examines the growth and instability of agricultural trade, including the relative dependence on exports and imports, as well as the foreign exchange market, the international monetary system, and the effects of foreign exchange rates on U.S. agricultural trade. Chapter 17 explores the rationale behind international trade as well as the beneficiaries of international trade. Finally, Chapter 18 focuses on agricultural trade policy and preferential trade agreements. This includes issues dealing with trade restriction and whether preferential trade agreements create or divert trade.

Each chapter concludes with a summary and a list of key terms. A "Testing Your Economic Quotient" section contains questions and problems to reinforce the key issues covered. Understanding the answers to these questions and problems will help students properly prepare for exams. References also are listed at the end of each chapter.

INSTRUCTOR ANCILLARIES

To access supplementary materials online, instructors need to request an instructor access code. Go to www.pearsonhighered.com/irc, where you can register for an instructor access code. Within 48 hours after registering, you will receive a confirming e-mail, including an instructor access code. Once you have received your code, go to the site and log on for full instructions on downloading the materials you wish to use.

ACKNOWLEDGMENTS

We wish to thank the many students who have given us comments and suggestions during the development phases of this and previous editions of the book. We also thank the following reviewers for their valuable feedback: Marlies Boyd, Modesto Junior College; Stephen King, Western Kentucky University; Pierre Boumtje, Southern Arkansas University; and Xiaowei Cai, California Polytechnic State University.

About the Authors

John B. Penson, Jr., holds the titles of Regents Professor and Stiles Endowed Professor of Agriculture in the Department of Agricultural Economics at Texas A&M University. He is also a senior scientist with the Norman Borlaug Institute for International Agriculture. Penson received a Ph.D. degree in agricultural economics from the University of Illinois. Penson has taught courses in Korea, Japan, Guatemala, Nicaragua, and Ecuador. His research has focused on the macroeconomics of agriculture and portfolio credit risk analysis. He has also conducted research in the Middle East and Eurasia.

Oral Capps, Jr., holds the titles of Executive Professor and Regents Professor in the Department of Agricultural Economics at Texas A&M University. He is a certified business economist and co-director of the Agribusiness, Food and Consumer Economics Research Center at Texas A&M University. He is also holder of the Southwest Dairy Marketing Endowed Chair. He received a Ph.D. in agricultural economics from Virginia Tech. He has received numerous teaching and research awards and is recognized internationally for his research in demand and price analysis.

C. Parr Rosson III is Professor and Department Head of the Agricultural Economics Department at Texas A&M University. He received his Ph.D. in agricultural economics from Texas A&M University. Rosson works in the areas of international trade and international marketing. He currently chairs the Education Committee of the Texas–Cuba Trade Alliance. He served on the Grains, Feed, Oilseeds and Planting Seeds Agricultural Trade Advisory Committee for the U.S. Trade Representative and U.S. Department of Agriculture from 2001 to 2015. He has conducted projects in Latin America, the Middle East, and Asia.

Richard T. Woodward is Professor in the Department of Agricultural Economics at Texas A&M University. His research is in the general area of environmental and resource economics. His recent research projects have focused on the use of transferable permits to address water quality and fishery challenges and problems of choice under uncertainty.

1

What Is Agricultural Economics?

Chapter Outline

Agricultural economics is an applied social science that deals with how producers, consumers, and societies use scarce resources in the production, marketing, and consumption of food and fiber products. In agricultural markets, the forces of supply and demand are at work. Credit: Brad McMillan/Cartoon Stock.

Agriculture certainly is among the most prominent sectors of any economy. Psalm 104 illustrates this point: "Bless the lord, O my soul, thou dost cause the grass to grow for the cattle, and plants for man to cultivate, that he may bring forth food from the Earth." Unequivocally, from biblical times agriculture has been a discipline worthy of study. We specifically are interested in the economic relationships inherent in the agricultural sector.

The roots of agricultural economics perhaps can be traced back to ancient Egypt, arguably to the first agricultural economist, Joseph. Joseph interpreted the dreams of the Pharaoh of Egypt and correctly predicted seven years of feast and seven years of famine.

What is agricultural economics? If you were to say "Agricultural economics is the application of economic principles to agriculture," you would be technically correct—but in a narrow context. This definition does not recognize the economic, social, and environmental issues addressed by the agricultural economics profession. To perceive agricultural economics as being limited only to the economics of farming and ranching operations would be incorrect. These operations account for only 2% to 4% of the nation's output. Actually, the scope of agricultural economics goes well beyond the farm gate to encompass a broader range of food- and fiber-related activities. When viewed from this broader perspective, the agricultural sector accounts for approximately 12% to 15% of the nation's output.

Before we define agricultural economics further, let us first examine the scope of economics and the role that agricultural economists play in today's economy. This examination will allow us to propose a more definitive answer to the question raised by the chapter title. A more in-depth assessment of the nation's food and fiber industry is presented in Chapter 2.

SCOPE OF ECONOMICS

Two frequently used clichés describe the economic problem: "You can't have your cake and eat it too" and "There's no such thing as a free lunch." Because we—individually or collectively—cannot have everything we desire, we must make choices. Consumers, for example, must make expenditure decisions with a budget in mind. Their objective is to maximize the satisfaction they derive from allocating their time between work and leisure, and from allocating their available income to consumption and saving, given current prices and interest rates. Producers must make production, marketing, and investment decisions with a budget in mind. Their objective is to maximize the profit of the firm, given its current resources and current relative prices. After considering the costs and benefits involved, society also must make choices on how to allocate its scarce resources among different government programs most efficiently.

Scarce Resources

The term *scarcity* refers to the finite quantity of resources that are available to meet society's needs. Because nature does not freely provide enough of these resources, only a limited quantity is available. **Scarce resources** can be broken down into the following categories: (1) natural and biological resources; (2) human resources; and (3) manufactured resources.

> **Scarce resources** can be divided into natural and biological resources, human resources, and manufactured resources.

Natural and Biological Resources   Land and mineral deposits are examples of scarce **natural resources**. The quality of these natural resources in the United States differs greatly from region to region. Some lands are incapable of growing anything in their natural state, and other lands are extremely fertile. Still other areas are rich in coal deposits or oil and natural gas reserves. In recent years, our society also has become aware of the increasing scarcity of fresh water, especially in the West. Whereas energy-related natural resources have represented critical scarce resources in recent decades, water could become *the* critical scarce natural resource in the near future. In addition to natural resources, scarce resources also include **biological resources** such as livestock, wildlife, and different genetic varieties of crops.

Human Resources   **Human resources** are services provided by laborers and management to the production of goods and services that also are considered scarce. Laborers, for example, provide services that, combined with scarce nonhuman resources, produce economic goods.[1] Workers in the automotive industry provide the labor input to produce cars and trucks. Farm laborers provide the labor input to produce crops and livestock. Labor is considered scarce even when the country's labor force is not fully employed. Laborers supply services in response to the going wage rate. Agribusinesses may not be able to hire all the labor services they desire at the wage they wish to pay.

Management, another form of human resource, provides entrepreneurial services, which may entail the formation of a new firm, the renovation or expansion of an existing firm, the taking of financial risks, and the supervision of the use of the firm's existing resources so that its objectives can be met. Without entrepreneurship, large-scale agribusinesses would cease operating efficiently.

Manufactured Resources   The third category of scarce resources is **manufactured resources** or, more simply, **capital**. Manufactured resources are machines, equipment, and structures. A product that has not been used up in the year it was made also is considered a manufactured resource. For example, inventories of corn raised but not fed to livestock or sold to agribusinesses represent a manufactured resource.

Scarcity is a relative concept. Nations with high per capita incomes and wealth face the problem of scarcity like nations with low per capita incomes and wealth. The difference lies in the degree to which resource scarcity exists and the forms that it takes.

Scarcity refers to the fixed quantity of resources that are available to meet societal needs.

Making Choices

Resource scarcity forces consumers and producers to make choices. These choices have a time dimension. The choices consumers make today will have an effect on how they will live in the future. The choices businesses make today will have an effect on the future profitability of their firms. Your decision to go to college rather than get a job today was probably based in part on your desire to increase your future earning power or eventual wealth, knowing what your earning potential would be if you did not attend college.

The choices one makes also have an associated **opportunity cost**. The opportunity cost of going to college now is the income you are currently foregoing by not getting a job now. The opportunity cost of a consumer taking $1,000 out of his or her savings account to buy a cell phone or other assorted technological devices is the interest income this money would have earned if left in the bank. An agribusiness firm considering the purchase of a new computer system also must consider the income it could receive by using this money for another purpose. The bottom line expressed in economic terms is whether the economic benefits exceed the costs, including foregone income. Simply put,

Opportunity cost refers to the implicit cost associated with the next best alternative in a set of choices available to decision makers.

---

[1] Goods and services produced from scarce resources also are scarce and are referred to as economic goods. Economic goods are in contrast to free goods, in which the quantity desired is available at a price of zero. Air has long been a free good, but pollution (a negative good), which makes the air unfit to breathe, is changing this notion in some areas.

opportunity cost is a concept associated with economic decisions. It refers to the implicit cost associated with the next best alternative.

To illustrate the concept of opportunity cost, consider the following hypothetical example. Suppose that RJR Nabisco has three alternatives for manufacturing snack foods:

Alternative 1: manufacture cookies alone and obtain a profit of $30 million.
Alternative 2: manufacture chips alone and obtain a profit of $25 million.
Alternative 3: manufacture both cookies and chips and obtain a profit of $35 million.

Because Alternative 3 offers the highest profit to RJR Nabisco, it is rational economically for the firm to adopt this choice and consequently manufacture both cookies and chips. However, in doing so, the firm foregoes Alternatives 1 and 2. The implicit cost associated with the next best alternative is to forgo a profit of $30 million. Thus, $30 million is the opportunity cost in this example.

Sometimes the choices we make are constrained not only by resource scarcity but also by noneconomic considerations. These forces may be political, psychological, sociological, legal, or moral. For example, some states have blue laws that prohibit the sale of specific commodities on Sundays. A variety of regulations exist at the federal and state levels that govern the production of food and fiber products, including environmental and food safety concerns. For example, specific chemicals are banned from use in producing and processing food products because of their potential health hazard. The Big Green movement in California in 1990 sought to ban the use of all agricultural chemicals that were shown to pose health hazards to laboratory animals. As another example, over the period February 2007 to August 2007, a nationwide recall of Peter Pan peanut butter took place due to its association with salmonella contamination. This product was not available in grocery stores for a period of 27 weeks.

Most resources are best suited for a particular use. For example, the instructor of this course is better qualified to teach this course than to perform open-heart surgery. By focusing the use of our resources on a specific task, we are engaging in **specialization**. With a given set of human and nonhuman resources, specialization of effort generally results in a higher total output. Individuals should do what they do comparatively better than others, given their endowment of resources. Some individuals might specialize in fields such as professional athletics, medicine, or law. Others might specialize in agricultural economics. States and nations may find it to their advantage to specialize in the production of coffee, rice, or computers and import other commodities for which their endowment of natural, human, and manufactured resources is ill-suited. As illustrated in Figure 1-1, Kansas has a surplus of wheat production but a shortage of orange production, while Florida has a surplus of orange production and a shortage of wheat production. Both states have a shortage of potato production, while Idaho has plenty to spare. Specialization in production provides the basis for trade among producers and consumers.

Choices in the allocation of resources made by society (a collection of individuals) might be quite different from the choices made by individual members of society. For example, all nations normally allocate some resources to military uses. Society as a whole must decide how best to allocate its resources between the production of civilian goods and services and the production of military goods, popularly referred to as the choice of "guns versus butter."

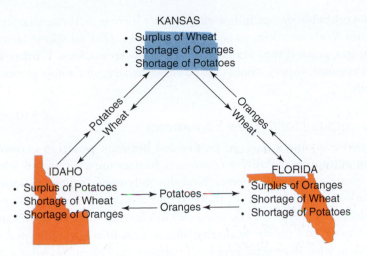

Figure 1-1
Specialization and resource allocation.

DEFINITION OF ECONOMICS

With the foregoing concepts of resource scarcity and choice in mind, we may now define the nature and scope of the field of economics as follows.

> **Economics** is a social science that deals with how consumers, producers, and societies choose among the alternative uses of scarce resources in the process of producing, exchanging, and consuming goods and services.

Microeconomics versus Macroeconomics

As with most disciplines, the field of economics can be divided into several branches. **Microeconomics** and **macroeconomics** are two major branches of economics. Microeconomics focuses on the economic actions of individuals or specific groups of individuals. For example, microeconomists are concerned with the economic behavior of consumers who demand goods and services and producers who supply goods and services, and the determination of the prices of those goods and services. Macroeconomics focuses on broad aggregates, such as the growth of the nation's gross domestic product (GDP), the gaps between the economy's potential GDP and its current GDP, and trade-offs between unemployment and inflation. For example, macroeconomists are concerned with identifying the monetary and fiscal policies that would reduce inflation, promote growth of the nation's economy, improve the nation's trade balance (exports minus imports), and reduce the national debt. Macroeconomics explicitly accounts for the interrelationships between the nation's labor, product, and money markets and the economic decisions of foreign governments and individuals.

Despite the differences between microeconomics and macroeconomics, there is no conflict between these two branches. After all, the economy in the aggregate is certainly affected by the events taking place in individual markets.

A word of caution: we must be careful when generalizing the aggregate or macroeconomic consequences of an individual or a microeconomic event. If not, we run the risk of committing a **fallacy of composition**, meaning that which is true in an individual situation is not necessarily true in the aggregate. For example, suppose Walt Wheatman adopts a new technology that doubles his wheat production. If the thousands of other wheat farmers in the United States and other wheat

Microeconomics is a branch of economics that focuses on the actions or behavior of individual agents or groups of agents.

Macroeconomics is another branch that centers attention on broad aggregates of the economy.

producers worldwide do not follow suit, Walt's income will rise sharply. It would be wrong for Walt or others to conclude, however, that all wheat farmers would achieve income gains if they also adopted this new technology. If other wheat producers did respond, supply would expand substantially, and wheat prices would fall dramatically.

Positive versus Normative Economics

Positive economics deals with what-is and what-would-happen-if questions.

Normative economics focuses on what-should-be or what-ought-to-be questions.

The study of economics also can be divided between **positive economics** and **normative economics**. Positive economics focuses on what-is and what-would-happen-if questions and policy issues. No value judgments or prescriptions are made. Instead, the economic behavior of producers and consumers is explained or predicted. For example, policymakers may be interested in knowing how consumers and producers would respond to a tax cut or alternatively to a tax hike. Or, policymakers may be interested in to what degree the problem of obesity may be mitigated if a notable tax is placed on sugar-sweetened beverages.

Normative economics focuses on determining "what should be" or "what ought to be." For example, policymakers might inquire as to which of several alternative policies *should be* adopted to maximize the economic welfare of producers and consumers. At the micro level, an automobile manufacturing plant might be interested in knowing the number of vehicles it *should be* producing to maximize profit.[2]

Alternative Economic Systems

An *economic system* can be defined as the institutional means by which resources are used to satisfy human desires; the term *institutional* refers to the laws, habits, ethics, and customs of the nation's citizens. **Capitalism** is a free market economic system in which individuals own resources and have the right to employ their time and resources however they choose, with minimal legal constraints from government. Prices signal the value of resources and economic goods. Under capitalism, as claimed by the Scottish economist and moral philosopher Adam Smith in his book *An Inquiry into the Nature and Causes of the Wealth of Nations* published

Adam Smith, the pioneer of capitalism.
Credit: GL Archive/Alamy Stock Photo.

---

2 For a more in-depth discussion of positive and normative economics, see Friedman (1974).

Thomas Piketty, the French economist who proposes redistribution of wealth through global taxation.   Credit: Rachel Torres/Alamy Stock Photo.

In 1776, individuals' efforts to maximize their own gains in a free market benefit society. One of the most important concepts of *The Wealth of Nations* is Smith's idea of the *invisible hand*. In his investigation as to why some countries are poor and stay poor, while other nations grow and prosper, Smith found that increases in productivity of <u>individuals</u> that come from their <u>individual</u> talents, the division of labor unhindered by government restrictions, and voluntary transactions in a free market result in rising prosperity. The "invisible hand of the market" is a metaphor conceived by Adam Smith to describe the self-regulating behavior of the marketplace.

Capitalism differs sharply from **socialism** or **communism**. Under socialism or communism, resources are generally collectively owned and the government decides how human and nonhuman resources are to be utilized across the various sectors of the economy. Prices largely are set by the government and administered to consumers and farmers. Winston Churchill noted that "socialism is a philosophy of failure, the creed of ignorance, and the gospel of envy; its inherent virtue is the equal sharing of misery" (www.brainyquote.com). *Capital in the 21st Century*, written in 2013 by the French economist Thomas Piketty, argues that inequality in wealth leads to stagnant or declining economic growth. To address this issue, Piketty proposes redistribution of wealth through a global tax.

A measurement of the equality or inequality in the distribution of wealth for any nation can be made by calculating the Gini coefficient, proposed in 1912 by the Italian statistician and sociologist Corrado Gini. A Gini coefficient close to 0 reflects complete equality in the distribution of wealth, while a Gini coefficient close to 1 reflects complete inequality. To provide perspective concerning this metric, since 2007, the Gini coefficient for the United States has risen from 0.463 to 0.477, a change of roughly 3%.

The United States has what is commonly referred to as a **mixed economic system**; that is, markets are not entirely free to determine price in some markets but are free in others. The government's intervention in the agricultural arena, for example, is well known. Loan guarantees to crop producers and guarantees to savings and loan depositors are forms of government intervention in the private sector. The government also controls numerous aspects of transportation, communications, education, and finance. Food assistance programs such as the Supplemental Nutritional Assistance Program (SNAP) and the Women, Infants, and Children (WIC) Program also are indicative of a mixed economic system.

DEFINITION OF AGRICULTURAL ECONOMICS

Because agricultural economics involves the application of economics to agriculture, we may define this field of study as follows:

> **Agricultural economics** is an applied social science that deals with how producers, consumers, and societies use scarce and natural resources in the production, processing, marketing, and consumption of food and fiber products.

WHAT DOES AN AGRICULTURAL ECONOMIST DO?

The application of economics to agriculture in a complex market economy such as that of the United States has a long and rich history. We can summarize this activity by discussing the activities of agricultural economists at the microeconomic level and at the macroeconomic level.

Role at Microeconomic Level

Agricultural economists at the micro level are concerned with issues related to resource use in the production, processing, distribution, and consumption of products in the **food and fiber system**. Production economists examine resource demand by businesses and their supply response. Market economists focus on the flow of food and fiber through market channels to their final destination and the determination of prices at each stage. Financial economists are concerned with issues related to the financing of businesses and the supply of capital to these firms. Resource economists focus on the use and preservation of the nation's natural resources. Other economists are interested in the formation of government programs for specific commodities that will support the incomes of farmers and provide food and fiber products to low-income consumers.

Role at Macroeconomic Level

Agricultural economists involved at the macro level are interested in how agriculture and agribusinesses affect domestic and world economies and how the events taking place in other sectors affect these firms and vice versa. For example, agricultural economists employed by the Federal Reserve System must evaluate how changes in monetary policy affect the prices of various food commodities. Macroeconomists with a research interest may use computer-based models to analyze the direct and indirect effects that specific monetary or fiscal policy proposals would have on the farm business sector. Macroeconomists employed by multinational food companies examine foreign trade relationships for food and fiber products. Others address issues in the area of international development.

Marginal Analysis

Economists frequently are concerned with what happens at the margin. A microeconomist may focus on how the addition of another input by a business, or the purchase of another product by a consumer, will change the economic well-being of the business and the consumer. A macroeconomist, on the other hand, may focus on how a change in the tax rate on personal income may change the nation's output, interest rates, inflation, and the federal budget deficit. The key word in this example is *change*, or, more specifically, how a change in price, quantity, and so on will affect other prices and quantities in the economy, and how this situation might change the economic well-being of consumers, businesses, and the economy as a whole. Many of the chapters to follow include a discussion of marginal analysis so as to better understand economic decisions made at the firm, household, or economy level.

Sonny Perdue, the current U.S. secretary of agriculture.   Credit: ZUMA Press, Inc./Alamy Stock Photo.

Zippy Duvall, the current president of the American Farm Bureau Federation.   Credit: Courtesy of Farm Bureau Federation.

Key agencies that agricultural economists deal with include the Economic Research Service (www.ers.usda.gov), the U.S. Department of Agriculture, and the American Farm Bureau Federation (AFBF) (www.fb.org), the voice of agriculture. The current U.S. secretary of agriculture is Sonny Perdue, and the current president of the AFBF is Zippy Duvall, a farmer from Georgia.

WHAT LIES AHEAD?

Chapter 2 gives an overview of the structure of the nation's food and fiber system and the important role it plays in the U.S. general economy. The remaining parts of the book can be summarized as follows:

- Part 2 focuses on understanding consumer behavior in the marketplace, particularly in explaining the demand for food and fiber products. Chapter 3 presents the theory of consumer behavior. Chapter 4 describes the conditions for consumer equilibrium and determination of market demand. Chapter 5 discusses the measurement and interpretation of demand elasticities.

- Part 3 changes the focus from the behavior of consumers to the behavior of producers of food and fiber products. Emphasis is placed on market equilibrium and market structures. Chapter 6 describes the measurement of production relationships, costs of production, and revenue. Chapter 7 describes the economics of input substitution and describes the economics of product substitution. Chapter 8 describes the determination of output and price under conditions of perfect competition. Finally, Chapter 9 describes the determination of output and price under conditions of imperfect competition.

- Part 4 examines the resource, environmental, and political setting in which producers and consumers of food and fiber products in the United States are immersed. Chapter 10 deals with resource and environmental economics. Chapter 11 focuses on the rationale for government intervention and outlines the development and application of income and price supports in the United States, primarily from the 1930s to the present.

- Part 5 switches attention to the macroeconomy—what makes it tick and the important links between the food and fiber system and the rest of the economy. Chapter 12 discusses product markets and national output. Chapter 13 also focuses on the tools of monetary and fiscal policy. Chapter 14 centers attention on business fluctuations, addressing consequences and policy applications. Chapter 15 concerns the macroeconomics of agriculture using information gleaned from Chapters 11 to 14.

- Part 6 draws attention to international linkages and to the global economy. Chapter 16 focuses on agriculture and international trade. It examines exchange rates and agricultural trade. Chapter 17 addresses the issue of why nations trade. Chapter 18 concerns agricultural trade policy and preferential trading arrangements.

Consequently, the book addresses seven different facets: (1) agricultural economics and the food and fiber sector; (2) consumer behavior; (3) business behavior and market equilibrium in perfectly competitive and imperfectly competitive environments; (4) resource and environmental economics; (5) government intervention in the food and fiber industry; (6) the macroeconomics of agriculture; and (7) international linkages primarily concerning trade, exchange rates, and trade policy.

Importantly, we wish to develop the understanding of and the ability to apply the economic principles that agricultural economists use to understand and predict individual and aggregate economic behavior and the impact of such behavior upon the well-being of society. In short, we plan to provide you the reader a *framework to think for yourself*, at least in conjunction with issues indigenous to economics.

SUMMARY

The purpose of this chapter is to define the field of agricultural economics as a subset of the general field of economics. The major points made in this chapter are summarized as follows:

1. Scarce resources are human and nonhuman resources that exist in a finite quantity. Scarce resources can be subdivided into three groups: (1) natural and biological resources; (2) human resources; and (3) manufactured resources.

2. Resource scarcity forces both consumers and farmers to make choices.

3. Most resources are best suited to a particular use. Specialization of effort may lead to a higher total output.

4. The field of economics can be divided into microeconomics and macroeconomics. Microeconomics focuses on the actions of individuals—specifically the economic behavior of consumers and farmers. Microeconomic analysis largely deals with the notion of partial equilibrium; events outside the market in question are assumed to be constant. Macroeconomics focuses on broad aggregates, including the nation's aggregate performance

as measured by gross domestic product (GDP), unemployment, and inflation. Macroeconomic analysis normally deals with the notion of general equilibrium; events in all markets are allowed to vary.

5. Positive economic analysis focuses on what-is and what-would-happen-if questions and policy issues. Normative economic analysis focuses on what-should-be or what-ought-to-be policy issues.

6. Capitalism, or free market economics, socialism, and communism represent alternative economic systems. The U.S. economy represents a mixed economic system. Some markets are free to determine price, and other market prices are regulated.

7. Agricultural economics is an applied social science that deals with how producers, consumers, and societies use scarce and natural resources in the production, processing, marketing, and consumption of food and fiber products.

8. Agricultural economists at the micro level are concerned with issues related to resource use in the production, processing, distribution, and consumption of products in the food and fiber system.

9. Agricultural economists involved at the macro level are interested in how agriculture and agribusinesses affect domestic and world economies and how the events taking place in other sectors affect these firms and vice versa.

KEY TERMS

Agricultural economics
Biological resources
Capital
Capitalism
Communism
Economics
Fallacy of composition
Food and fiber system

Human resources
Macroeconomics
Management
Manufactured resources
Microeconomics
Mixed economic system
Natural resources
Normative economics

Opportunity cost
Positive economics
Scarce resources
Scarcity
Socialism
Specialization

TESTING YOUR ECONOMIC QUOTIENT

1. Land, labor, and capital are examples of what three types of scarce resources?

   a.

   b.

   c.

2. An agribusiness firm may undertake three alternatives:

   Alternative 1: buy cane sugar and manufacture various sugars and sweets, making a profit of $10 million;

   Alternative 2: buy wheat and produce bread, rolls, and pastries, making a profit of $15 million; or

   Alternative 3: buy corn and produce Tex-Mex foods, making a profit of $12 million.

   a. Which alternative should this agribusiness firm undertake? Why?

   b. The opportunity cost associated with these three choices is $_____ million.

3. a. Concern has been expressed on the part of Congress and the President about what should be the optimal tax rate for those individuals who make more than $500,000 per year. This issue corresponds to what branch of economics?

   b. What branch of economics is concerned with the effects of food safety (e.g., *E. coli*) on consumer demand for beef (i.e., what-if types of questions)?

   c. What branch of economics is concerned with the rate of inflation and the unemployment rate?

   d. What branch of economics deals with the consumption expenditures of AGEC 105 students at Texas A&M University?

4. To economists, the word *marginal* means _____.

5. Circle the correct answer. The U.S. economy represents what kind of economic system?

   a. Capitalistic

   b. Socialistic

   c. Communistic

   d. Mixed

6. _____ is an applied social science that deals with how producers, consumers, and societies use scarce resources in the food and fiber sector.

7. Circle the correct answer. Economic reasoning that is true for one individual but not for society as a whole is referred to as

   a. specialization.

   b. fallacy of composition.

   c. opportunity cost.

   d. normative economics.

8. The Belford family owns a farm near San Angelo, Texas. Three alternatives exist for how to use the farm:

   Alternative 1. Grow cotton. Cotton yield would be 500 pounds per acre. The price of cotton is $0.96 per pound and production expenses are $285 per acre.

   Alternative 2. Grow wheat. Wheat yield would be 50 bushels per acre. The price of wheat is $7.25 per bushel and production expenses are $210 per acre.

   Alternative 3. Lease out the acres. The Belfords' neighbor, Auld McDonald, will pay $200 per acre for leasing, but the Belfords would still have expenses of $40 per acre.

Based on this information, answer the following:

a. Which alternative should the Belfords undertake? Why?

b. Given your answer to the previous question, what is the Belfords' opportunity cost per acre?

c. What is the total economic cost per acre for your answer?

9. Most resources are best suited for a particular use. For example, climate and other conditions in Florida allow resources to be used in orange production in lieu of wheat or potato production. What economic concept deals with this issue?

10. A Gini coefficient of _____ indicates perfect inequality in the distribution of wealth.

11. What economist was a champion of capitalism and referred to the "invisible hand of the market"?

12. What economist proposed a redistribution of wealth through a global tax?

REFERENCES

Friedman M: *Essays in positive economics*, Chicago, 1974, University of Chicago Press.

Piketty, T: *Capital in the 21st century*, August 2013, Belknap Press.

Smith, A: *An inquiry into the nature and causes of the wealth of nations*, first published in 1776.

GRAPHICAL ANALYSIS

In many of the chapters to follow, students must understand the construction and interpretation of graphs. Given the emphasis in this book on graphical analysis, we provide a tutorial on this subject. We begin with the construction of a graph from the numbers in a table documenting the relationship between two variables.

Constructing a Graph

Two variables can be related in different ways. For example, there is a direct relationship between yields and fertilizer usage (at least over some relevant range). That is to say, the greater the amount of fertilizer applied, the higher the yield. In more general terms, the increase in one variable may be associated with an increase in another variable. Two variables can also be inversely related. As the price of gasoline increases, individuals will find ways to reduce their consumption of this product, all other factors invariant. Here, an increase in one variable is associated with a decrease in another variable. Finally, students will encounter instances later in this book in which the relationship between two variables is mixed. For example, consider the relationship between yields and rainfall. Yields will increase sharply as we move from a situation of no rainfall to some normal amount. Beyond this level of rainfall, however, yields may actually begin to decline as a result of farmers not being able to get into the fields at the proper time, low-lying areas being washed out, and so on.

Table 1-1
TWO RELATED VARIABLES: PRICE AND QUANTITY

Price per Pair ($)	Quantity Sold during the Week	Location on Graph
9	20	A
8	30	B
7	40	C
6	50	D
5	60	E
4	70	F

To illustrate how to graph two related variables, let us assume that a local farm input supply dealer has noted a relationship between the price charged for work gloves and the number of pairs of work gloves sold during the week (see Table 1-1). The data in Table 1-1 should suggest to you that there is an inverse relationship between the price of a pair of work gloves and the number of pairs sold. As the price decreases, there is an increase in the quantity sold.

The price–quantity relationship in Table 1-1 can be viewed as coordinates on a graph. In economics, it is customary to put the dollar values (price in this instance) on the vertical, or Y, axis and quantity on the horizontal, or X, axis. Figure 1-2 shows the location of these price–quantity coordinates on a graph. Point A, for example, represents the observation that 20 pairs of gloves will be sold if the price per pair is $9. Assume for the moment that the sales of work

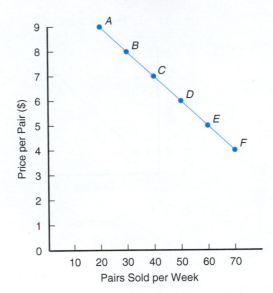

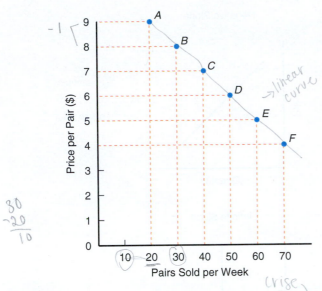

Figure 1-2
Graphing relationship between price and quantity.

gloves are perfectly divisible; that is, we can sell one-fourth or one-eighth of a work glove. This division allows us to have a quantity relationship at every possible price between the $4 and $9 range cited in Table 1-1 and also allows us to connect points *A* through *F* with a solid line. This line is normally referred to as a *linear curve* by economists, although this line does not curve at all. The term *nonlinear curve* refers to the situation wherein the relationship among respective variables is not linear.

Slope of a Linear Curve

An important feature of a curve to an economist is its slope, or the ratio of the change in the vertical axis to the change in the horizontal axis (rise over run). To illustrate the calculation of the slope for a linear curve, let us return to the price–quantity relationship observed for work gloves in Table 1-1. The slope of the linear curve is found by dividing the change in the values on the *Y*, or vertical, axis by the corresponding change in the values on the *X*, or horizontal, axis. As we move from point *A* to point *B* on this curve, the price per pair of work gloves falls by $1 because we moved from $9 a pair to $8 a pair. The corresponding change in quantity of gloves sold per week was 10 pairs, or 30 pairs minus 20 pairs. The slope of this curve therefore would be

slope = Change in price ÷ Change in quantity

= −$1.00 ÷ 10 pairs

= −$0.10 per pair

Thus, the slope of this linear curve at *all* points along this curve is $0.10. A specific property of a linear curve (which you should prove to yourself by examining other points along the curve) is that its slope is the same between any two points (i.e., its slope is constant).

Because economists often discuss basic demand and supply relationships in terms of the slopes of these curves, you must understand the difference between a positive slope, a negative slope, a zero slope, and an infinite slope. Each of these slopes is illustrated in Figure 1-3.

Positively and negatively sloped curves can take on a variety of shapes. Figure 1-4, for example, shows that a positively sloped linear curve (also called a ray) starting from the origin at a 45-degree angle has a slope of 1.0. At all points along this line, the quantity of *Y* is exactly equal to the quantity of *X*. A line with a positive slope of 0.5 will be flatter than the line having a positive slope of 1.0. Finally, lines with a slope of greater than 1.0 will be steeper than either of the first two curves.

Figure 1-4 suggests the following conclusion: the greater (smaller) a positive slope, the steeper (flatter) a linear curve will be. The opposite is true for negatively sloped linear curves.

Slope of a Nonlinear Curve

Although the slope of a linear curve is constant over the entire range of the curve, the slope of a nonlinear curve is not. A nonlinear curve, in fact, can exhibit a positive,

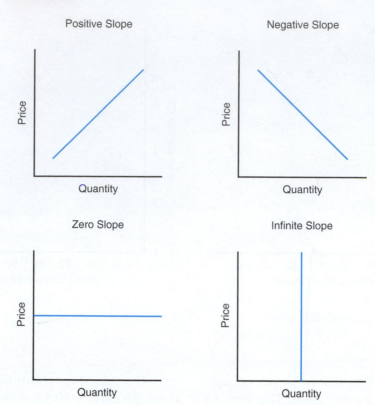

Figure 1-3

Alternative slopes of linear curves.

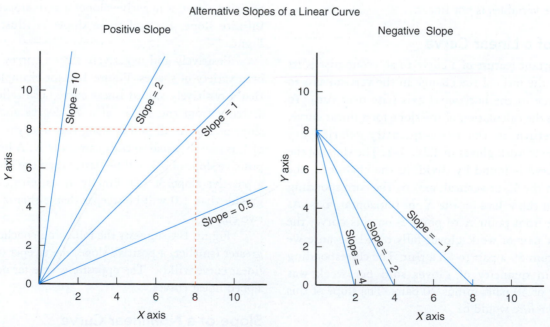

Figure 1-4

A ray from the origin (zero units of Y and zero units of X) with a 45-degree angle will have a slope of 1. Linear curves with a positive (negative) slope of less than 1 will be flatter (steeper), while linear curves with a slope greater than 1 will be steeper (flatter).

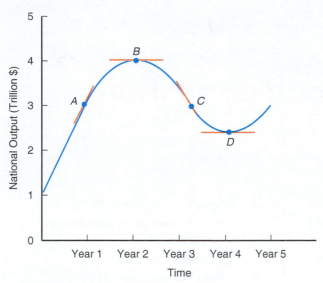

Figure 1-5
Slope of a nonlinear curve.

negative, or zero slope. Consider the nonlinear curve presented in Figure 1-5, which shows a time path for a business cycle over a period of time.

The slope at specific points along a nonlinear curve is calculated by computing the slope of a linear curve tangent to the nonlinear curve at these points. The slope at point *A* in Figure 1-5 is positive, indicating a positive growth in the economy at that point. The slope at point *B* is zero, indicating no change in the nation's output during the period. The slope of point *C* is negative, indicating a negative growth in the economy or a recession. At some point in time, the economy will bottom out (point *D*) and begin a period of positive economic growth.

The U.S. Food and Fiber Industry

Chapter Outline

The Economic Research Service (ERS) of the U.S. Department of Agriculture describes the term *food system* as a complex network of farmers and the industries that link them. Those links include makers of farm equipment and chemicals as well as firms that provide services to agribusinesses, such as providers of transportation and financial services. The system also includes the food marketing industries that link farms to consumers, and which include food and fiber processors, wholesalers, retailers, and food service establishments (Food Market Structures: Overview, Economic Research Service, USDA).

Agriculture is important in the development of any economy. Farmers and ranchers today are highly integrated partners in the U.S. economy. Their links to other sectors in the domestic economy include not only the markets to which they sell their output but also the financial markets from which they borrow funds, the labor markets from which they hire labor and seek off-farm employment, and the manufactured input markets from which they purchase chemicals, fertilizers, and equipment.

INDICES

Before embarking on the description and trends associated with the food and fiber industry, it is necessary to first understand the notion of how economists report measures of economic activity. Trends in output, productivity, and prices are commonly reported through the aid of *index* values. An index is nothing more than a percentage comparison from a fixed point of reference or benchmark. By comparing output and prices of apples in various years, for example, economists can describe changes in apple prices and apple production relative to the benchmark or base period.

To illustrate, consider Table 2-1. Milk production from 1995 to 2015 ranged from 154,006 million pounds in 1996 to 208,633 million pounds in 2015. Milk prices received by dairy producers during the same period ranged from $12.11 per hundred weight (2002) to $23.97 per hundred weight (2014). Let 2000 be the base period or benchmark. The selection of the base period or benchmark is arbitrary but important. To calculate the **output index** associated with milk production for any given year, simply divide the milk production in that year by the production level in the base period. So, to arrive at the output index of 0.93 for 1995, we divide 155,292 million pounds (production in 1995) by 167,393 million pounds (production in the base period). In the same way, to obtain the output index of 1.14 for 2008, we divide 189,992 million pounds (production in 2008) by 167,393. Similar calculations are made and exhibited in Table 2-1 to obtain the price indices of milk.

The index of the base period is always either 1 or 100, and the choice of the base period is arbitrary. If the base period changes, the corresponding set of indices also changes. Indices are unitless measures. Interpretation of the indices is quite important. For example, in Table 2-1, the price index of 0.98 in 2002 is interpreted as follows: relative to 2000 (the base period), milk prices in 2002 were lower by 2.0%. By the same token, milk prices in 2011 were higher by 63% relative to 2000. So, in short, indices provide a straightforward way to make comparisons.

Trends in economic activity are commonly reported through the use of index values. An *index* is a percentage comparison from a fixed point of reference. Indices commonly used in practice include the Consumer Price Index and the Index of Prices Received (or Paid) by farmers.

Table 2-1
OUTPUT AND PRICE INDICES FOR MILK, 1995 TO 2015

Year	Milk Production (Million Pounds)	Output Index (Base Year 2000)	All Milk Price ($/cwt)	Price Index (Base Year 2000)
1995	155,292	0.93	12.74	1.03
1996	154,006	0.92	14.88	1.21
1997	156,091	0.93	13.34	1.08
1998	157,262	0.94	15.50	1.26
1999	162,589	0.97	14.36	1.17
2000	167,393	1.00	12.32	1.00
2001	165,332	0.99	14.97	1.22
2002	170,063	1.02	12.11	0.98
2003	170,394	1.02	12.52	1.02
2004	170,934	1.02	16.05	1.30
2005	176,929	1.06	15.13	1.23
2006	181,782	1.09	12.88	1.05
2007	185,654	1.11	19.13	1.55
2008	189,992	1.14	18.33	1.49
2009	189,334	1.13	12.83	1.04
2010	192,848	1.15	16.26	1.32
2011	196,245	1.17	20.14	1.63
2012	200,366	1.20	18.51	1.50
2013	201,218	1.20	20.05	1.63
2014	206,046	1.23	23.97	1.95
2015	208,633	1.25	17.08	1.39

Source: National Agricultural Statistics Service, USDA.

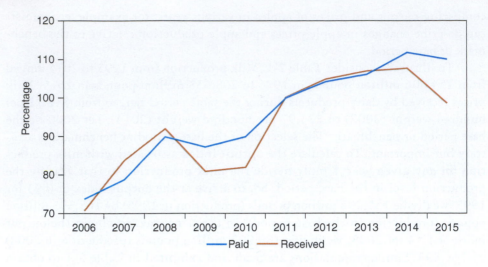

Indices commonly used in practice include the Consumer Price Index, the Wholesale Price Index, and the Index of Prices Received (or Paid) by Farmers. Perhaps the most visible price index today is the Consumer Price Index (CPI). Price indices are useful in that they are measures of inflation. Inflation or its opposite, deflation, affects prices, interest rates, expenditures, and disposable income. Consider the following graph above, which relates to the Index of Prices Received and Paid from 2006 to 2015. The base period is 2011 = 100. In 2015, the Index of Prices Received was 98.5, meaning that relative to 2011, prices received by farmers were lower by 1.5%. In 2015, in similar fashion, prices paid by farmers were higher by 10% as the index for prices paid was 110.

Nominal values refer to economic measures for which no adjustments to inflation have been made.

Nominal (current) dollar values refer to economic measures for which no adjustments to inflation (or deflation) have been made. Suppose that your annual income in 1980 was $25,000 and that at present it is $50,000. Your nominal income today is twice that of 1980. Are you twice as well off today as in 1980? The answer is not necessarily because prices of goods and services that you buy also have changed. Most prices of goods and services have increased since 1980, although they may not have necessarily doubled. To arrive at the actual change in purchasing power of your income from 1980 to the present, some adjustments for price changes need to be made.

Real values are inflation-adjusted values.

Real or constant dollar values refer to economic measures for which adjustments to inflation (or deflation) have been made. To derive the real value, economists use the following:

$$\text{Real value} = \frac{\text{Nominal value}}{\text{Price index}} \tag{2.1}$$

To illustrate, consider Table 2-2. Total expenditures for food away from home are exhibited for the years 1980 through 2014. Nominal expenditures ranged from $120.3 billion in 1980 to $731.3 billion in 2014. The corresponding CPI ranged from 0.824 in 1980 to 2.367 in 2014. Using formula (2.1), the associated real expenditures ranged from $144.0 billion (1981) to $308.9 billion (2014). In this way, adjustments for inflation are made. The $308.9 billion figure is in terms of dollars related to the base period of the CPI, 1982 to 1984. This figure corresponds to the **inflation-adjusted** total food-away-from-home expenditures in 2014 in terms of 1982 to 1984 dollars.

Table 2-2
NOMINAL AND REAL TOTAL FOOD-AWAY-FROM-HOME EXPENDITURES

Year	Nominal Expenditures (Million Dollars)	CPI (Base Period 1982 to 1984 = 1.00)	Real Expenditures (Million Dollars)
1980	120,296	0.824	145,990
1981	130,914	0.909	144,020
1982	139,776	0.965	144,846
1983	150,883	0.996	151,489
1984	161,046	1.039	155,001
1985	168,831	1.076	156,906
1986	181,695	1.096	165,780
1987	196,557	1.136	173,026
1988	214,396	1.183	181,231
1989	228,449	1.240	184,233
1990	244,975	1.307	187,433
1991	256,337	1.362	188,206
1992	263,442	1.403	187,770
1993	278,451	1.445	192,700
1994	291,002	1.482	196,358
1995	307,297	1.524	201,638
1996	311,582	1.569	198,586
1997	333,387	1.605	207,718
1998	349,976	1.630	214,709
1999	365,853	1.666	219,600
2000	391,086	1.722	227,111
2001	403,406	1.771	227,784
2002	411,571	1.799	228,778
2003	446,356	1.833	243,511
2004	466,035	1.888	246,841
2005	507,604	1.953	259,910
2006	541,795	2.016	268,748
2007	570,527	2.073	275,218
2008	586,703	2.153	272,505
2009	583,135	2.145	271,858
2010	601,383	2.181	275,737
2011	635,180	2.249	282,420
2012	669,720	2.296	291,690
2013	697,605	2.330	299,401
2014	731,258	2.367	308,939

Source: Economic Research Service, U.S. Department of Agriculture, http://www.ers.usda.gov/Data/ FoodMarketIndicators.

WHAT IS THE FOOD AND FIBER INDUSTRY?

The term *agriculture* means different things to different people. Some might think solely of farmers and ranchers when they use this term; others might think of agribusiness firms such as Tyson Foods, HEB, and McDonald's. In recent years, many agricultural economists have referred to the **food and fiber industry** when describing the agricultural sector. The food and fiber industry includes farms, ranches, and agribusinesses.

The food and fiber system—from the farmer to the consumer—is one of the largest sectors in the U.S. economy. This system typically produces output valued at roughly 12% to 15% of the nation's output (Economic Research Service, USDA).

The food and fiber system comprises the economic activities of the farms and the firms that assemble, process, and transform raw agricultural commodities into final products to U.S. and to foreign consumers. The food and fiber system accounts for roughly 12% to 15% of the U.S. gross domestic product (GDP).

Historically, one out of every six jobs in the U.S. economy has been tied in one way or another to the food and fiber industry. The system includes all economic activities supporting farm production, such as machinery repair and fertilizer production, food processing and manufacturing, transportation, and wholesale and retail distribution of food and apparel products.

As Figure 2-1 suggests, the food and fiber system encompasses the activities of the farm input supply sector, the farm sector, the processing and manufacturing sector, and the wholesale and retail trade sector. In addition to farms and ranches captured in the farm sector, the food and fiber system includes such firms as

- John Deere, DeKalb Seed, Ralston-Purina, and others that supply goods and services to farmers and ranchers,
- Anheuser-Busch, Del Monte Foods, Cargill, and others that utilize raw agricultural products in fiber manufacturing and food processing operations, and
- Ahold, Wal-Mart, Kroger, Sysco, Supervalu, and others that distribute finished food and fiber products at the wholesale level, the retail level, and the food service level.

In short, the food and fiber system comprises the economic activities of the farms and the firms that assemble, process, and transform raw agricultural commodities into final products for distribution to U.S. and foreign consumers.

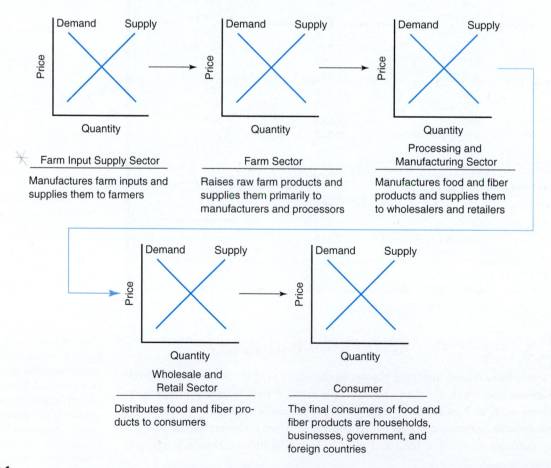

Figure 2-1

The food and fiber industry consists of the farm input supply sector, the farm sector, the processing and manufacturing sector, and the wholesale and retail sectors. These sectors are linked together by a series of markets in which farmers purchase production inputs and sell their raw products to processors and manufacturers, which, in turn, create food and fiber products that are distributed to customers in wholesale and retail markets.

Table 2-3
VALUE ADDED TO THE U.S. ECONOMY BY THE AGRICULTURAL SECTOR FOR 2015

Component Accounts	2015 ($ Billion)
Value of crop production	185.0
Food grains	12.1
Feed crops	58.0
Cotton	5.0
Oil crops	36.9
Fruits and tree nuts	30.5
Vegetables	19.2
All other crops	27.2
Value of livestock production	190.0
Meat animals	96.2
Dairy products	35.4
Poultry and eggs	47.0
Other farm income	28.8
Value of agricultural sector production	426.1
Net farm income	56.4

Source: Economic Research Service, U.S. Department of Agriculture, Farm Income and Wealth Statistics.

Simply put, the food and fiber system includes all economic activities that support farm production and the conversion of raw products to goods for consumption—farm inputs, farm production, food processing and manufacturing, transportation, wholesale and retail trade, distribution of food and apparel, and eating establishments.

In Table 2-3, we present figures for the value added to the U.S. economy by the agricultural sector for 2015. The value of crop production was $185.0 billion, and the value of livestock production was $190.0 billion over this period. Revenues from other farm income were calculated to be $28.8 billion. Overall, the value of agricultural sector production was $426.1 billion. Net farm income in 2015 amounted to $56.4 billion.

CHANGING COMPLEXION OF FARMING

The nature of farming and ranching operations in the U.S. farm sector has changed dramatically during the post–World War II period. The focus of this section is on the collective structure and performance of farmland ranches and on the changing complexion of farming activities in the United States. We can assess these attributes by examining recent trends in the physical structure, productivity, profitability, and financial structure of farms in general.

Physical Structure

An examination of the changing physical structure of the farm sector must necessarily focus on things such as the number and size of farms and ranches, their ownership and control, and the ease of entry into the farm sector.

Number and Size of Farms   A trend toward fewer but larger farms has been occurring. The number of farms has declined from 6.8 million in 1935 (the peak) to about 2.1 million currently (U.S. Department of Agriculture). Between 1935 and 1974, the number of farms dropped by two-thirds from 6.8 million to 2.5 million. Since 1974, farm numbers have been more stable, especially since the 1990s. As illustrated in Figure 2-2A, the number of U.S. farms ranged from 2.1 million to 2.2 million over the period 1990 to 2015.

The average farm size during the post–World War II period doubled, from about 200 acres to more than 400 acres per farm. As shown in Figure 2-2B, over the period 1990 to 2015, the average farm size ranged from 418 acres to 464 acres. Currently, the average farm size is close to 440 acres. National averages can, however, give a misleading picture of the physical structure of the farm operation. About 60% of U.S. farms had sales less than $10,000, and these farms accounted for only 3% of the value of agricultural production. At the other extreme are farms with sales in the millions. These million-dollar farms make up less than 2% of all U.S. farms, but they account for a sizeable share of the value of production.

It is noteworthy that the share of total farm receipts earned by the 50,000 largest farms (2% of the total number of U.S. farms) has been increasing during the last 35 years. Large-scale farms currently account for 10% of U.S. farms, but they are responsible for 75% of the value of production. Arguably, concentration of production may be more of a critical issue in the discussion of farm structure than the decline in the number of farms.

As exhibited in Figure 2-2C, the average value per acre of farm real estate on a nominal basis has risen almost monotonically from $627 in 1990 to $2,605 in 2014. Farm real estate values have increased roughly four times in the last 25 years.

Specialization, Diversification, Organization, and Contracting

U.S. farms tend to be specialized rather than diversified. About half of U.S. farms produce just one commodity. Smaller farms are the most likely to produce one commodity, but three-fourths of the farms with sales of at least $500,000 produce no more than three commodities.

Most of the operators of farms with sales less than $50,000 also work off the farm or are retired. Retired operators and operators reporting a major occupation other than farming made up roughly 60% of all farms. Yet, these farms accounted for only 13% of the value of production. Census data show that farmers have been combining off-farm work with farming to some extent since the 1930s. Additionally, census data show that the average age of farm operators increased from 48 in 1940 to 54 in 1997. In 2004, about 27% of farm operators reported their age as 65 or older, up from 17% in 1968. Like their nonfarm counterparts, many of today's farm households are dual-career households. For those operators who report a nonfarm major occupation or work off-farm, the health of the local economy, nonfarm job growth, and the level of nonfarm wages may be as important as farm programs.

Concern has been expressed over a perceived increase in the number of corporate farms and a corresponding decline in family farms. Data from the Census of Agriculture show that family-owned farms are not losing their share of U.S. agriculture to nonfarm corporations. At present, most U.S. farms are family farms. U.S. farms are mostly organized as individual operations, but farms organized as partnerships and corporations accounted for nearly 40% of the value of production in the agricultural sector.

Noteworthy trends in the changing structure of the farm sector are the declining number of farms and the rising average size of farms. Also, the share of total farm receipts earned by the largest farms has been increasing during the last 30 years. U.S. farms tend to be specialized rather than diversified.

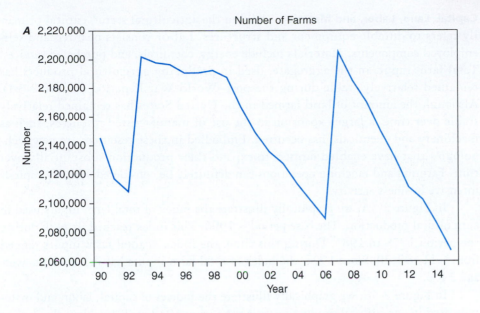

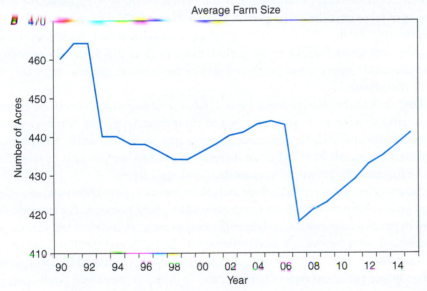

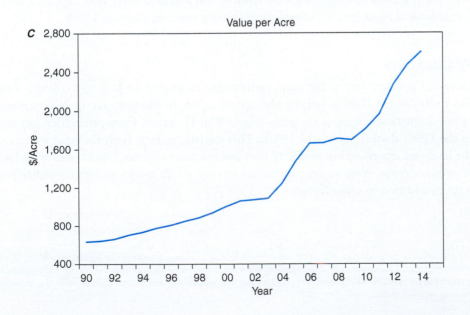

Figure 2-2

Two significant trends in the changing structure of the farm sector are the declining number of farms and the rising average size of farms. (A) The number of farms in this country has declined from nearly 6 million in 1950 to roughly 2.1 million currently. (B) The average size of farms has roughly doubled during this period.

(*Source:* U.S. Department of Agriculture.)

Capital, Land, Labor, and Materials   Within the agricultural sector, **capital** primarily refers to durable equipment and structures. **Labor** consists of hired and self-employed components. **Materials** include energy, chemicals, and purchased services. Total farm input, in the aggregate, used in producing agricultural products has remained relatively stable during the post–World War II period (Figure 2-3A). Although the amount of **land** farmed in the United States has remained relatively stable over time, a large expansion in the use of manufactured resources, such as machinery and chemicals, has occurred. Embodied in these resources are new technologies that have enabled farmers to expand their production substantially over time. Farming and ranching operations can definitely be considered a highly capital-intensive business activity.

In Figure 2-3A, we graphically illustrate the index of total farm inputs used in agricultural production. The base period is 1996. This index reached its peak during the period 1978 to 1983. During this time, the index of total farm inputs ranged from 1.11 to 1.20. Since 1996, the index of total farm inputs has increased on average 2.8%.

In Figure 2-3B, we graphically illustrate the indices of capital, labor, and materials used in agricultural production over the period 1948 to 2013. Note the decline in the use of labor over the entire period. From 1948 to 1980, the use of capital inputs was on the rise. From 1980 to 1993, capital inputs declined and, subsequently, since 1994 they have leveled off. Over the period, from 1948 to 2013, the use of materials such as chemicals, energy, and purchased service has been on the rise, especially from 1998 to the present.

This shift in the relative use of capital, labor, and materials reflects the changing relative productivity of these resources and their relative prices. Improvements in farm machinery and variable production inputs such as hybrid seeds, fertilizers, and agricultural chemicals have led to an increase in production per unit of labor input and have fostered the growth of large-scale, specialized farms.[1]

Advances in machinery technology and achievements in farm chemicals have affected regional production patterns and once-conventional cropping practices. For example, farmers used to practice crop rotation and diversification to conserve their soil and control pests. Farm chemicals were increasingly used by farmers in the 1970s and 1980s to grow one crop exclusively year after year. Disease-control techniques also have allowed livestock farms to specialize in one particular type of livestock and to utilize confinement production practices. Many groups concerned with the environment and food safety have urged the adoption of limited input (i.e., low-chemical) production practices since the 1990s.

The total quantity of inputs used in producing raw agricultural products has remained relatively stable over the past 60 years. In the farm sector, there has been an expansion in the use of manufactured inputs and a decline in labor use.

Productivity

Increased productivity is the main contributor to growth in U.S. agriculture. **Productivity**, or the level of output per unit of input, in the farm sector has increased rather dramatically during the post–World War II period. Crop production per acre in the 1990s doubled since the 1950s. This statistic reflects both the productivity of the land and the changing productivities and amounts of other inputs used with land to produce crops. Year-to-year variations in crop yields reflect unusual weather patterns in addition to other factors.

---

[1] Specialization in the production of farm commodities has become increasingly apparent during the post–World War II period. Such specialization has come about primarily as a result of the development of capital-intensive production technologies and the proliferation of government programs that have reduced the need for farm diversification as a method of reducing exposure to risk.

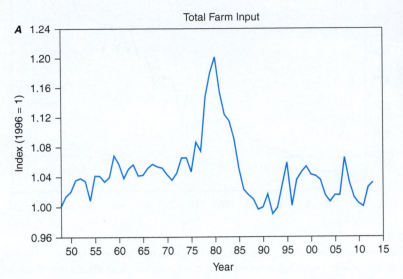

Figure 2-3A
Index of total farm inputs used in agricultural production, 1948 to 2013 (1996 = 1.00).

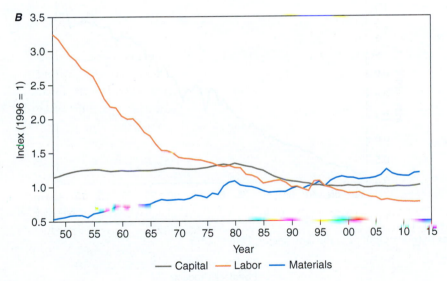

Figure 2-3B
Indices of capital, labor, and materials used in agricultural production, 1948 to 2013 (1996 = 1.00).

Figure 2-4 shows the annual trend in output relative to input use in agriculture during 1948 to 2013. The ratio of output to input in the farm sector, termed productivity, increased noticeably from 1948 to 2013. Agricultural productivity rose at an average annual rate of nearly 4.0% over this period.

As shown in Figure 2-5A, total U.S. agricultural output increased at an average annual rate of 4.2% over the period 1948 to 2013. Growth rates in crop and livestock output averaged 4.3% and 3.4%, respectively (see Figures 2-5B and 2-5C). The level of aggregate input use remained relatively unchanged.

Recent years have seen the development of several biotechnologies that enhance the productivity of farming and ranching operations, such as bovine somatotropin (BST). This hormone, which is administered to dairy cows by daily

The ratio of output to input or productivity in the farm sector has increased dramatically over the past 60 years. Total U.S. agricultural output has increased at an average annual rate of nearly 4% since 1948. The level of aggregate input use has remained relatively unchanged.

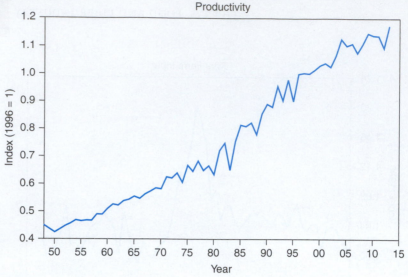

Figure 2-4

Index of agricultural productivity, 1948 to 2013 (1996 = 1.00). The productivity, or level of farm output per unit of farm input, has increased sharply during the post–World War II period. One reason for this rising productivity has been the technological advances embodied in farm inputs.

(*Source:* United States Department of Agriculture.)

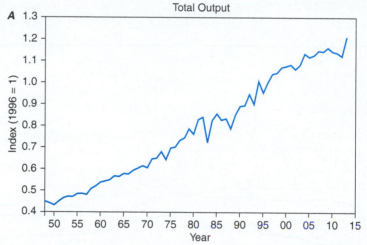

Figure 2-5A

Index of total output from the farm sector, 1948 to 2013.

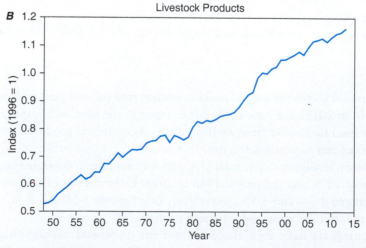

Figure 2-5B

Index of total output associated with livestock products, 1948 to 2013.

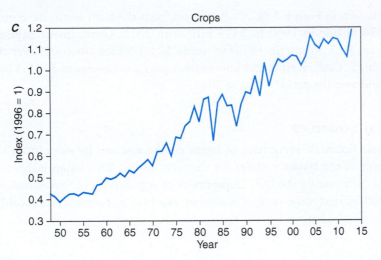

Figure 2-5C
Index of total output associated with crops, 1948 to 2013.

injections, increases milk production per cow. U.S. farmers have adopted genetically engineered (GE) crops widely since their introduction in 1996. Soybeans and cotton genetically engineered with herbicide-tolerant traits have been the most widely and rapidly adopted GE crops, followed by insect-resistant cotton and corn.

Profitability

Gross income from farming activities and related production expenses were relatively stagnant from the 1960s to the early 1970s. Beginning in 1972, however, gross income from farming activities, which consists of cash receipts from marketing crops and livestock and government payments to farmers, has risen, albeit somewhat erratically, while production expenses have behaved in a similar manner.

Net farm income is defined as

$$\text{Net farm income} = \text{Gross farm income} - \text{Production expenses} \qquad (2.2)$$

and gross farm income is defined as

$$\begin{aligned}\text{Gross farm income} = \ &\text{Cash receipts from farm marketings} \\ &+ \text{Government payments} \\ &+ \text{Other income from farm sources} \qquad (2.3)\end{aligned}$$

As depicted in Figure 2-6A, production expenses began to increase persistently after 1973 due in part to the rising prices of petroleum and rising interest rates on farm debt. As exhibited in Figure 2-6B, nominal net farm income, with some upswings and downturns, generally has been on the rise, especially since 2010. In 2014, net farm income reached at 90.5 billion, but a year earlier net farm income was $123.3 billion. The peak of net farm income from 1949 to 2014 was $123.3 billion. When adjustments for inflation are made, by dividing net farm income by the Consumer Price Index, real net farm income peaked in 1973 and then declined for the most part until 2002. Since 2002, albeit with some peaks and valleys, real net farm income has recovered, especially since 2010. Over the period 1960 to 2014, real net farm income was at its lowest point in 1983 at $14.4 billion.

Gross farm income has increased, albeit somewhat erratically, and production expenses have behaved in similar fashion. Inflation-adjusted or real net farm income has grown modestly over the past 30 years. Equity associated with the farm sector has been on the rise over the past decade. The debt-to-asset ratio and the debt-to-equity ratio associated with the farm sector have been relatively stable over the past 10 years.

As shown in Figure 2-6C, cash receipts from the farm sector have grown from almost $34 billion in 1960 to $377 billion in 2015. Government payments have ranged from $0.5 billion (in 1974) to nearly $24.4 billion (in 2005) over the period 1960 to 2015. Cash receipts and government payments averaged $165.3 billion and $8.3 billion over the period 1960 to 2015.

Financial Structure

The overall **financial structure** of farms can be assessed by examining the major components of the **balance sheet** for the farm sector. This balance sheet, which is published each year by the U.S. Department of Agriculture (*Economic Indicators of the Farm Sector*), indicates the value of real estate **assets** (e.g., farmland and buildings) and

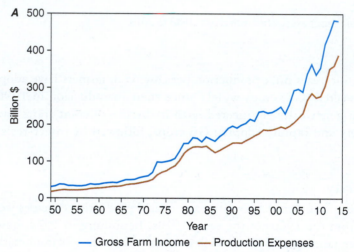

Figure 2-6A

Gross farm income and production expenses, 1960 to 2014.

(*Source:* U.S. Department of Agriculture.)

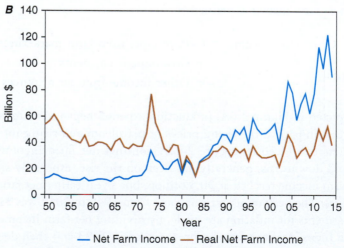

Figure 2-6B

Net farm income and real net farm income, 1960 to 2014.

(*Source:* U.S. Department of Agriculture.)

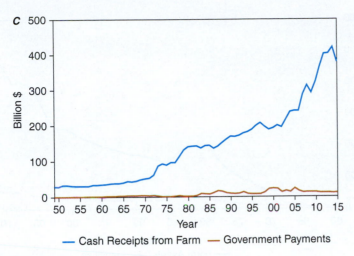

Figure 2-6C

Cash receipts from the farm sector and government payments, 1960 to 2015.

(*Source:* U.S. Department of Agriculture.)

non–real estate assets (e.g., machinery, trucks, and crop and livestock inventories) on farms and the financial assets (e.g., checking account balances and savings accounts) and liabilities of U.S. farms. **Equity**, or net worth, is given by

$$\text{Equity} = \text{Value of real estate assets} + \text{Value of non-real estate assets} + \text{Financial assets} - \text{Liabilities} \tag{2.4}$$

Trends in these balance sheet items over the period 1960 to 2015 are presented in Figure 2-7. Examining Figure 2-7A, we see that the change in the value of farm real estate represents the major component of the change in the total value of all farm assets. Since 1986, real estate assets associated with the farm sector have grown

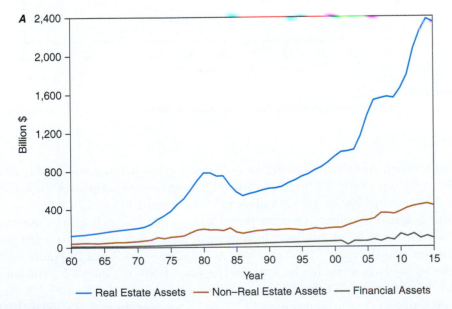

Figure 2-7A

Value of real estate assets, non–real estate assets, and financial assets associated with the farm sector, 1960 to 2015 (billion dollars).

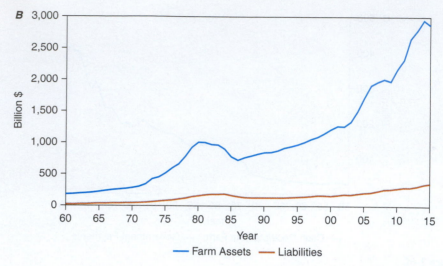

Figure 2-7B

Assets and liabilities associated with the farm sector, 1960 to 2015 (billion dollars).

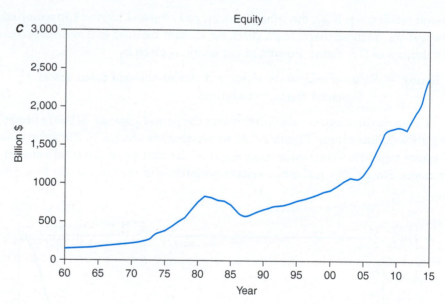

Figure 2-7C

Equity associated with the farm sector, 1960 to 2015 (billion dollars).

rather steadily. As exhibited in Figure 2-7*B*, farm assets have grown steadily since 1986, totaling nearly $2.9 trillion in 2015. Farm sector liabilities grew from $22.4 billion in 1960 to $188.8 billion in 1984. From 1985 to 1992, farm sector liabilities actually declined. Over the period 1993 to 2015, farm liabilities rose from $134.3 billion to $364.2 billion. In a nutshell, farm assets greatly exceeded farm liabilities over the period 1960 to 2015. As a result, as exhibited in Figure 2-7*C*, equity in the farm sector has been on the rise since 1986, reaching $2.5 trillion in 2015. From Figure 2-7*D*, the debt-to-asset ratio ranged from 11.3% to 22.2% while the debt-to-equity ratio ranged from 12.7% to 28.5% over the period 1960 to 2015. Both ratios reached their peak in 1985. Since 2000, the debt-to-asset ratio

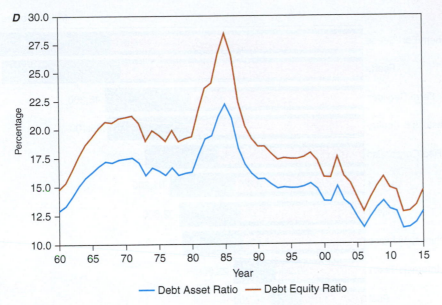

Figure 2-7D
Debt-to-asset ratio and debt-to-equity ratio associated with the farm sector, 1960 to 2015 (percent).

and the debt-to-equity ratio have not changed appreciably. From 2000 to 2015, the debt-to-asset ratio and the debt-to-equity ratio averaged 12.8% and 14.7%, respectively.

OTHER SECTORS IN THE FOOD AND FIBER INDUSTRY

From the wide fluctuations in net farm income from 1980 to the present, it is not hard to believe that farms are directly affected by the economic health of the rest of the economy, both domestic and foreign. In turn, the other sectors in the food and fiber industry are dependent on a healthy farm sector. Farm input suppliers, such as John Deere Company, the DuPont Chemical Company, and the Dow Chemical Company, are dependent on a healthy farm sector for a strong market for their products. Similarly, food processors and manufacturers are dependent on the farm sector for a steady stream of inputs to their production processes. Let us look more closely now at some of the characteristics of the other firms involved in the U.S. food and fiber industry.

Farm Input Suppliers

Farm input suppliers provide U.S. farmers and ranchers with the inputs they need to produce crops and livestock. There are six broad categories of farm input suppliers: (1) the feed manufacturing industry, (2) the fertilizer industry, (3) the agricultural chemical industry, (4) the farm machinery and equipment industry, (5) the hired farm labor market, and (6) farm lenders.

The relative importance of individual categories of farm production expenses for 2014 is illustrated in Figure 2-8. This figure indicates that 17% of total farm production expenses go toward the purchase of fertilizers and agricultural chemicals. Feed, farm services, and livestock and poultry purchased for feeding operations

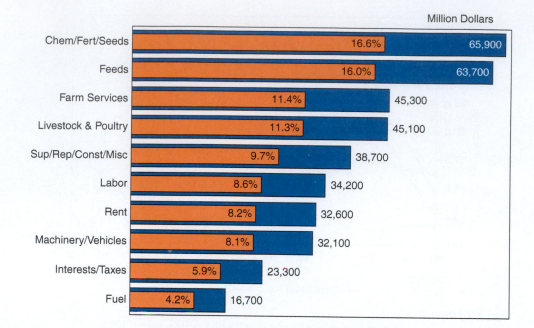

Figure 2-8

Major input expenditures associated with the farm sector, 2014.

(*Source:* USDA NASS, August 2015.)

constitute roughly 39% of total farm production expenses; wages paid to labor, rent, and machinery/vehicles amount to approximately 25% of total farm production expenses; and interest payments to banks and other farm lenders and taxes (nearly 6%) as well as fuel (slightly more than 4%) also represent notable input expenditure categories.

Food Processors, Wholesalers, and Retailers

The food marketing channel is primarily composed of the farm sector, the processing sector, the wholesaling sector, and the retailing and food service sector. The ultimate goal of the food marketing system is to deliver food produced on farms to consumers. The food retailing and food service sector not only is an important component of the food marketing channel but also is vital to the U.S. economy. In 2014, U.S. consumers spent $1,459.1 billion on food: $737.8 billion on food at home and $731.3 billion on food away from home (Economic Research Service, USDA). This figure corresponds to roughly 8% of the U.S. **gross domestic product** for 2014. At-home food outlets include conventional supermarkets, superstores, supercenters, membership clubs, combination (food and drug) stores, natural and organic outlets, limited-assortment stores, convenience stores, dotcoms, and gasoline stations. Away-from-home food outlets include fast-food, limited-menu, and full-menu establishments.

The business of food retailing and food service is undergoing salient change. According to the Food Marketing Institute, the singular force driving this change is the consumer. The popular press is replete with characterizations of the "time-starved consumer," the "nutrition-conscious or health-conscious consumer," and the "environment-conscious consumer." The food retailing and food service sectors provide significant service to consumers. In 2014, over 80% of the U.S. food dollar went toward value-added services and material transportation, processing, distribution, labor, packaging, and energy.

At the turn of the 20th century, Americans bought virtually all their food as ingredients or in raw form to be prepared for meals eaten at home. A century later, consumers are buying nearly half their food at restaurants and take-out establishments. Also, an increasing number of meals eaten at home are fully or partially prepared by outside sources. In 1930, the share of the food dollar for at-home consumption was almost 80% and for away-from-home consumption, roughly 20%. In 1990, these shares were 57.0% for at-home consumption and 43% for away-from-home consumption. In 2014, these shares were almost evenly split between at-home and away-from-home consumption. According to Figure 2-9, the restaurant industry's share of the food dollar in the United States in 1955 was 25%, but in 2014, this share rose to 47%. Bottom line, Americans are spending a greater share of the food dollar in away-from-home premises rather than for at-home consumption.

Driving this trend is the increase in dual-income families, time-starved consumers, and consumers who lack any culinary expertise. As a result, consumers decide either to eat away from home or to buy an increasing portion of their food in a ready-to-eat or ready-to-heat form.

Additionally, the percentage of income spent for food has declined over time (see Figure 2-10). In 1929, the first year data of this type were recorded, 23.4% of disposable income (i.e., income after taxes) was spent for food. Over the period

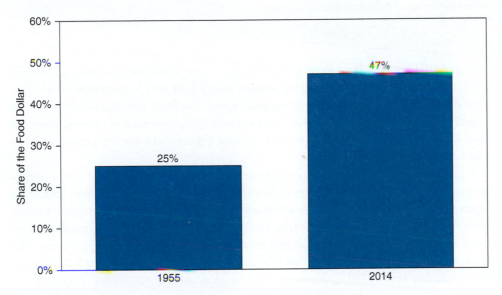

Figure 2-9
Restaurant industry's share of the food dollar in the United States in 1955 and 2014.
(*Source:* Published by National Restaurant Association.)

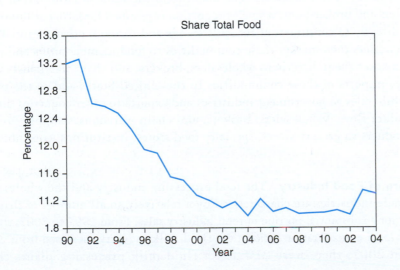

Figure 2-10
Percentage of disposable personal income spent on food, 1990 to 2014.

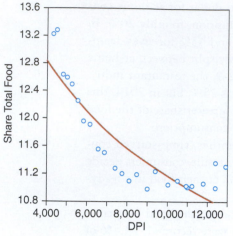

Figure 2-11

Illustration of Engel's Law, data from 1990 to 2014.

Note: Share of total food corresponds to the percentage of disposable income spent on food. DPI refers to disposable personal income (billion dollars).

Ernst Engel.   Credit: INTERFOTO/Alamy.

1990 to 2014, the percentage of income spent for food ranged from 11.0% to 13.3%. This percentage tapered off almost every year and is currently at 11.3%. The decline in the percentage of income spent for food also is consistent with Engel's Law; as disposable personal income rises, the proportion of income spent on food declines. This relationship is evident from Figure 2-11 for the period 1990 to 2014, the past 25 years.

A decline in the percentage of income spent for food generally reflects a highly developed economy in which there is money to spend for personal services and other discretionary items. Some of these additional services ordinarily are purchased along with food, which largely explains why the percentage of income spent for food away from home has not fallen as has the percentage of income spent for food at home. The food-at-home market now includes a larger service component through packaging and foods that require little preparation by the consumer.

Value-Added Process

Figure 2-12 illustrates the value added to the product as it flows through the food and fiber industry to consumers. The majority of the value of farm output goes to assemblers and brokers; only a small proportion is retained for farm consumption or is sold directly to consumers at roadside stands and through other means. Brokers and assemblers then market these commodities to food manufacturers and processors or market them directly to wholesalers, brokers, and chain warehouses. Along the way, imports of these commodities to the United States add to the product flow, while sales to government industries and exports to other countries diminish the product flow. Wholesalers, brokers, and chain warehouses supply processed food products to grocery stores, specialty food stores, institutions, consumers, and the military.

Structure of Food Industry   The food processing industry and the wholesale and retail trade industries are characterized by a relatively small number of firms that account for a substantial portion of total industry sales. From 1997 to 2007, the four-firm national concentration ratio in the fluid milk industry increased from 21% to 46%. In 2007, there were 21% fewer fluid milk processing plants than in

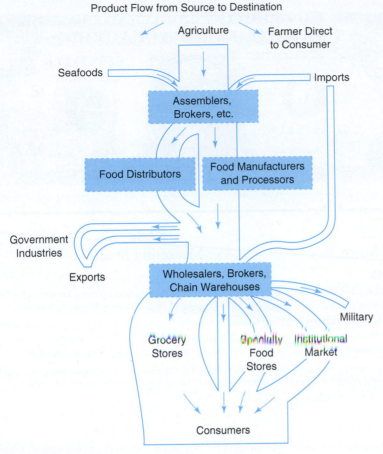

Figure 2-12
This figure illustrates the channels through which food and fiber products flow. The width of these channels reflects their relative magnitude. Almost all raw agricultural products flow to brokers and other middlemen firms that either pass them on to wholesalers or sell them to manufacturers and processors.

1997, processing 26% more milk per plant. Structural changes in fluid milk processing are occurring at the same time as rapid consolidation in milk production. From 1997 to 2007, 43% fewer farms produced over twice as much milk per farm. Consolidation in beef and pork slaughter has been of special interest to policy officials given the historically high and growing rates of concentration. For example, the four largest steer and heifer slaughter firms increased their share of slaughter to 85% in 2010, after remaining between 78% and 81% between 1998 and 2009. We will discuss the ramifications of concentration in the food industry in Chapter 9.

The Marketing Bill   The **marketing bill** for food is defined as the portion of food expenditures associated with the activities of firms beyond the farm gate. As shown in Figure 2-13, the marketing bill represents close to 83% of every dollar spent on food. Labor expenses, packaging expenses, transportation costs, and advertising expenses of food processors, wholesalers, and retailers are among the major categories of expenditures made by these firms. For example, nearly 40 cents of every dollar spent on food represents labor expenses incurred beyond the farm gate.

The marketing bill is the difference in dollars between the farm value and consumer expenditures for food produced on U.S. farms. Consumer expenditures include spending at grocery stores, restaurants, and institutions. The marketing bill pays for

The U.S. food markets are a network of processors, wholesalers, retailers, and food service purveyors. The ultimate goal of the food marketing system is to deliver food produced on farms to consumers. The marketing bill, defined as the difference between the farm value and consumer expenditures for food, has increased rather steadily since the early 1950s. The farm value share of consumer expenditures is roughly 17% currently, while the marketing bill share is about 83% at present.

17.2¢
Farm Share

82.8¢
Marketing Share

Figure 2-13
The marketing bill is the value of food expenditures contributed by firms beyond the farm gate. The farm value of total food expenditures currently is 17%. The marketing bill for food is defined as the portion of food expenditures associated with the activities of firms beyond the farm gate. Labor expenses, packaging expenses, transportation costs, and advertising expenses of food processors, wholesalers, and retailers are among the major categories of expenditures.

Note: Includes food eaten at home and away from home. Other costs include property taxes and insurance, accounting and professional services, promotion, bad debts, and many miscellaneous items.

(*Source:* Siamphoto/Fotolia.)

all of the major functions performed by the food industry, processing, wholesaling, transporting, and retailing.

Labor costs overshadow all other cost components of the marketing bill. The industry's next highest costs are for packaging, transportation, advertising, and energy. Recent trends such as high energy costs and the rising demand for more convenient packaging have increased all these expenses. Changes in the marketing bill may result from changes in price, product mix, product quantity, and the quality of marketing services. The bill includes both at-home and away-from-home markets.

Based on information from the National Farmers Union in 2016, the share of the food dollar received by farmers varies depending on the commodity. For example, the retail value of a pound of bacon is $4.39. But the hog farmer receives $0.78, a share of 17.8%. The retail value of a gallon of fat-free milk is $3.89. The dairy farmer receives $1.25, a share of 32.1%. Finally, the retail value of bread is $2.79, but the wheat farmer receives $0.11, a share of 3.9%.

Fiber Manufacturers

Cotton is the single most important textile fiber in the world, accounting for nearly 80% of natural fiber use and more than one-third of total fiber demand worldwide. Demand for cotton products by consumers has grown notably since the end of the 1990s, both in the United States and in foreign countries.

The cotton and textile industries are particularly important components of the U.S. food and fiber system. As a percentage of the gross domestic product, personal consumption expenditures on clothing and accessories are slightly more than 3%. The U.S. cotton industry accounts for more than $40 billion in products and services annually, generating nearly a half million jobs from the farm level to the textile mill level of the marketing chain.

The marketing bill for food is defined as the portion of food expenditures associated with the activities of firms beyond the farm gate. Labor expenses, packaging expenses, transportation costs, and advertising expenses of food processors, wholesalers, and retailers are among the major categories of expenditures.

Shippers and Handlers

The transportation of food and fiber products—when commodities move from U.S. farms and ranches to processors and manufacturers and then on to wholesalers and retailers—is an extremely important dimension of the U.S. food and fiber industry. Millions of tons of food products flow annually over 200,000 miles of railroad tracks, 3 million miles of intercity highways, and 26,000 miles of improved waterways.

The transportation of fresh fruits and vegetables is of particular importance because of the perishability of the product. Trucks are replacing railroads as the mode of shipping fresh fruits and vegetables. Piggybacks, or trucks containing fresh fruits and vegetables traveling on railroad flatbed cars, represent a small component of total fruit and vegetable shipments.

Importance of Export Markets

The volume of agricultural exports is important not only to the producers of the commodities marketed but also to the firms that move these commodities through marketing channels to foreign ports. These firms provide storage and transportation facilities and marketing services, which add value to commodities as they move overseas. The volume of agricultural exports also is important because it offsets, at least in part, imports of nonagricultural products and thus lowers the U.S. trade deficit. The United States historically provided approximately 70% of the world's coarse grain exports and 65% of the world's soybean exports.

World events influence domestic consumers and producers. Growth in wheat production by Argentina, Australia, Canada, and the European Community increases supplies on world markets, depresses wheat prices, and lowers revenue received by U.S. wheat producers. Domestic consumers of bread and other bakery products benefit from this situation eventually. Other world events, including World Trade Organization negotiations, a drive for food security in traditional food-importing nations, and trends in foreign exchange rates and world weather shocks also influence world trade flows and market shares and affect U.S. consumers and producers. We will discuss international linkages in subsequent chapters.

SUMMARY

The purpose of this chapter is to acquaint you with the structure and performance of the farm sector during the post–World War II period and its role in the U.S. food and fiber industry. The major points made in this chapter are summarized as follows:

1. The U.S. food and fiber industry consists of different groups of business entities called sectors, which are in one way or another associated with the supply of food and fiber products to consumers. In addition to the **farm sector**, this industry consists of firms that supply manufactured inputs to farms and ranches, firms that process raw food and fiber products, and firms that distribute food and fiber products to consumers.

2. Among the physical structural changes taking place in the farm sector during the post–World War II period is the trend toward fewer but larger farms. We have also seen a tremendous expansion in the use of manufactured inputs, such as machinery and chemicals, and a decline in labor use. Rising capital requirements in general during the period have increasingly represented a barrier to entry for would-be farmers.

3. Although the total quantity of inputs used in producing raw agricultural products has remained relatively stable during the post–World War II period, the total quantity of output has increased substantially. These results, taken together, imply an increase in productivity, or the ratio of output to inputs.

4. **Gross farm income** has increased, albeit somewhat erratically, during the post–World War II

period, and production expenses have behaved in similar fashion. The result is a highly variable level of profits, or **profitability**, from one year to the next.

5. The financial structure of the farm sector during the post–World War II period shows that financial assets represent a considerably smaller portion of total farm assets. Real estate assets are the major component of farm asset.

6. The U.S. food marketing sector is the network of processors, wholesalers, retailers, and restaurateurs that market food from farmers to consumers. Approximately 83% of the personal consumption expenditures on food went to pay for activities taking place beyond the farm gate.

7. Along the flow of products from farmers to processors and eventually on to consumers, middlemen play a vital role. Classifications of middlemen firms include merchant middlemen firms, agent middlemen firms, speculative middlemen firms, processing and manufacturing firms, and facilitative organizations.

8. In recent times, the number of **mergers** and acquisitions in food industries has increased sharply

relative to historical levels. Consequently, food industries have become more concentrated. **Concentration** is particularly high in industries marketing products such as breakfast cereals, beer, candy, and soft drinks.

9. The food processing industry and the wholesale and retail trade industries are characterized by a relatively small number of firms that account for a substantial portion of total industry sales. Although aggregate concentration has increased, the number of food marketing companies has remained relatively constant.

10. Farmers and ranchers get approximately 17% of each dollar spent on food. This share varies considerably by commodity. The remaining portion goes to food processors, wholesalers, and retailers. The major categories of expenditures include labor, packaging, transportation, and advertising.

11. The transportation of food and fiber products along the marketing chain is an extremely important component. The storing and exporting of nonperishable commodities also is an important dimension of marketing agricultural commodities.

KEY TERMS

Assets
Balance sheet
Capital
Concentration
Equity
Farm sector
Financial structure
Food and fiber industry

Gross domestic product
Gross farm income
Inflation-adjusted
Labor
Land
Marketing bill
Materials
Mergers

Net farm income
Nominal dollar values
Output index
Price index
Productivity
Profitability
Real dollar values

TESTING YOUR ECONOMIC QUOTIENT

Circle the correct answer.

1. The percentage of disposable income currently spent on food in the United States is
   a. less than 5.
   b. between 5 and 10.
   c. between 10 and 15.
   d. between 15 and 20.
2. Bill Toney, a hobby farmer from Virginia, has a net worth of $16 million. He has assets of $30 million and liabilities of $ _____ million.
3. The portion of food expenditures associated with the activities of firms beyond the farm gate is known as the _____.

Circle the correct answer.

4. Suppose that the index of prices received by farmers for 2015 was 0.97 and the base year of this index was 2000. Then,
   a. relative to 2015, farm prices were 97% higher in 2000.
   b. relative to 2000, farm prices were 3% lower in 2015.
   c. relative to 2015, farm prices were 3% lower in 2000.
   d. relative to 2000, farm prices were 97% higher in 2015.

5. Since World War II, there have been increases in productivity in the agricultural sector. This trend is due in part to
   a. the substitution of capital and materials for labor.
   b. increases in farm size.
   c. the decline in the numbers of farms.
   d. all of the above.
6. List the five sectors of the food and fiber industry:
   a.
   b.
   c.
   d.
   e.
7. The major component of the marketing bill for food is _____.
8. On average, U.S. farmers get approximately _____ cents of the dollar spent for food.
9. If your nominal income for 2012 was $75,000 and the consumer price index for 2012 was 2.5, what was your real income for 2012? _____
10. Develop the output indices of the Nouveau Cattle Slaughtering Plant (base year 2014).

Year	Pounds of Cattle Slaughtered	Output Index
2013	180,000	_____
2014	250,000	_____
2015	200,000	_____

11. Which of the following is true? Circle all that apply.
   a. Currently, the food and fiber sector accounts for about 12% to 15% of gross domestic product (GDP).
   b. Off-farm income is important to agricultural producers today.
   c. Twenty percent of farmers produce 80% of the agricultural output in the food and fiber industry.
   d. The food and fiber industry is responsible for one out of every ten jobs.
12. Today, the number of farms in the United States is roughly in the order of _____ million.
13. The following information pertains to a farm in the Rio Grande Valley.

Cash receipts from farm marketings	$500,000
Receipts of government payments	$100,000
Other income from farm sources	$50,000
Production expenses	$300,000
Value of real estate assets	$10,000,000
Value of non–real estate assets	$1,000,000
Financial assets	$5,000,000
Liabilities	$8,000,000

a. Net farm income for this operation is $ _____.
b. The equity for this operation is $ _____.
14. The output produced in bushels and the price of this output ($/bushel) for a farmer over the last 3 years are as follows:

Year	Output	Price ($)	Consumer Price Index	Production Expenses ($)
2010	70,000	3.20	1.10	120,000
2011	75,000	2.90	1.20	140,000
2012	80,000	2.80	1.15	135,000

Let the base year be 2011:
   a. For the year 2010, what is the output index? Interpret this measure.
   b. For the year 2012, what is the price index? Interpret this measure.
   c. Assume that the only source of income for this farmer is from the production of this output. What is the best year in terms of real income for this farmer?
15. Suppose your nominal income in 1995 was $24,000. Suppose, too, that the Consumer Price Index for 2012 was 2.5 and that the base year for this index was 1995. How much nominal income would you need in 2012 in order to match the spending power of your $24,000 in 1995?
16. Verify the calculations associated with the output and price indices in Table 2-1.
17. Verify the calculations associated with real total food-away-from-home expenditures in Table 2-2.
18. The term *real* as opposed to *nominal* means that economists are making adjustments for _____.
19. Assume that for John Paxton, a soybean producer from Iowa, the only source of farm income is from the production of soybeans. Paxton produced 100,000 bushels of soybeans in 2014, receiving $8 per bushel. Assuming this producer had production expenses of $300,000, and assuming the CPI for 2014 was 2.00, his real farm income for 2014 was
   a. $250,000.
   b. $500,000.
   c. $800,000.
   d. can't tell; insufficient information.

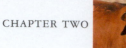

Use the table below to answer questions 20 and 21 dealing with the output and price of corn for 2013 and 2014.

Year	Output (in Bushels)	Price per Bushel
2013	10,000	$4.00
2014	12,000	$5.00

20. Relative to 2013, the price of corn for 2014 was

   a. higher by 20%.

   b. lower by 25%.

   c. lower by 20%.

   d. higher by 25%.

21. Which of the following statements is *true*? (Let 2013 be the base year.)

   a. The output index for 2013 is less than 1.

   b. The output index for 2014 is 1.25.

   c. The output index for 2014 is 1.20.

   d. The output index for 2014 is 0.83.

REFERENCES

Economic Report of the President, Washington, D.C., published annually, U.S. Government Printing Office.

U.S. Department of Agriculture: *Economic indicators of the farm sector, income and balance sheet statistics* (ERS-USDA Statistical Bulletin Series), Washington, D.C., published annually, U.S. Government Printing Office. www.ers.usda.gov. internet site.

Theory of Consumer Behavior

The biological process of photosynthesis, in which the addition of light to a plant's environment results in plant growth, can be thought of in a stimulus–response context. The stimulus is the addition of light, and the response is the plant growth. This process can be studied in a controlled environment using sophisticated measuring devices.

Economic behavior also can be thought of in a stimulus–response context. For example, a fall in the price of ice cream acts as a stimulus, causing consumers to purchase more ice cream. These purchases can be measured and recorded. In most respects, however, the similarities end here. The complex process of photosynthesis can be examined and studied directly, but most economic behavior processes cannot be. In fact, this example illustrates the distinction between the natural sciences (e.g., biology, chemistry, and physics) and the social sciences (e.g., economics). Most economic behavior processes cannot be studied in a controlled environment.

We can examine the technical relationships of converting inputs to outputs in a production process, but we cannot observe the process of connecting the economic stimulus to an economic decision. Why does Robbin purchase more ice cream than Willis when both face the same prices and have the same income? The most prominent economic theories of consumer behavior assume that consumers are rational and seek to maximize their satisfaction while staying within their budget. In this chapter, we discuss consumer theory and how it can be used to understand the behavior of consumers.

UTILITY THEORY

Consumers typically face a broad set of choices when allocating their income among food and nonfood goods and services. Historically, considerable attention has been given to the development of a theoretical framework that will help us understand the choices consumers make. In the following discussion, we will assume that consumers are rational individuals who maximize their satisfaction or utility.

Consumers are offered a broad array of items in U.S. supermarkets. How do consumers decide which products to purchase?   Credit: Tyler Olson/Fotolia.

Total Utility

A consumer purchases a good or service because of the satisfaction he or she expects to receive. Early researchers of consumer behavior argued that utility was cardinally measurable.[1] They also argued that the utility derived from a given commodity is independent of the utility derived from other commodities. For example, the latter belief suggests that the consumer can determine the utility of taco consumption independently from hamburger consumption. Total utility, accordingly, then would be equal to the total utility derived from each of the individual commodities. The psychological units of satisfaction derived from consumption are generally referred to as **utils**.

$$\text{total utility} = (\text{quantity of hamburgers} \times \text{quantity of pizza}) \qquad (3.1)$$

> We assume that consumers are rational and maximize their satisfaction or utility. A utility function is an algebraic expression that allows us to rank consumption bundles.

A **utility function** is an algebraic expression that allows us to rank a consumption bundle by the total utility or satisfaction it provides. The Rolling Stones released in June 1965 their single "I Can't Get No Satisfaction." This song dealt in part with the lack of utility or satisfaction derived from commercialization. **Consumption bundles** refer to particular combinations of goods being considered. The utility function describes the **total utility** derived from consuming a particular bundle. Consequently, utility or satisfaction is a function of consuming individual commodities.

To clarify the meaning and use of the utility function, consider a consumer who has the following utility function (although it is highly unlikely the consumer is aware of this mathematical representation of his or her utility function):

---

[1] The term **cardinally measurable** centers attention on the attempt to quantify the amount of satisfaction obtained from consumption. On the other hand, the term **ordinally measurable** implies only a ranking of the amount of satisfaction in some sort of ordered fashion.

Table 3-1

EXAMPLE OF TOTAL UTILITY DERIVED FROM THE CONSUMPTION OF HAMBURGERS AND PIZZA

Bundle	Quantity of Hamburgers	Pizza	Total Utility
A	3	10	30
B	4	7	28
C	2	15	30

If consumption bundle *A* consists of 3 hamburgers and 10 slices of pizza per week, the consumer with a utility function such as Equation 3.1 would derive a total utility of 30 from the consumption of this bundle (i.e., 3×10). This bundle, two other bundles of consumer goods, and the subsequent total utility they provide are summarized in Table 3-1.

If we wanted to know whether bundle *B*, which consists of 4 hamburgers and 7 pizza slices per week, is preferred, not preferred, or indifferent to bundle *A*, we know from Equation 3.1 that the utility this consumer derives from consuming bundle *B* would be 28 (i.e., 4×7). Therefore, this consumer would prefer bundle *A* to bundle *B* because the utility provided by bundle *A* (30) is greater than the utility provided by bundle *B* (28). Suppose that bundle *C* consists of 2 hamburgers and 15 slices of pizza. The utility derived from consuming this bundle also would equal 30. Therefore, this consumer would be indifferent between bundles *A* and *C*.

The notion of a utility function may seem mysterious. In fact, it is hard to imagine a consumer thinking in terms of a specific utility function when purchasing goods and services, as suggested by Equation 3.1. Yet, the concept of satisfaction that the utility function expresses is the foundation of consumer economic analysis.

Marginal Utility

If utility is measurable, it is appropriate to question how total utility changes as a greater (or lesser) amount of a particular good is consumed. The change in total utility, associated with a specific change in the consumption of a commodity, is referred to as **marginal utility (MU)**. As stipulated in Chapter 1, in economics, the term *marginal* is synonymous with the word *change*. To illustrate this, the marginal utility of hamburgers is shown in Equation 3.2, where Δ indicates the change in a value.

$$\text{MU}_{\text{hamburgers}} = \frac{\Delta \text{utility}}{\Delta \text{hamburgers}} \tag{3.2}$$

This measure constitutes the change in utility associated with a change in the consumption of hamburgers. This value will always be greater than zero only if we assume that the consumer's appetite never becomes totally satiated. This value will fall as hamburger consumption increases; similarly, this value will rise as hamburger consumption decreases.

To illustrate the notion of marginal utility, assume that the data in Table 3-2 reflect the utility of Sue Shopper regarding hamburger consumption. The first column in this table indicates the quantity of hamburgers Sue purchases and consumes per week. The second column represents her total utility associated with each specific consumption level. The third column presents the corresponding levels of marginal utility. Note that each successive increment of hamburgers increases utility by a smaller amount. When consumption of hamburgers increases from 2 to 3, utility

The marginal utility of a good refers to the change in utility or satisfaction due to the change in consumption of that good. The law of diminishing marginal utility is one of the few laws in economics. This law stipulates that as the consumption of a good increases, the associated marginal utility declines.

Table 3-2
CALCULATION OF MARGINAL UTILITY FOR SUE SHOPPER

Quantity of Hamburgers Consumed per Week	Total Utility	Marginal Utility
1	20	10
2	30	9
3	39	8
4	47	7
5	54	6
6	60	5
7	65	4
8	69	3
9	72	2
10	74	0
11	74	−4
12	70	

increases by 9 utils. When consumption of hamburgers increases from 8 to 9, utility increases by only 3 utils. When marginal utility is zero, total utility is maximized. In addition, marginal utility can be negative at higher levels of hamburger consumption. Utility actually decreases by 4 utils as hamburger consumption increases from 11 to 12. Figure 3-1 shows the shape of the total and marginal utility curves associated with the data presented in Table 3-2.

Law of Diminishing Marginal Utility

Is it clear why Sue's marginal utility declines when her consumption increases? If you consume one hamburger and then another, the second hamburger gives you less satisfaction than the first. Because there is so much truth to this notion, it has been given law-like status. The **law of diminishing marginal utility** suggests that as consumption per unit of time increases, marginal utility decreases. Think about the consumption of your favorite beverage. The first beverage generates a certain level of satisfaction in terms of quenching your thirst. But additional beverages provide less satisfaction than previously. In fact, after a certain point, additional consumption of your favorite beverage likely will result in dissatisfaction! Often the law of diminishing marginal utility can be witnessed during weekends of tailgating at college or professional football games.

Does it seem logical to assume that the marginal utilities provided by different commodities are independent? Put differently, would the utility you derive from hamburger consumption depend on the amount of soft drinks, french fries, and tacos you consume? Because most people would answer yes to this question, we must consider the consumer's consumption of all other goods and services before we can fully understand what influences consumer behavior.

INDIFFERENCE CURVES

Cardinal measurement for utility is unreasonable and unnecessary. Cardinality implies that society can add utils like it can add distances. Instead, utility can be viewed as being ordinally measurable—that is, as a personal index of satisfaction in which the magnitude is used only to rank consumption bundles. In this light, we do not need to actually record or measure the number of utils of satisfaction associated with a particular bundle of goods.

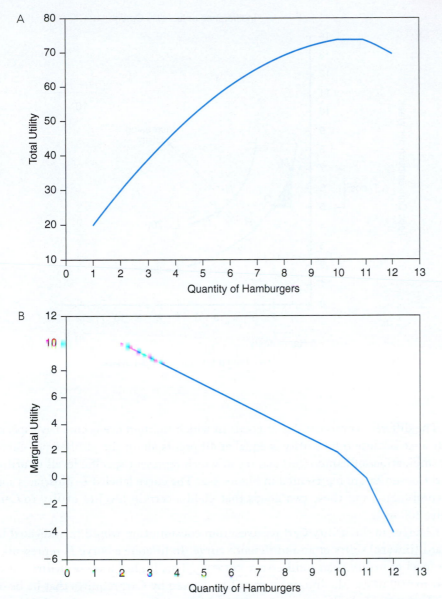

Figure 3-1

Total utility continues to increase as the number of hamburgers consumed increases, at least up to 11 hamburgers. At this point, total utility is maximized. Beyond 11 hamburgers, total utility decreases (A). Marginal utility declines as Sue increases her consumption of hamburgers (B).

Modern consumption theory dismisses the notion that utility is cardinally measurable and instead measures utility in ordinal terms. All we really need to know is that bundle N is preferred to bundle M; we do not need to know by how much.

Concept of Isoutility

The basic building block of modern consumption theory is the notion of an **isoutility curve**, which accounts for substitution in consumption for two products. The term *iso* is of Greek origin and means "equal."[2] An isoutility curve often is referred to as an **indifference curve**. A consumer is indifferent to consumption bundles that yield an equal level of satisfaction or utility.

The indifference curve or isoutility curve represents the combination of consumption bundles that provide a consumer a given level of satisfaction. The slope of the indifference curve is the marginal rate of substitution.

---

[2] For example, consider an isosceles triangle, a triangle with two equal sides.

Figure 3-2

An indifference curve represents all combinations of two goods that yield an equal level of satisfaction or utility. In other words, the total utility derived from consumption is equal at all points along an indifference curve. The utility associated with consuming 7 tacos and 1 hamburger per week (point M) by Carl Consumer is equal to the utility associated with consuming 5 tacos and 2 hamburgers (point Q). Indifference curve I₇ represents a higher level of utility than curve I₂. Why? Bundle P on curve I₇ corresponds to approximately the same number of tacos but more hamburgers than bundle Q.

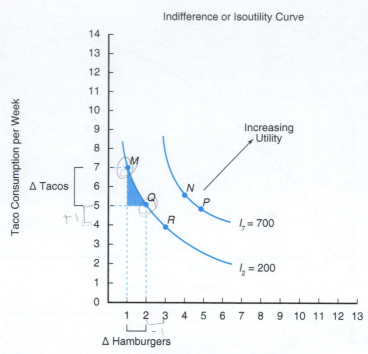

The different combinations of goods in which indifference occurs have special significance because total utility is equal at all points along the indifference curve. The combinations of hamburgers and tacos, which represent specific levels of utility to Carl Consumer, are represented in Figure 3-2. The curve labeled I_2 illustrates specific combinations of these two goods that yield a certain level of utility to Carl, namely, 200 utils.

Changes in the utility Carl receives from consumption would be indicated by outward (inward) shifts of an indifference curve. Indifference curve I_7 represents a *higher* level of utility than indifference curve I_2. This higher level of utility corresponds to 700 units. Maximization of utility derived by Carl requires that he be on the highest possible indifference curve. Carl prefers indifference curve I_7 to I_2 because the amount of utility expressed by I_7 exceeds the amount of utility expressed by I_2. But recall we do not need to know how much utility in terms of utils.

Marginal Rate of Substitution

To maintain a constant level of utility, one must change consumption of one commodity to obtain additional consumption of another; that is, a consumer may substitute one commodity for another to maintain a constant level of utility. The rate at which the consumer is willing to substitute one good for another is called the **marginal rate of substitution**. The marginal rate of substitution of hamburgers for tacos, for example, represents the number of tacos Carl is willing to give up for an additional hamburger to maintain the same level of satisfaction, or, in mathematical terms, as

$$\text{marginal rate of substitution of hamburgers for tacos} = \frac{\Delta \text{tacos}}{\Delta \text{hamburgers}} \quad (3.3)$$

The marginal rate of substitution associated with moving from point M to Q in Figure 3-2 would be −2 (i.e., −2 1). Carl is willing to give up 2 tacos for 1 additional hamburger. If he instead moved from point Q to R, we see that the marginal rate of substitution would be -1 (i.e., $-1 \div 1$).

The marginal rate of substitution represents the slope for any specific segment of an indifference curve for two goods. Importantly, this slope is negative, meaning that the marginal rate of substitution is negative. Hence, this concept indeed represents a trade-off of one commodity for another. We also can equate the marginal rate of substitution of hamburgers for tacos in Equation 3.3 with the ratio of their marginal utilities, or

$$\frac{\Delta \text{tacos}}{\Delta \text{hamburgers}} = \frac{\text{MU}_{\text{hamburgers}}}{\text{MU}_{\text{tacos}}} \tag{3.4}$$

Thus, for Carl Consumer, the cutback in taco consumption times the marginal utility of tacos is identical to the increase in hamburger consumption times the marginal utility of hamburgers.

Why does the marginal rate of substitution fall as we move down the indifference curve? In Figure 3-2, the marginal rate of substitution fell from -2 to -1 when Carl moved down the indifference curve. When Carl consumed 7 tacos (point M), he was willing to give up 2 tacos to eat 1 more hamburger (a movement from point M to Q). When Carl consumed 5 tacos (point Q), he was willing to give up only 1 taco to receive 1 more hamburger (a movement from point Q to R). Perhaps the most intuitive explanation we can offer at this point relies on the law of diminishing marginal utility. As taco consumption falls, its marginal utility rises. As hamburger consumption increases, its marginal utility falls. Thus, the marginal rate of substitution *falls* as one moves *down* an indifference curve (e.g., increasing hamburger consumption and reducing taco consumption).

THE BUDGET CONSTRAINT

We often hear the phrase "I wanted to purchase it, but I just could not afford it." This phrase portrays that we are all faced with what economists call a **budget constraint**; that is, purchases by a consumer cannot exceed his or her income. If consumption decisions are made as a household, income should include all forms of family income. It should also *exclude* tax obligations to reflect the **disposable income** of the household.[3] We must discuss the budget constraint using a unit of time, such as the maximum expenditures per day, per week, and so on.

If all other factors remain constant, when the disposable income of a consumer increases, the percentage of income spent for food decreases. For a poor consumer, a greater percentage of income is used to purchase food. This observation is commonly referred to as **Engel's Law**, previously encountered in Chapter 2. In the United States, we have a relatively high per capita national income, and we spend a relatively small percentage of our total consumption expenditures, roughly 10%, on food. In less developed countries, a greater share of disposable personal income is spent on food relative to the United States. Figure 3-3 illustrates Engel's Law, using information from the household portion of the 1987–1988 Nationwide Food Consumption Survey. Engel's Law states that the greater the weekly income, the lower the proportion of income spent on food. Each point in this graph corresponds to a particular household, a total of approximately 4,000.

The total expenditures made by a consumer on a number of items can be determined by multiplying the total quantity of each good or service purchased by its respective price and then totaling the value of all purchases. For example, suppose

The budget constraint defines the feasible set of consumption choices facing a consumer. This constraint depends upon the prices of goods in question and the income available to a consumer.

---

[3] *Disposable income* is defined as income after taxes.

Figure 3-3
Scatter plot of weekly income and total food budget share.

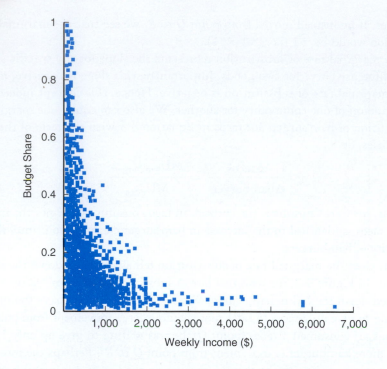

Carl Consumer had a specific amount of money to spend on food eaten away from home per week. If he limited this consumption to purchases of hamburgers and tacos, Carl's total expenditure would be equal to the price of hamburgers times the quantity of hamburgers he consumed *plus* the price of tacos times the quantity of tacos he consumed, or

$$\left\{ \begin{matrix} \text{price of} \\ \text{hamburgers} \end{matrix} \times \begin{matrix} \text{quantity of} \\ \text{hamburgers} \end{matrix} \right\} + \left\{ \begin{matrix} \text{price of} \\ \text{tacos} \end{matrix} \times \begin{matrix} \text{quantity of} \\ \text{tacos} \end{matrix} \right\}$$
$$= \begin{matrix} \text{income spent on} \\ \text{food eaten away from home} \end{matrix} \tag{3.5}$$

This budget constraint limits Carl's consumption of hamburgers and tacos to no more than the total income allocated to their consumption. Consequently, the budget constraint is dependent on the prices of the respective items as well as the amount of income to be spent.

When the budget constraint is graphically depicted, it is referred to as the *budget line.* The slope of Carl's budget line is equal to the negative of the price ratio, or

$$\text{slope of budget line} = -\frac{\text{price of hamburgers}}{\text{price of tacos}} \tag{3.6}$$

which suggests that the budget constraint will become *steeper* (flatter) as the price of hamburgers rises (falls) relative to the price of tacos (see Figure 3-4D). Similarly, the budget constraint will become steeper (flatter) as the price of tacos falls (rises) (see Figure 3-4C).[4]

To illustrate, suppose that Carl has $5 a week to divide between the consumption of tacos and hamburgers. Tacos cost $0.50 each and hamburgers cost $1.25 each. Some of the combinations of tacos and hamburgers Carl can afford with a weekly

---

[4] The slope of the budget line can be derived mathematically by rearranging Equation 3.5 to read

$$\text{quantity of tacos} = \frac{\text{income}}{\text{price of tacos}} - \left[\frac{\text{price of hamburgers} \times \text{quantity of hamburgers}}{\text{price of tacos}} \right]$$

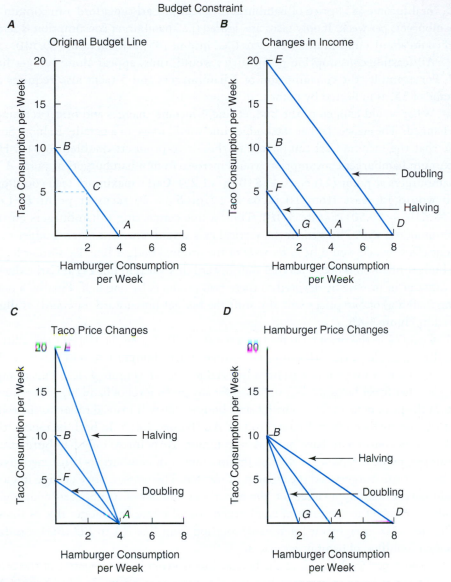

Figure 3-4

Let the line connecting points A and B represent the original budget constraint or budget line. This line suggests that Carl Consumer could spend his entire weekly budget of $5 to buy 4 hamburgers costing $1.25 each, 10 tacos costing $0.50 each, or some combination of these two food items that appears along line AB.

budget of $5 appear in Table 3-3. This budget constraint is illustrated graphically in Figure 3-4A. We know from Equation 3.5 that if just hamburgers are desired, taco consumption would be zero, and the quantity of hamburgers consumed would be 4

Table 3-3
EXAMPLE OF BUDGET CONSTRAINT

Tacos ($0.50 Each)	Hamburgers ($1.25 Each)	Expenditure ($)
10	0	5
5	2	5
0	4	5

(i.e., total income [$5]/price of hamburgers [$1.25]). Carl can afford a maximum of 4 hamburgers per week. If only tacos are desired (i.e., hamburger consumption is zero), Carl could afford a maximum of 10 tacos (i.e., income [$5]/price of tacos [$0.50]).

All feasible consumption possibilities would, thus, appear along budget line *AB*. For example, the consumption of 2 hamburgers and 5 tacos also requires an income of $5, as indicated by point *C* in Figure 3-4A.

What would happen to the budget line if income changes and prices remained unchanged? The answer is that the budget line would move in a parallel fashion. Suppose that the income Carl can devote to these two products doubled to $10. His maximum hamburger consumption would increase from 4 hamburgers at point *A* to 8 hamburgers at point *D* (i.e., 8 = $10 ÷ $1.25). Carl's maximum taco consumption would increase from 10 tacos at point *B* to 20 tacos at point *E* (i.e., 20 = $10 ÷ $0.50) (Figure 3-4B). Thus, a line connecting 8 hamburgers on the horizontal axis with 20 tacos on the vertical axis would represent a new budget constraint (*DE*), which would lie *to the right* of the original budget line. By similar logic, the budget line would take a parallel shift inward (leftward) to line *FG* if Carl reduced the amount of income he devoted to these two products by one-half. Finally, a doubling (halving) of *both* prices will also shift the budget line inward (outward) as illustrated in Figure 3-4B.

Changes in the price ratio for two products will change the slope of the budget line. For example, if the price of tacos doubles, Carl's budget line will rotate to the left (a counterclockwise rotation) from line *AB* to line *AF* (Figure 3-4C). This change suggests that fewer tacos can be purchased for any given level of hamburger consumption. If the price of tacos falls in half, Carl's budget line will instead rotate to the right (a clockwise rotation) from line *AB* to line *AE* (Figure 3-4C). In both instances, the budget lines continue to have point *A* in common. At point *A*, only hamburgers are consumed; point A represents the maximum number of hamburgers consumed given the price of hamburgers and the amount of income. Therefore, a price change in tacos would have absolutely no effect on the maximum number of hamburgers consumed. Similarly, changes in the price of hamburgers would rotate the budget line as shown in Figure 3-4D. A rightward (leftward) rotation from line *BA* to *BD* (*BG*) signifies halving (doubling) of hamburger prices.

To summarize, the slope of the budget line is given by the negative of the price ratio. This ratio also is constant because the respective prices do not change. An increase (decrease) in income will shift the budget line outward to the right (inward to the left) from the origin. This shift will be parallel in nature as long as the price ratio does not change. A change in the ratio of the two product prices, however, will alter the slope of the budget line.

Importantly, the slope of the budget line is not only equal to the negative of the price ratio. This slope is also equal to (Δtacos/Δhamburgers) = ($MU_{hamburgers}$/MU_{tacos}) as stipulated by Equation 3.4. This relationship will play a key role in Chapter 4 in regard to the concept of consumer equilibrium.

SUMMARY

The major points made in the chapter are summarized as follows:

1. The budget constraint represents the amount of income the consumer has to commit to consumption in the current period. A proportional change in all prices and income has *no effect* on the budget constraint. For this reason, economists argue that only relative price changes matter. When presented graphically, the budget constraint is frequently referred to as the budget line. The slope of the budget line, which tells us the rate of exchange between two goods as their prices change, is given

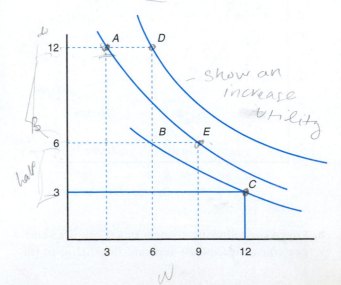

by the negative of the price ratio. A change in relative prices will change the slope of the budget line. Finally, an increase (decrease) in income will shift the budget line to the right (left). Changes in income do not affect the slope of the budget line.

2. We assume that consumers are rational and maximize their satisfaction, or utility. Thus, consumers are assumed to be able to rank all their choices.

3. Early researchers of consumer behavior argued that utility could be measured. The term *utils* was used as a unit of measure. A hamburger might yield 10 utils, a soda 4 utils, and so on. Marginal utility describes the change in utility or utils as more of a good is consumed and is thought to diminish as consumption increases. This phenomenon is known as the law of diminishing marginal utility.

4. Today no one really believes that utility can be measured in utils. Instead, utility is thought of in the context of a personal index of satisfaction. The magnitude of this index (or function) serves to order the consumption bundles or combinations of goods the consumer faces.

5. All consumption points that provide the same utility form an isoutility or indifference curve. Increases (decreases) in utility are indicated by a shift in an indifference curve to the right (left). The negative of the slope of this curve is known as the marginal rate of substitution (MRS). This rate indicates the willingness of the consumer to substitute one good for another. The decline in the willingness of the consumer to substitute one good for another as one moves down an indifference curve indicates the existence of the principle or law of diminishing marginal utility.

KEY TERMS

Budget constraint
Cardinal measure
Consumption bundles
Disposable income
Engel's Law

Indifference curve
Isoutility curve
Law of diminishing marginal
   utility
Marginal rate of substitution

Marginal utility
Ordinal measure
Total utility
Utility function
Utils

TESTING YOUR ECONOMIC QUOTIENT

1. Based on the following table, graph the total utility curve of Robbin Denison for buffalo wings. Then, calculate the marginal utility between each point and plot the corresponding graph. Why is the slope of the MU curve negative?

ROBBIN'S UTILITY FUNCTION	
Number of Wings	Total Utility
A 1	30
B 2	58
C 3	84
D 6	150
E 12	222
F 24	306

2. Robbin likes a beverage (Coca-Cola, of course) with her wings. There is only one place to buy this combination, Wings 'n' Suds. Using the following graph, let the number of Coca-Colas be on the vertical axis and the number of wings be on the horizontal axis. Label the graph.

a. Robbin has her choice of getting 12 bottles of Coca-Cola and 3 wings or 3 bottles of Coca-Cola and 12 wings, free of charge. Which bundle will she choose? Why?

b. Which would Robbin choose if she could have either 12 bottles of Coca-Cola and 6 wings or 6 bottles of Coca-Cola and 9 wings?

c. Calculate the MRS between points A and E.

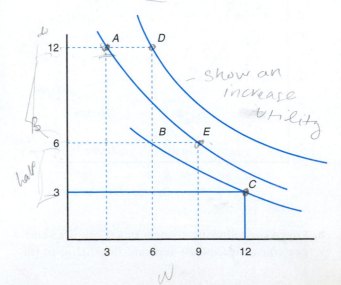

3. After careful scrutiny, Robbin budgets $12/week for Wings 'n' Suds. Consider the following:

   a. Graph and label the axes to show how much of each good Robbin is *able* to buy, if the price of a buffalo wing is $0.50, while the price of Coca-Cola is $1.50.

   b. In order to attract more customers to Wings 'n' Suds, management decides to lower the price of Coca-Cola to a buck. Show what happens to Robbin's budget line compared to (a).

   c. Instead of (b), assume there is a sudden shortage of buffalo wings available. Now, Wings 'n' Suds has to raise the price of a wing to $3. Show how this event changes Robbin's budget line relative to (a).

   d. Robbin, after winning $1,000 in the Texas lottery, decides that she can now spend $40/week at Wings 'n' Suds. Show how this event changes Robbin's budget line relative to (a).

   e. What combination of wings and Coca-Colas *should* Robbin buy in (a), (b), (c), and (d)?

4. Given the following set of indifference curves, calculate the marginal rate of substitution between the following:

   a. Points *A* and *B*.

   b. Interpret this measure.

   c. Which combination yields the highest level of satisfaction? Circle the correct answer(s).

      i. 7 tacos, 1 hamburger

      ii. 2 tacos, 5 hamburgers

      iii. 5 tacos, 7 hamburgers

      iv. 7 tacos, 5 hamburgers

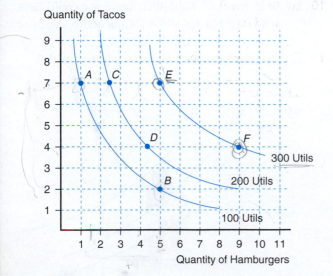

5. Given the following changes in a consumer's budget constraint, please indicate in writing to the

right of each graph what caused the budget constraints to change.

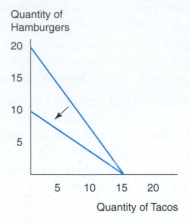

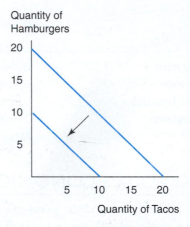

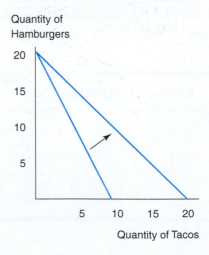

6. Assume you have interest in only two goods: food and environmental quality. That is, these goods are the only ones that provide utility to you. Consider the following graph:

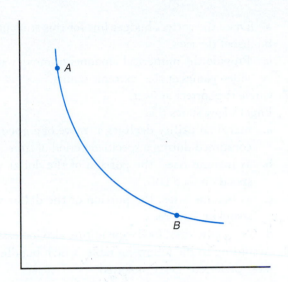

Let point *A* correspond to 80 units of food and 20 units of environmental quality.

Let point *B* correspond to 50 units of food and 30 units of environmental quality.

a. Label the axes. What is the technical name of the curve mentioned earlier?

b. Calculate how many units of food you are willing to give up in order to obtain one more unit of environmental quality so as to maintain the same level of satisfaction.

7. Suppose Glenn Gibbs (a native of Manchester, England) has an income of $30. He derives satisfaction from the consumption of tea and biscuits. The price of tea is $3.00 per cup and the price of biscuits is $0.50/unit.

a. Graphically construct the budget line for this situation. Label your axes carefully.

b. Now, suppose the price of tea increases to $5.00 per cup and Glenn's income remains at $30. Assuming the price of biscuits remains at $0.50/unit, redraw the budget line to reflect this situation.

8. Given the following data

Quantity of Sodas	Total Utility	Marginal Utility
1	20	—
2	25	?(a)
6	37	?(b)

a. MU between 1 and 2 is _____.

b. MU between 2 and 6 is _____.

9. What caused the budget constraint to change? Circle the correct answer.

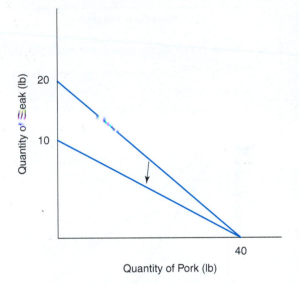

Quantity of Pork (lb)

a. The price of steak rose.

b. The price of steak fell.

c. The expenditure on steak and pork rose.

d. The price of pork fell.

10. In question 9, if expenditure on steak and pork equals $100, what is the price per pound of pork? _____

11. A rational consumer maximizes his or her satisfaction or _____.

12. The marginal utility of a good (e.g., ice cream) declines with increases in the consumption of that good. This phenomenon is referred to as the _____.

13. a. For a representative consumer, 4 wings and 3 bottles of Dr. Pepper generate the same utility as 6 wings and 2 bottles of Dr. Pepper. How many wings must the consumer give up in order to get one more bottle of Dr. Pepper? _____

b. What is the technical name associated with the trade-off in (a)? _____

14. What is the name associated with the following graph? _____

15. a. What is the name associated with the following diagram? _____

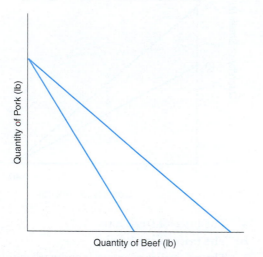

b. What happened to shift the curve above? _____

16. Suppose that a college student can spend $50 on entertainment. This student derives satisfaction only from watching movies and playing video games. The price of a movie is $5 and the price of a video game is $2.

a. Draw the correct budget line for this student.
b. Label the axes.
c. Provide the numerical amounts of movies and video games at the "extreme points."

17. Circle the correct answer.
Engel's Law states that
a. marginal utility declines as more of a good is consumed during a specified period of time.
b. as income rises, the portion of the dollar we spend on food falls.
c. as income rises, the portion of the dollar we spend on food rises.
d. as income rises, food expenditure also increases.

18. According to the following table, which bundle is most preferred? Least preferred?

Bundle	Number of Wings	Bottles of Coca-Cola	Total Utility
A	3	3	18
B	6	4	48
C	9	2	36
D	12	1	24

19. When total utility is at a maximum, marginal utility is _____.

20. We do not need to actually measure the level of satisfaction derived by a consumer from the consumption of goods. We only need a ranking among the alternative consumption bundles. Thus, utility is a(n) _____ concept.

21. The mathematical representation of the satisfaction a consumer derives from a bundle of goods is called _____ the function.

22. The graph of alternative consumption bundles that provide a consumer a given level of satisfaction is called a(n) _____ curve.

23. True or False. Circle the correct answer.
a. Total utility may never be negative.
b. Marginal utility may never be negative.
c. The marginal rate of substitution is always negative.
d. The law of diminishing marginal utility states that total satisfaction declines as more of a good is consumed.

24. Suppose a college student can spend $30 on Subway sandwiches and Chick-Fil-A sandwiches. The price of a Subway sandwich is $5 and the price of a Chick-Fil-A sandwich is $3. Which of the following diagrams describes the correct budget line for this student?

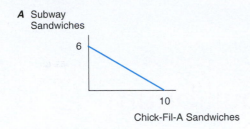

A Subway Sandwiches

6

10

Chick-Fil-A Sandwiches

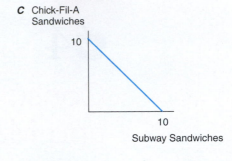

C Chick-Fil-A Sandwiches

10

10

Subway Sandwiches

B Subway Sandwiches

10

6

Chick-Fil-A Sandwiches

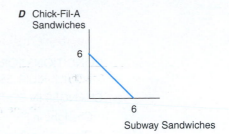

D Chick-Fil-A Sandwiches

6

6

Subway Sandwiches

✎4

Consumer Equilibrium and Market Demand

In this chapter, we shall merge the concepts of budget constraint and marginal utility presented in Chapter 3 to determine consumer equilibrium and market demand. Our goal is to understand how consumers will react to changes in prices and income when deciding about consuming specific commodities. The example of tacos and hamburgers employed in Chapter 3 will continue to be used when we examine the concept of consumer equilibrium.

This chapter begins with a discussion of the equilibrium conditions for any individual consumer. The notion of equilibrium corresponds to the point at which the consumer maximizes his or her satisfaction subject to the budget constraint. Changes in this equilibrium when the prices of various goods change or as the consumer's income changes also are discussed. The chapter closes with a focus on the law of demand for consumers in general, including the impact that tastes and preferences have upon market demand, and with an approach to measuring changes in the economic well-being of consumers when market conditions change.

CONDITIONS FOR CONSUMER EQUILIBRIUM

If there were no budget constraint, the consumer would move toward consuming all goods and services at a point at which marginal utility of each good or service is zero.[1] With the presence of a budget constraint, consumer decisions can be thought of as the process of choosing products so that the total utility derived from consumption is maximized, subject to the amount of the budget constraint. The key assumption is that consumers try to *maximize* their satisfaction or utility.

---

[1] Going beyond this condition implies a reduction in total utility because marginal utilities are subsequently negative.

Preferences and affordability are the economic forces behind consumer purchases.
Credit: naka/Fotolia.

Let us initially characterize the conditions for maximization of utility graphically. Consider the budget constraint and the set of indifference curves for Carl Consumer in Figure 4-1. If his utility is to be maximized subject to the budget constraint, it necessarily follows that we find the point on the budget line that yields the *highest* utility.

This situation occurs in Figure 4-1 at point A. Why? Point C on the I_1 curve would not exhaust the income allocated for consumption of these two goods. Carl can increase his utility to 200 by moving from point C on the I_1 indifference curve to point A on the I_2 indifference curve. Although point B on the I_4 indifference curve represents higher utility than point A, it requires more income than Carl has available. Therefore,

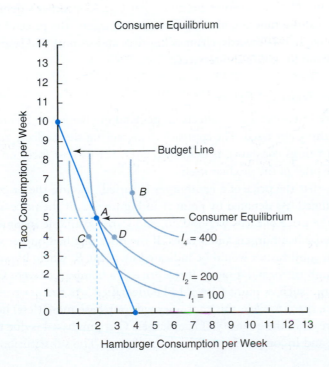

Figure 4-1

The rational consumer is said to be in equilibrium with regard to the consumption of two goods when the slope of the budget line (the budget constraint) is equal to the slope of the indifference curve. In the case of Carl Consumer, this equality occurs at point A. The indifference curve I_2 represents the highest utility curve attainable, given the nature of the budget constraint. At point A, where the budget line is tangent to the indifference curve I_2, Carl would purchase 5 tacos and 2 hamburgers. The total cost of these items completely exhausts his $5 weekly fast-food budget.

Consumer equilibrium refers to the condition wherein the slope of the budget line and the slope of the indifference curve are equal. Also, the marginal utilities per dollar spent on the goods in question must be equal. At the consumer equilibrium point, we realize the quantity of goods or services that maximizes total utility subject to the budget constraint.

point *A*, which consists of 5 tacos and 2 hamburgers, represents Carl's **consumer equilibrium**, conditional on the price of hamburgers, the price of tacos, and the income or outlay Carl wishes to spend. There would be economic incentives for him to move to point *A* if he were located anywhere else on this graph. Only at point *A* is the slope of the budget line equal to the slope of the indifference curve. This situation epitomizes the shortest law in economics, namely, that incentives matter!

We can express the conditions for consumer equilibrium in mathematical terms. We see that consumer equilibrium between tacos and hamburgers for Carl Consumer is reached when the absolute values of these slopes are equal, or when

$$\frac{\text{price of hamburgers}}{\text{price of tacos}} = \frac{MU_{\text{hamburgers}}}{MU_{\text{tacos}}} \qquad (4.1)$$

The first term in Equation 4.1 represents the rate at which the market is willing to exchange tacos for hamburgers. This ratio coincides with the negative of the slope of the budget line. The second term in this equation represents the rate at which the consumer is willing to exchange tacos for hamburgers, given **tastes and preferences**. This ratio coincides with the slope of the indifference curve or the marginal rate of substitution. When the equality expressed in Equation 4.1 holds, the consumer is in equilibrium. We also can use Equation 4.1 to restate the condition for consumer equilibrium as

$$\frac{MU_{\text{hamburgers}}}{\text{price of hamburgers}} = \frac{MU_{\text{tacos}}}{\text{price of tacos}} \qquad (4.2)$$

Equation 4.2 suggests that at equilibrium, the marginal utility derived from the last dollar spent on each good is equal. The equilibrium condition expressed in Equation 4.2 can be expanded to include all goods and services purchased by the consumer.

CHANGES IN EQUILIBRIUM

The condition previously stated for consumer equilibrium suggests that the demand for each good is influenced by consumer income and all prices. Indeed, there is a causal economic relationship that suggests that Carl Consumer's demand for hamburgers (or tacos) is a function of the price of hamburgers, the price of tacos, and his available income. In other words, changes in prices and income will lead to changes in consumer demand for goods and services.

Changes in Product Price

Economists are interested in the effects on demand of changes in income and prices, holding everything else fixed. The approach to measuring these effects, *ceteris paribus* (i.e., all other things constant), is examined later. Let us first focus on the effects of changes in the price of the product itself.

Suppose that the price of a hamburger is varied, leaving the price of tacos and income unchanged. As denoted by point *A* in Figure 4-2, if the price of hamburgers were $1.25, the price of tacos were $0.50, and Carl's available income were $5, he would consume 2 hamburgers and 5 tacos. If the price of a hamburger were to fall to $1, Carl's new equilibrium would be indicated by point *B*, where he would prefer to purchase 3 hamburgers and 4 tacos. If the price of a hamburger were varied further, other equilibria (such as point *C*) would be identified. As the price of hamburgers declines (rises), *ceteris paribus*, this consumer will purchase more (less) hamburgers.

The increase in the quantity of hamburgers Carl purchased is due to the **substitution effect** and **income effect** of the price change. The substitution effect occurs

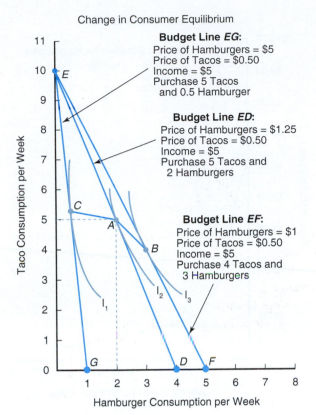

Figure 4-2

The consumer's equilibrium position in consumption can change as the budget line changes (see also Figure 3-1). Beginning with Carl Consumer's equilibrium at point A, where $5 is spent on tacos and hamburgers, we see that 5 tacos and 2 hamburgers are bought. If the price of hamburgers were to fall to $1, the budget line would rotate to the right and the new equilibrium would occur at point B. If the price of hamburgers rose to $5, equilibrium would occur at point C. The line joining points C, A, and B is called the price–consumption curve because it shows the amounts of hamburgers and tacos Carl would consume as the price of hamburgers changes.

when the price of hamburgers declines relative to the price of tacos. Carl will substitute hamburgers for tacos because the price of hamburgers has declined in relation to the unchanged price of tacos. Yet, simultaneously, when the price of hamburgers fell, Carl's real income increased. The real income effect of a decrease in the price of hamburgers means that the consumer can buy more hamburgers or tacos (or both) even though actual income has not changed.

The line joining points C, A, and B in Figure 4-2 is called the **price–consumption curve**. This curve shows the quantities of hamburgers and tacos Carl will consume as the price of hamburgers changes. If the price were continuously varied, a full set of prices and quantities of tacos and hamburgers could be identified.

If we graphed this series of prices and quantities of hamburgers associated with points C, A, and B, we would get a downward-sloping **demand curve**, which would indicate the quantity of hamburgers demanded by Carl at alternative price levels, holding all other factors constant. The demand curve typifies the amounts of a commodity consumers are willing and able to purchase at each possible price during some specific time in a specific market, all other factors held constant. Demand may change over time, and a careful specification of demand will include the time period for which it applies and the specified market. Demand for the same commodity may be different, not only in different time periods but also in different markets.

A demand curve is a schedule that shows, holding all other factors invariant, the inverse relationship between the price of a commodity and the amount of the commodity consumed. Factors affecting consumer demand are the price of the product, the price of other products, disposable income, and tastes and preferences.

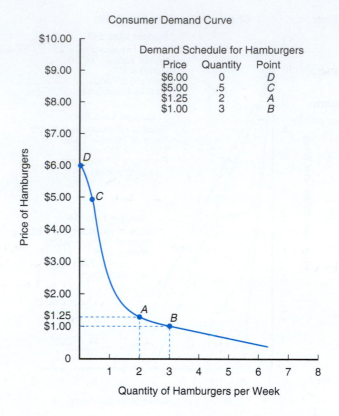

Figure 4-3

If we hold the price of tacos constant at $0.50 and the budget constraint constant at $5, we can derive Carl Consumer's demand curve for hamburgers by examining the quantities of hamburgers consumed. For example, point A here corresponds to point A in Figure 4-2, where 2 hamburgers were purchased at a price of $1.25. If the price were to fall to $1, we see in both figures that 3 hamburgers would be desired (point B). Point C in both figures also indicates that Carl would consume only 1 hamburger every other week if the price rose to $5. Finally, at a price of $6 per hamburger, Carl would purchase no hamburgers. A line connecting these and other price–quantity combinations represents the consumer demand curve. The associated demand schedule of prices and quantities corresponding to this figure also is given.

The consumer demand curve for hamburgers in Carl Consumer's example is illustrated in Figure 4-3. This demand curve shows that if the price of a hamburger were $1.25, Carl would demand 2 hamburgers (see point *A*). Point *A* in Figures 4-2 and 4-3 therefore represents the same quantity demanded. The same relationship holds for points *B* and *C* in these two figures also. Finally, as suggested by Figure 4-3, if the price of hamburgers rose to $6 or above, the quantity of hamburgers demanded by Carl would fall to zero (see point *D*). The demand curve and the price–consumption curve are really two sides of the same coin. The demand curve represents price and quantity pairs in equilibrium. Price is on the *Y*, or vertical, axis in this graph and quantity is on the *X*, or horizontal, axis. The price–consumption curve constitutes the preliminary step to a representation of the demand curve. In this curve, equilibrium price and quantity pairs are identified using the tangencies of the indifference curves and budget lines. This curve then is transformed into a demand curve.

Changes in Other Demand Determinants

We have learned that the demand for hamburgers is a function of the price of hamburgers, the price of tacos, and the level of income. A demand curve shifts when changes in income and other prices occur. There are two major economic "shifters" of a consumer demand curve: (1) changes in the consumer's disposable income and (2) changes in the prices of other goods and services. Both changes are studied in a *ceteris paribus* context, which assumes that all other factors in the economy remain constant during the time period in question.

Change in Income   When prices are held fixed and income is varied, the effects that changes in income have on consumer demand for a particular product can be assessed. This impact has led economists to classify goods into two categories: (1) normal goods and (2) inferior goods. They are defined as follows:

- **Normal goods** are those goods for which a rise (fall) in income will lead to increased (decreased) consumption. Examples of normal goods are gasoline, housing, and steak.
- **Inferior goods** are those goods for which a rise (fall) in income will lead to decreased (increased) consumption. In the past, margarine has been considered to be an inferior good, while butter has been considered a normal good. When income rises, consumers tend to eat more butter and less margarine. Riding the bus has always been considered an inferior good relative to other modes of transportation. When income increases, consumers purchase their own car, take a taxi, or fly.

In Figure 4-4, hamburgers and tacos were both normal goods when Carl Consumer's available income increased from $5 to $6. This rise in available income increased the equilibrium consumption of hamburgers from 2.0 units (point *A*) to 2.6 units (point *B*). Taco consumption would change from 5.0 to 5.5 units. However, if income were to rise to $8, taco consumption would fall to roughly 4.0 units and hamburger consumption would rise close to 5.0 units (point *C*). Carl obviously has a strong preference for hamburgers when his income increases. Although tacos were initially considered a normal good when income expanded, they became an inferior good to Carl at a higher income level.

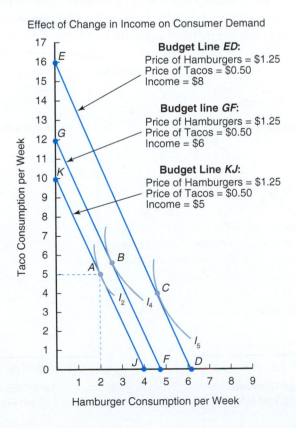

Effect of Change in Income on Consumer Demand

Budget Line *ED*:
Price of Hamburgers = $1.25
Price of Tacos = $0.50
Income = $8

Budget line *GF*:
Price of Hamburgers = $1.25
Price of Tacos = $0.50
Income = $6

Budget Line *KJ*:
Price of Hamburgers = $1.25
Price of Tacos = $0.50
Income = $5

Figure 4-4

An increase in income shifts the budget line in a parallel fashion to the right. If Carl Consumer's fast-food budget increased from $5 to $6 per week, he could attain a higher indifference curve (I_4) and new equilibrium position (point B). If his fast-food budget rose to $8, his equilibrium position would shift to point C. The nature of these changes in demand when income changes can be further studied, determining if both products are normal or inferior goods.

The Engel curve is the schedule that shows how many units of a good the consumer will purchase at different income levels, all other factors constant. From this relationship, goods may be classified as normal or inferior.

If income varied continuously and we recorded the equilibrium choices, such as *A*, *B*, and *C* in Figure 4-4, we could plot a curve relating income to consumption of a good. Such a curve is called an **Engel curve**, named after the 19th-century German statistician Ernst Engel. Recall from Chapters 2 and 3 that *Engel's Law* suggests that when consumer income increases, the proportion of income spent on food decreases, *ceteris paribus*.

A different Engel curve exists for each commodity. In any analysis of Engel curves, income is on the vertical axis and quantity consumed is on the horizontal axis. The Engel curves for hamburgers and tacos, based on this example, are illustrated in Figure 4-5*A* and *B*. Figure 4-5*A* shows that this particular consumer will want more hamburgers per day as income increases, which suggests that this good is a normal good. Figure 4-5*B* shows that tacos are a normal good over the *AB* segment of the Engel curve, but inferior over the *BC* segment. Here, over this segment, taco consumption per day falls when income rises.

The demand curve for hamburgers likely will shift if income changes, *ceteris paribus*. In the case of a normal good or a luxury good, a rise in income leads to a **rightward shift in the consumer's demand curve**. In the case of an inferior good, a rise in income leads to a **leftward shift in the demand curve**, *ceteris paribus*.

Changes in Other Prices   A demand for a product also may shift if the prices of other products change. If hamburgers and tacos are **substitutes**, a rise in the price of tacos may cause you to eat more hamburgers and fewer tacos. Your demand curve for hamburgers would therefore shift to the right when the price of tacos rises with no corresponding change in the price of hamburgers. A decline in the price of tacos would cause the demand curve to shift to the left, again if hamburger and tacos are substitutes.

Instead, if we consider the relationship between soft drinks and hamburgers, we may find that a rise in the price of soft drinks leads to fewer hamburgers consumed. In this case, hamburgers and soft drinks are said to be **complements**. Complements are goods that generally are consumed together. The demand curve for hamburgers would shift to the left when the price of soft drinks rises. Furthermore, a decline in the price of soft drinks would shift the demand curve for hamburgers to the right. Finally, a rise in the price of salt may cause no change in hamburger consumption. If so, hamburgers and salt are said to be **independent goods**.

Figure 4-5

An Engel curve depicts the relationship between Carl Consumer's income and his consumption of hamburgers and tacos.

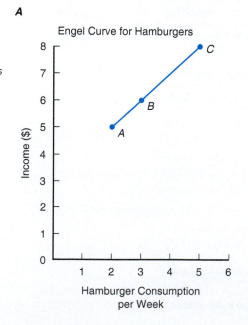

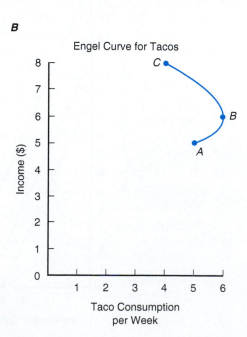

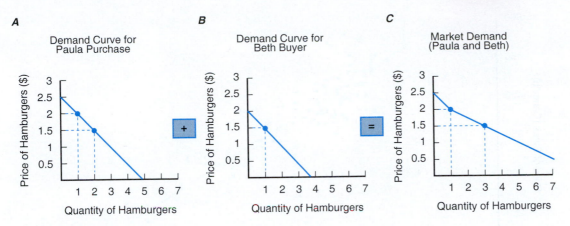

Figure 4-6

The market demand for hamburgers or any other good or service is given by the summation of the quantities demanded by individual consumers at a particular price. Assume that Paula Purchase and Beth Buyer are the only two consumers in the economy. At a price of $2.50, neither consumer would want to purchase hamburgers; therefore, the market demand would be zero. If the price were $2.00, however, Paula would want to purchase one hamburger. Because Beth would still defer from hamburger consumption, the market demand curve reflects a quantity of one. This process then is repeated for lower price levels until the entire market demand curve is revealed (C).

THE LAW OF DEMAND

Market Demand

The concept of demand applies to a single individual or firm and to any number of individuals or firms. The economy is composed of a myriad of consumers who make expenditures on many goods. Thus, the sum of all relevant consumers comprises market demand. Consequently, it is important to distinguish demand curves for individuals from demand curves for the *market* of consumers.

To illustrate the concept of market demand, suppose that there are only two consumers in a market, Paula Purchase and Beth Buyer. The demand curves for each are represented in Figure 4-6A and B. The **market demand curve** is the horizontal summation of the two individual demand curves as shown in Figure 4-6C. At a price of $2.00, the quantity demanded by Paula would be 1 hamburger; Beth would purchase none at this price. The combination of a price of $2.00 and a quantity of 1 hamburger represents one point on the market demand curve. If the price of hamburgers were $1.50, the market demand would be 3 hamburgers (i.e., two by Paula and one by Beth). By varying the price in this way, a continuous market demand curve can be constructed, as shown in Figure 4-6C.[2]

The effects of a change in price, a change in other prices, or a change in income apply at the market level like they do for an individual consumer. If tacos and hamburgers are substitute goods for both consumers, a rise in the price of tacos will shift each consumer's demand curve for hamburgers to the right. Thus, the market demand curve for hamburgers in Figure 4-6C will shift to the right also. Similar reasoning applies to changes in income. If the good is a normal good, increases in individual consumer incomes imply rightward shifts of both the individual *and* market demand curves. We have drawn demand curves sloping downward, suggesting that consumers demand more at a lower price. This situation occurs with such regularity that it is referred to as the law

The market demand curve is the horizontal summation of all individual demand curves at given market prices. As a general law, demand curves are presumed to be downward sloping. Movement along a demand curve when the price of the good changes is referred to as a change in the quantity demanded. A shift in the demand curve resulting from a change in prices of other goods, income, population, and/or tastes and preferences is referred to as a change in demand.

[2] Some specific cases exist in which horizontally adding individual curves may not lead to the market demand curve. Suppose that the demand curve for consumer A depends on the demand by consumer B. One cannot simply add the individual demand curves *independently* to obtain the market demand curve. In our discussion, we rule out this possibility.

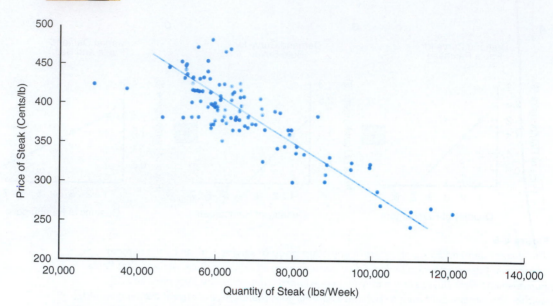

Figure 4-7
Scatter plot of the demand curve for steak.

of demand.[3] In other words, as a general rule or law, demand curves have a negative or downward slope. An example of a market demand curve is exhibited in Figure 4-7. This relationship corresponds to the demand curve for steak in the Bryan–College Station, Texas. It is based on actual data using price and purchase information from area grocery stores. Each dot in Figure 4-7 represents the price of steak in a particular week and the corresponding amount of steak purchased for consumption.

Interpretation of Market Demand

Knowledge of the properties of the market demand for a particular good (e.g., food) or service (e.g., air travel) is extremely important to businesses participating in these markets. It helps businesses define their production schedules so they can meet demand expectations. Consumers in general also have an interest in the nature of the market demand curve, because it will ultimately have a major effect on the market price of the good or service. Policymakers also have an interest in market demand, both from a commodity policy perspective and from a macroeconomic perspective.

Before discussing the role that noneconomic factors such as tastes and preferences play in market demand, let us distinguish the difference between a **change in the quantity demanded** and a **change in demand**.

Change in Quantity Demanded   When consumers change the amounts purchased because the price of the product changes, a change in the quantity demanded occurs. For example, as Figure 4-8 illustrates, if the price of a product falls from P_A to P_B, the quantity demanded will increase from Q_A to Q_B. Similarly, an increase in the price of the product from P_B to P_A would cause the quantity demanded to decline from Q_B to Q_A. The movement between points A and B on the D_1 demand curve thus represents a change in the quantity demanded.

Change in Demand   Although a change in the price of a product will result in movement along its demand curve, changes in the prices of substitutes and complements or

---

[3] Is there a logical or empirical possibility of an upward-sloping demand curve? The answer to this question is *yes*. When this situation occurs, these goods are called Giffen goods. Although we will not develop the argument here, normal goods will never be Giffen goods. Indeed, a good must be inferior for it to be a Giffen good. In reality, such goods are extremely rare. Even if some individuals have upward-sloping demand curves, the sum over all individuals would nevertheless yield a downward-sloping market demand curve.

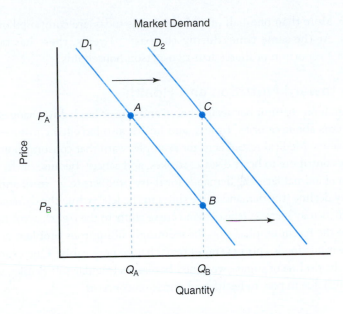

Figure 4-8
The increase in quantity from Q_A to Q_B can be the result of a change in the quantity demanded as price falls from P_A to P_B (movement from point A to B on the D_1 demand curve), or a change in demand (movement from point A on the D_1 demand curve to point C on the D_2 demand curve).

changes in consumer income will cause the demand curve to shift. An increase in income will shift the market demand curve from D_1 to the right at D_2, assuming the good in question is a normal good. This situation suggests that quantity Q_B would now be purchased at price P_A (designated by point C on Figure 4-8) instead of quantity Q_A purchased at price P_A (designated by point A on Figure 4-8). A decrease in income would have the opposite effect. The effects on quantity here are due to a shift in the demand curve, or more succinctly in economic parlance, a change in demand.

Similarly, if the price increases of a good that is considered a substitute, the demand curve for the good in question will shift to the right, *ceteris paribus*. The movement from point A to C in Figure 4-8 may be due to a rise in the price of the competing good. This situation also reflects a change in demand as opposed to a change in the quantity demanded. Examples of competing (substitute) goods are quite common, such as Tropicana and Minute Maid orange juice, Pepsi and Coca-Cola soft drinks, and Scott and Bounty paper towels.

TASTES AND PREFERENCES

Demand determinants include own-price, prices of related goods, and incomes of consumers. These determinants are economic factors of demand. However, noneconomic determinants also exist, namely, the composition of the population, attitudes toward nutrition and health, and attitudes toward food safety, lifestyles, technological forces, and advertising. Typically, such determinants carry the label **tastes and preferences**.

Composition of the Population

The composition of the population plays a role in the demand for particular commodities. The proportion of persons in the over 65-year-old age group is on the rise. The number of Americans aged 65 and over has more than doubled in the last three decades. The age shift of the population accounts in part for the decline in whole milk consumption as well as for the increase in consumption of fruits and vegetables. Also, the changing racial and ethnic composition of the population is of substantial importance to the domestic food market. The fastest growing ethnic groups are Hispanics and Asians. The increase in demand for Cajun, Mexican, and Southern-style foods may be due in part to racial and ethnic elements.

Finally, the distribution of households of various sizes influences the domestic food market (Senauer, Asp, & Kinsey, 1991). Single-person households have increased

Changes in tastes and preferences refer to changes in the composition of the population, attitudes toward nutrition and health, and attitudes toward food safety, lifestyles, technological forces, and advertising.

dramatically. More than one-half of all households today are composed of only one or two persons. At the same time, during the past 30 years, there has been a steady decline in the proportion of more-than-two-person households.

Attitudes toward Nutrition and Health

Medical research on the link between tobacco and cancer has led to changes in smoking habits. Concerns about calories, fitness, and health also have led consumers to change their eating habits. Medical researchers, for example, warn that consumption of too much red meat may contribute to heart disease, strokes, and cancer. Because of the emphasis on the reduction of animal fats, the demands for dairy products (e.g., milk and cheese) and red meats may decline (the demand curve shifts to the left), while the demand for poultry and fish products may increase (the demand curve shifts to the right), *ceteris paribus*. Additionally, with the recent emphasis on the seemingly ubiquitous problem of obesity, the demand for various foods and beverages may change noticeably. One example centers attention on the decline of sugar-sweetened beverages (regular soft drinks, sport drinks, and fruit drinks) due in part to health and nutrition concerns.

Food Safety

Food safety issues currently are a major concern with consumers. Such issues pertain to pesticide and herbicide residues in foods, antibiotics and hormones in poultry and livestock food, irradiation, nitrates in food, additives and preservatives, sugar, and artificial coloring. Three examples are the case of salmonella in poultry, peanut butter, and seafood; the chemical daminozide, sold under the name Alar and used on apples; and the outbreak of *Escherichia coli* bacteria in meat-packing plants. The impact of food safety concerns typically shifts the demand curve for a commodity to the left, *ceteris paribus*. Another example is the presence of bovine spongiform encephalopathy (BSE), more popularly known as mad cow disease.

Lifestyles

Changes in lifestyles are evident in the United States. Fashions, particularly for clothing and automobiles, serve as excellent examples of this phenomenon. The demand curve for items that are in fashion shifts to the right, all other factors invariant; the demand curve for goods that are not in fashion shifts to the left, all other factors invariant. Moreover, there has been an increasing trend for both spouses to work outside the home, resulting in less time for food preparation. One of the major social trends of the last several decades has been the increase of women in the labor force. Concomitantly, enormous growth has occurred in the number of fast-food restaurants and the increase in the demand for food away from home in general. Also, because of this social trend, there exists a rise in the demand for prepared foods for at-home use. In this example, not only is there a budget constraint, but also there is a time constraint.

Technological Forces

New technology in household food preparation, especially microwave ovens, and concurrent innovations in food processing continue to decrease the time needed for meal preparation at home. Most households in the United States own a microwave oven. Consequently, consumers want their food preparation to be easy and quick. During the past few decades, a myriad of convenience foods, particularly frozen items, ready-to-serve items, and mixes, have been introduced into the marketplace. Convenience presently is a major attribute in food products, especially in catering to the time-starved consumer.

Improvements and developments in processing and marketing also have contributed to the popularity of some foods. The development of single-serving, boxed fruit juices as well as an increased variety of blends has spurred the consumption of fruits. Improvements in processing techniques have permitted the production of shortening and margarine made entirely from vegetable oils. Cane and beet sugar consumption have been affected by the development of high-fructose corn syrup.

Advertising and Promotion

The impacts of advertising and promotion cannot be overlooked. Generic advertising and promotion campaigns are presently in full swing for fluid milk, citrus products, cotton, yogurt, cheese, butter, beef, pork, and lamb. The "Got Milk" advertising campaign for milk, "The Other White Meat" campaign for pork, and the "Fabric of Our Lives" campaign for cotton are testimonials. Food-related or beverage-related advertising is a multibillion-dollar industry in the United States. The purpose of generic advertising campaigns is to persuade consumers to buy particular products and to increase the demand for the products. Branded advertising and promotional effects attempt to make the demand curve more steep (less responsive to price) by developing brand loyalties.

CONSUMER SURPLUS

An examination of the demand curve presented in Figure 4-9 reveals that the consumer pays $6 each for 5 units of this product. The consumer, however, was willing to pay $10 for 1 unit, $9 for 2 units, and so on. Therefore, although the consumer actually pays $6 per unit to consume 5 units, he or she was willing to pay much more to purchase a smaller quantity.

If one pursues this line of reasoning, it can be argued that the area *ABC* in Figure 4-9 is a measure of the excess the consumer was willing to pay to consume 5 units of this product. This difference between willingness to pay and the amount

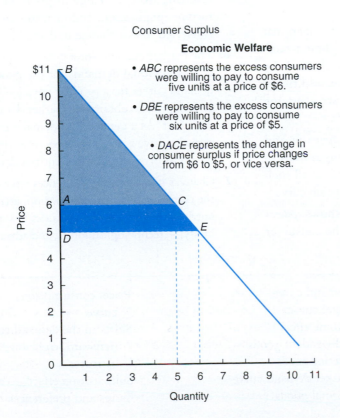

Consumer Surplus

Economic Welfare

- *ABC* represents the excess consumers were willing to pay to consume five units at a price of $6.

- *DBE* represents the excess consumers were willing to pay to consume six units at a price of $5.

- *DACE* represents the change in consumer surplus if price changes from $6 to $5, or vice versa.

Figure 4-9

The change in the economic welfare of consumers can be approximated by the concept of consumer surplus, tantamount to the difference between what the consumer was willing to pay for a product and what the consumer had to pay.

Consumer surplus is the difference between willingness to pay and the amount actually paid. Consumer surplus operationally is the area below the demand curve but above the market price.

actually paid is referred to as **consumer surplus**. Although not without controversy, it is a "bread and butter" tool of economic policy analysis.[4]

Using this approach, consumer economic well-being would increase if the price of a product fell from $6 to $5 per unit. The amount of this gain would be equal to area *ACED*. Thus, a federal government policy that resulted in cheaper food prices would increase consumer economic well-being. A loss in consumer surplus equal to area *ACED* would occur if the market price were to rise from $5 to $6. The concept of consumer surplus allows economists to place a monetary value on the gains or losses attributed to price changes. In this example, the change in consumer surplus is equal to the area of *BDE* (equal to $18) minus the area of *ABC* (equal to $12.50). Consequently, the change in consumer surplus is $5.50. The demand curve permits the calculation of consumer surplus. As such, the demand curve is valuable not only to firms but also to government policymakers. Often agricultural, government, macroeconomic, or firm policies result in changes of prices. In these cases, changes in consumer surplus represent the economic benefits to consumers of particular policy changes. The concept of consumer surplus will be discussed further in later chapters, when we discuss changing supply and demand conditions as well as policy alternatives.

---

[4] The issue is whether area *ABC* represents the consumer's actual willingness to pay for the privilege of purchasing the commodity at $6 or whether area *ABC* represents only an approximation of it.

SUMMARY

The major points made in this chapter are summarized as follows:

1. Factors affecting consumer demand are
   - price of the product,
   - price of other products,
   - disposable income of consumers, and
   - tastes and preferences.

2. The consumer is at equilibrium when the marginal rate of substitution or slope of the highest attainable indifference curve is equal to the slope of the budget line, or, alternatively, when an additional dollar spent on each good would return the same marginal utility per dollar.

3. Each tangency point between the budget line and the indifference curve when price changes leads to a new consumer equilibrium. The locus of all such tangencies is called the price–consumption curve.

4. A demand curve is a schedule that shows, *ceteris paribus*, how many units of a good the consumer

is willing and able to buy at different prices for that good during some specified time in a specified market.

5. Movement along a demand curve when the price of the good changes is referred to as a change in the quantity demanded; a shift in the demand curve resulting from a change in prices of other goods, income, population, and/or tastes and preferences is referred to as a change in demand.

6. The market demand for a good is equal to the sum of all individual demands for the good. Individual demand curves for a product are added horizontally at each price to obtain the market demand curve for a product. As a general law, demand curves are presumed to have a negative slope, reflecting the inverse relationship of prices and quantities demanded.

7. Changes in tastes and preferences refer to changes in the composition of the population, attitudes toward nutrition and health, and attitudes toward food safety, lifestyles, technological forces, and advertising.

KEY TERMS

Ceteris paribus
Change in demand
Change in quantity
   demanded
Complements
Consumer equilibrium
Consumer surplus

Demand curve
Engel curve
Income effect
Independent goods
Inferior goods
Market demand curve
Normal goods

Price–consumption
   curve
Shifts in the demand curve
   (leftward, rightward)
Substitutes
Substitution effect
Tastes and preferences

TESTING YOUR ECONOMIC QUOTIENT

1. Betty and Wilma wish to buy blouses and jeans. Betty has a budget of $600, while Wilma has a budget of $1,200. Using the indifference curve analyses given in the following table, determine prices for the blouses and jeans. For which commodity can you derive a demand curve? Derive this curve for Betty and Wilma. Finally, derive the market demand curve, assuming that they are the only consumers.

		Quantity of Blouses	Pairs of Jeans	Price of Blouses	Price of Jeans
Betty	A				
	B				
Wilma	A				
	B				

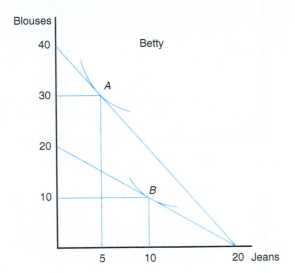

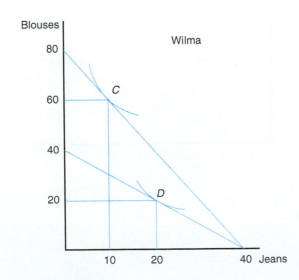

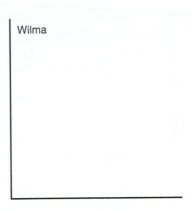

2. The budget for this consumer is $48.
   a. What is the price per pound of steak?
   b. What is the price per pound of chicken?
   c. List two true statements centered on point A regarding equilibrium and regarding utility and budget (many answers are possible):
      i. _____.
      ii. _____.

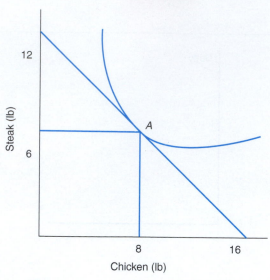

Steak (lb)

12

6

8   16

Chicken (lb)

4. Derive separate Engel curves for products A and B from the following indifference curve analysis. Assume P_A = $1.50 and P_B = $2.25. Are these goods normal or inferior?

See Indifference Curves That Follow:

	Quantity of A	Quantity of B	Income
PT_1			
PT_2			
PT_3			

3. a. Illustrate a demand curve for Chobani Yogurt. Label your axes appropriately.

b. The graph below depicts a _____ demand curve.

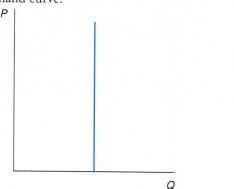

P

Q

c. The graph below depicts a _____ demand curve.

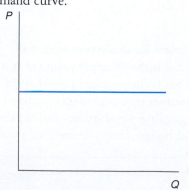

P

Q

Engel Curve for Product A

Engel Curve for Product B

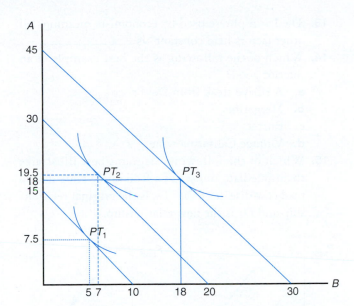

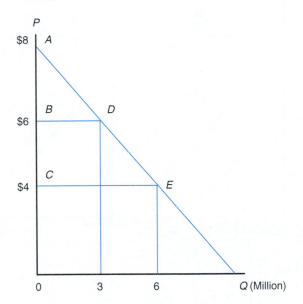

5. The graph below is a market demand curve for steaks.

a. Clearly label the consumer surplus associated with a market price of $6 using the letters on the graph from the previous page.

b. Suppose the market price decreases to $4. Are consumers better off or worse off because of this decrease in market price? Using the letters on the previous graph, label the change in consumer surplus.

c. Calculate the dollar amount of the consumer surplus in (a). Be careful of your units of measurement.

d. Calculate the dollar amount of the change in consumer surplus in (b). Be careful of your units of measurement.

6. a. The graph below is a demand curve for shrimp from Galveston Bay. Suppose that the Environmental Protection Agency finds that there are toxic levels of a pesticide in Galveston Bay. Show graphically what happens to the demand curve for shrimp in light of this information.

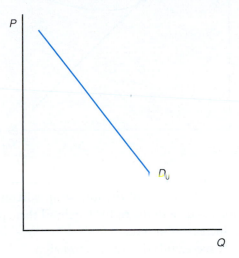

b. Is the situation described in (a) a change in quantity demanded or a change in demand? Circle your answer.

7. Consider the following diagram.

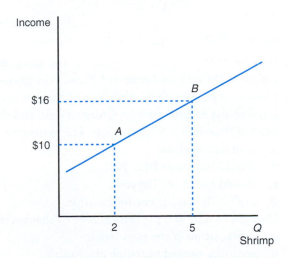

a. This diagram pertains to a(n) _____ curve.

b. Shrimp is a(n) _____ good.

8. According to the following graph, the consumer is said to be in _____ at point *A*.

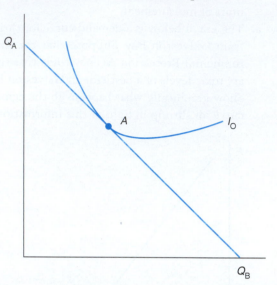

9. Assume that the good in question is a normal good. Which of the following will shift the demand curve to the right? Circle all that apply.
   a. Increase in income.
   b. A successful advertising campaign.
   c. Decrease in income.
   d. The American Medical Association reports that the consumption of this product would lower cholesterol levels by 25%.
   This rightward shift corresponds to a change in _____.

For questions 10, 11, and 12, circle the correct answer.

10. Assume that for a given consumer, the marginal utility of Dreyer's Ice Cream is 120, and the price of Dreyer's is $4 per gallon. Also, assume that the marginal utility of Blue Bell Ice Cream is 160, and the price of Blue Bell is $5 per gallon. This consumer
    a. is in equilibrium.
    b. should buy more Blue Bell.
    c. should buy more Dreyer's.
    d. can't tell; insufficient information.

11. Changes in taste and preferences refer to changes in
    a. composition of the population.
    b. attitudes toward nutrition and health.
    c. attitudes toward food safety.
    d. all of the above.

12. The connection of all tangency points between budget lines and indifference curves is called the
    a. Engel curve.
    b. price–consumption curve.
    c. demand curve.
    d. none of the above.

13. The Latin phrase used by economists meaning "all other factors held constant" is _____.

14. Which of the following is the best example of an inferior good?
    a. A ribeye steak from Del Frisco's
    b. Margarine
    c. Butter
    d. Vintage California wine

15. Which of the following diagrams best illustrates the immediate impact of a food safety scare due to mad cow disease? (*Note:* D_0 is the original relationship and D_1 is the new relationship.)

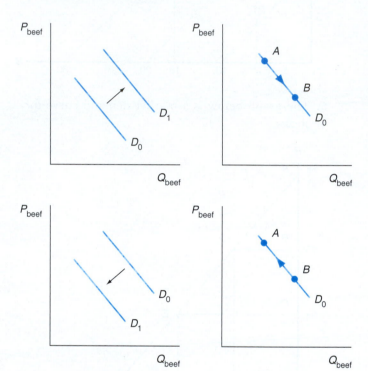

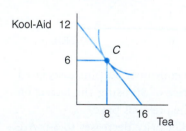

16. The demand curve for a product shifts to the right. Which of the following statement(s) is/are plausible explanation(s) for this situation?
    a. The price of a competing product decreased.
    b. A successful television advertising campaign was launched by the manufacturer of the product.
    c. Assuming the commodity in question is a normal good, the income available to the consumer increased.
    d. b and c.

17. Individual demand schedules for three customers from the local H-E-B store in the purchase of ice cream are given below:

	Customers		
	A	B	C
Price/Quart	Quart/Week	Quart/Week	Quart/Week
$1.75	2	9	5
$1.65	4	11	7
$1.60	5	12	8

What is the quantity demanded of ice cream at a price of $1.65 assuming that the only consumers in this market are customers A, B, and C?

a. 16
b. 22
c. 25
d. Can't tell; insufficient information

18. Consider the diagram to the right. At point C, which of the following statement(s) is (are) true?

a. The optimal purchase for this consumer is 8 glasses of Kool-Aid and 6 glasses of tea.
b. The marginal rate of substitution is –3/4.
c. $\dfrac{MU_{Kool\text{-}Aid}}{P_{tea}} = \dfrac{MU_{tea}}{P_{Kool\text{-}Aid}}$
d. All of the above.

19. In light of the diagram associated with question 18, if the budget for this consumer is $48, then

a. at point C, the consumer is said to be in equilibrium.
b. the price per glass of Kool-Aid is $4.
c. the price per glass of tea is $3.
d. all of the above.

REFERENCE

Senauer B, E Asp, and J Kinsey: *Food trends and the changing consumer*, St. Paul, 1991, Eagen Press.

5

Measurement and Interpretation of Elasticities

Chapters 3 and 4 discuss consumer response to a decline in a product's price by purchasing more of that product. In fact, economists are so sure of this inverse relationship between price and quantity demanded that it is referred to as the law of demand. In Chapter 4, we learned that the market demand curve for a commodity shifts to the right or the left when consumers respond to changes in other prices, income, and other factors.

What is left unsaid thus far is the degree of consumer responsiveness to change in prices and incomes. Estimates of the degree of responsiveness are expressed in what economists refer to as elasticities. The concept of demand elasticity was put forward

Alfred Marshall. Credit: INTERFOTO/Alamy.

by British economist Alfred Marshall, the 19th-century pioneer of microeconomic theory. The purpose of this chapter is to discuss the measurement of specific, widely used concepts of elasticities and provide actual estimates of these elasticities and their meaning to economic analyses.

OWN-PRICE ELASTICITY OF DEMAND

Economists compare the change in quantity demanded and the change in the price of a good in percentage terms. This comparison is formulated as a ratio and is called the **own-price elasticity of demand**.[1] The own-price elasticity of demand measures the sensitivity to changes in the price of particular products. The own-price elasticity of demand is defined as

$$\text{own-price elasticity of demand} = \frac{\text{percentage change in quantity}}{\text{percentage change in price}} \qquad (5.1)$$

The percentage change in the quantity of hamburgers demanded, for example, is equal to the change in hamburgers divided by the average quantity of hamburgers consumed during the period. The percentage change in the price of hamburgers is equal to the

As is the case for most products, the consumption of fluid milk is affected by the price of fluid milk. In fact, consumption and price are inversely related.   Credit: Ryan McVay/Photodisc/Getty Images.

---

[1]The elasticity of demand presented in this chapter is an arc elasticity that applies to discrete changes in price. When the changes approach zero, a point elasticity of demand can be defined.

change in the price of hamburgers divided by the average price of hamburgers during this period.

To illustrate the calculation of this elasticity, assume that your consumption of hamburgers drops from 3 hamburgers to 2 hamburgers when the price increases from $1.00 to $1.25 per hamburger. The average quantity over this range would be equal to 2.5 (i.e., [2 + 3]/2), while the average price would be $1.125 (i.e., [$1.25 + $1]/2). The own-price elasticity of demand in this case would be

$$\text{own-price elasticity of demand} = \frac{(Q_A - Q_B) \div ([Q_A + Q_B] \div 2)}{(P_A - P_B) \div ([P_A + P_B] \div 2)}$$

$$= \frac{(2 - 3) \div 2.5}{(\$1.25 - \$1.00) \div \$1.12} = -1.8 \qquad (5.2)$$

in which Q_A and P_A represent the quantity and price after the change, and Q_B and P_B represent the quantity and price before the change. Thus, a 1% fall (rise) in the price of a hamburger will increase (reduce) quantity demanded by 1.8%. Often the minus sign is ignored (i.e., we might simply say that the own-price elasticity is 1.8). The minus sign indicates that the demand curve is indeed downward sloping. The minus sign also reflects the law of demand.

We may simplify Equation 5.2 with some algebraic manipulation:

$$\text{own-price elasticity of demand} = \frac{\Delta Q}{\Delta P} \times \frac{\overline{P}}{\overline{Q}} \qquad (5.3)$$

where

$$\Delta Q = Q_A - Q_B; \ \Delta P = P_A - P_B; \ \overline{P} = \frac{P_A + P_B}{2}; \ \text{and} \ \overline{Q} = \frac{Q_A + Q_B}{2}$$

This formula, given by either Equation 5.2 or 5.3, measures average price elasticity between two points on the demand curve and is technically called the **arc elasticity**. Differential calculus permits the determination of price elasticity at a specific point on the demand curve. This measure, dealing with infinitesimal changes, is called the **point elasticity**. Because demand curves slope downward to the right, the measure of own-price elasticity is always negative. The effects of a change in the price of a good on the demand for this good are summarized in Table 5-1.

When the price elasticity of demand for a good exceeds one (in absolute value), we call the response **elastic**; that is, the percentage change in quantity demanded exceeds the percentage change in price. If the price elasticity of demand is equal to one, the curve would represent a **unitary elastic** demand. When the price elasticity of demand for a good is less than one (in absolute value), the demand is called **inelastic**. The percentage change in the quantity demanded is less than the percentage change in the product price.

If the demand curve were perfectly flat, or horizontal, it would represent a **perfectly elastic demand**. If the demand curve were perpendicular to the horizontal axis, or completely vertical, it would represent a **perfectly inelastic demand** (see Figure 5-1A and B).

Along the demand curve, the elasticity may be changing. Consider the case of the linear demand curve for a hypothetical product illustrated in Figure 5-2. The demand response by the consumer is elastic along the upper portion of the curve. We see this result from the elasticities, calculated using Equation 5.1, that are presented in column 6 of Table 5-2. The demand response is unitary elastic at the midpoint of this curve, inelastic to the right of this point, and elastic to the left of this point.

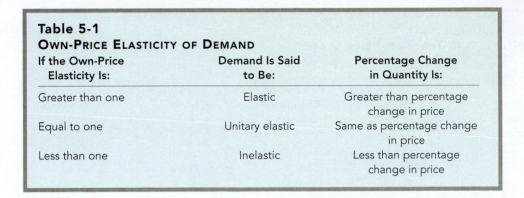

Table 5-1
OWN-PRICE ELASTICITY OF DEMAND

If the Own-Price Elasticity Is:	Demand Is Said to Be:	Percentage Change in Quantity Is:
Greater than one	Elastic	Greater than percentage change in price
Equal to one	Unitary elastic	Same as percentage change in price
Less than one	Inelastic	Less than percentage change in price

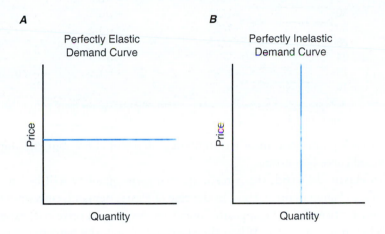

Figure 5-1
(A) A perfectly elastic demand curve is parallel to the horizontal axis. (B) A perfectly inelastic demand curve is parallel to the vertical axis.

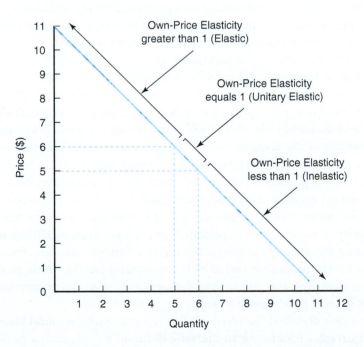

Figure 5-2
Graphical illustration of consumption expenditures and the own-price elasticity data given in Table 5-2.

Why does this elasticity change along a linear demand curve, although the change in quantity divided by the change in price is constant? The ratio of price to quantity is continuously changing as we move down the demand curve. In fact, the ratio of price to quantity approaches zero when price approaches zero. Therefore, in the case of a linear demand curve, we can conclude that the own-price elasticity of demand falls (rises) when the product price falls (rises). Note that the elasticity is

Table 5-2
CONSUMPTION EXPENDITURES AND THE OWN-PRICE ELASTICITY

(1)	(2)	(3)	(4)	(5)	(6)
Price ($)	Quantity Demanded	Total Expenditure, (1) × (2) ($)	Percentage Change in Quantity, % Δ (2)	Percentage Change in Price, % Δ (1)	Own-Price Elasticity, (4) ÷(5)
11	0	0	+1/0.5	−1/10.5	−21.00
10	1	10	+1/1.5	−1/9.5	−6.33
9	2	18	+1/2.5	−1/8.5	−3.40
8	3	24	+1/3.5	−1/7.5	−2.14
7	4	28	+1/4.5	−1/6.5	−1.44
6	5	30	+1/5.5	−1/5.5	−1.00
5	6	30	+1/6.5	−1/4.5	−0.69
4	7	28	+1/7.5	−1/3.5	−0.47
3	8	24	+1/8.5	−1/2.5	−0.29
2	9	18	+1/9.5	−1/1.5	−0.14
1	10	10			

The own-price elasticity of demand measures the percentage change in the quantity demanded of a good, given a 1% change in price. This measure is negative, reflective of the law of demand. If this elasticity is greater than one, the demand for a good is termed elastic; if this elasticity is less than one, the demand for a good is said to be inelastic. The reciprocal of the own-price elasticity is known as the price flexibility. Price flexibility measures the percentage change in price attributed to a 1% change in quantity demanded.

different at each point on a linear (straight-line) demand curve, but the slope of the linear demand curve is constant.

With **elastic demand**, the percentage change in quantity will be greater than the percentage change in price; thus, in the case of elastic demand, consumer expenditures rise when prices fall. The opposite conclusion holds for a price rise; expenditures will fall when price increases. When the elasticity is one, the percentage change in quantity will be equal to the percentage change in price. There would be no change in consumer expenditures when price changes. The percentage change in quantity will be less than the percentage change in price if demand is inelastic; thus, consumer expenditures fall (rise) when price falls (rises).

Note that in Table 5-2 total expenditures made by the consumer would be $18 if the price were equal to $9. Total expenditure would be $28 if the price fell to $7. Total expenditures would have risen by $10 (i.e., $28 − $18) if the price were to fall by $2 (i.e., $9 − $7). This relationship will always hold whenever the change in price takes place in the elastic portion of the demand curve. If the price were to fall from $4 to $2, total expenditures would fall by $10 (i.e., $28 − $18). This change in total expenditures takes place in the inelastic portion of the demand curve. The opposite conclusion holds for a price rise; a rise in price raises (lowers) expenditure if demand is inelastic (elastic).

From the perspective of a business firm, it is important to obtain the degree of price sensitivity of its customers. A principal goal of most firms is to increase revenue/profits. Raising the price of the product to increase firm revenue only makes sense if the demand for the product is inelastic. If the demand for the firm's product were elastic (price sensitive), then the appropriate strategic pricing strategy would be to lower prices in order to maximize revenue.

The concept of elasticities often is a key input in making sound business decisions. To illustrate, a firm facing an **inelastic demand** for its product could increase price and raise revenues at the same time. The rise in price would lead to increases in profits because the higher price would reduce the quantity sold, cutting total costs. In another case, top management may operate under the assumption that the brand name of its product is so strong that it could raise price without any serious impact on sales. However, if the demand for this branded product is elastic, the appropriate course of action would be to cancel the planned price increase and offer instead a price reduction, perhaps through a discount coupon.

Further, suppose that the own-price elasticity is known along with the percentage change in quantity demanded. Then the corresponding percentage change in price can be calculated. Price flexibility measures the percentage change in price attributed to a 1% change in quantity demanded. The reciprocal of the own-price elasticity is known as the **price flexibility**.

INCOME ELASTICITY OF DEMAND

As noted earlier, it is useful to assess the effects of changes in income on changes in quantity demanded in percentage terms. This measure is called the **income elasticity** of demand. The income elasticity of demand measures the sensitivity to changes in income. The income elasticity of demand is defined as

$$\begin{array}{c} \text{income} \\ \text{elasticity} \\ \text{of demand} \end{array} = \frac{\text{percentage change in quantity}}{\text{percentage change in income}} \qquad (5.4)$$

Consequently, the income elasticity of demand is a measure of the responsiveness of the quantity of a good purchased due to changes in income, all other factors constant. In Figure 4-5 (A, B), the income elasticity of demand for hamburgers over the segment AB of the Engel curve for hamburgers is equal to

$$\begin{array}{c} \text{income} \\ \text{elasticity} \\ \text{of demand} \end{array} = \frac{(Q_A - Q_B) \div ([Q_A + Q_B] \div 2)}{(I_A - I_B) \div ([I_A + I_B] \div 2)}$$

$$= \frac{(2 - 3) \div ([2 + 3] \div 2)}{(5 - 6) \div ([5 + 6] \div 2)} = 2.20 \qquad (5.5)$$

Therefore, a 1% increase in income leads to a 2.2% increase in the demand for hamburgers. An income elasticity greater than one implies that a 1% increase in income will cause consumption to rise more than 1%. Goods with income elasticities greater than one are called luxuries by economists. When the income elasticity is less than one but greater than zero, the good is called a **necessity**. Both necessities and luxuries are *normal goods*. When the income elasticity is negative, the good is referred to as an *inferior good* (see Table 5-3). Most foods and beverages are classified as necessities, and many nonfood products, such as furniture, a physician's services, and recreation, are considered **luxuries** by economists.

Again, we may simplify Equation 5.5 as follows:

$$\begin{array}{c} \text{income} \\ \text{elasticity} \\ \text{of demand} \end{array} = \frac{\Delta Q}{\Delta I} \times \frac{\bar{I}}{\bar{Q}} \qquad (5.6)$$

where

$$\Delta Q = Q_A - Q_B; \ \Delta I = I_A - I_B; \ \bar{I} = \frac{I_A + I_B}{2}; \ \text{and} \ \bar{Q} = \frac{Q_A + Q_B}{2}$$

According to Tomek and Robinson (1981), "The income elasticity for food in the aggregate, as well as for many individual products, is thought to decrease as incomes increase." Income elasticities will typically change over different income levels, and this change can be positive or negative. When incomes rise, *ceteris paribus*, demand increases for foods such as beef, poultry, shellfish, fresh fruits, and vegetables but decreases for other foods such as sugar, potatoes, eggs, and breakfast cereal (see Blaylock & Smallwood, 1986). In the domestic market, most foods have small, positive income elasticities. Consequently, large increases in income are necessary to generate substantial increases in consumption.

The income elasticity of demand measures the percentage change in the quantity demanded of a good given a 1% change in income. If this elasticity is negative, the good in question is classified as an inferior good. If this measure is positive but less than one, the good in question is labeled a necessity. If this measure is greater than or equal to one, the good in question is labeled a luxury good. Both necessities and luxuries are normal goods.

Table 5-3
INCOME ELASTICITY CLASSIFICATIONS

If the Income Elasticity Is:	The Good Is Classified as:
Greater than one	A luxury and a normal good
Less than one but greater than zero	A necessity and a normal good
Less than zero	An inferior good

CROSS-PRICE ELASTICITY OF DEMAND

The cross-price elasticity of demand measures the change in the quantity demanded for one good in light of a 1% change in the price of another good. If this elasticity is positive, the goods in question are classified as substitutes. If this elasticity is negative, the goods in question are classified as complements. If this elasticity is equal to zero, the goods in question are classified as independent.

We can measure the effects of changes in the price of tacos on the demand for hamburgers by calculating the **cross-price elasticity** of demand as

$$\text{cross-price elasticity of demand} = \frac{\text{percentage change in quantity of humburgers}}{\text{percentage change in price of tacos}} \tag{5.7}$$

$$= \frac{(Q_{HA} - Q_{HB}) \div ([Q_{HA} \div Q_{HB}] \div 2)}{P_{TA} - P_{TB} \div ([P_{TA} \div P_{TB}] \div 2)} \tag{5.8}$$

in which Q_H refers to the quantity demanded of hamburgers and P_T refers to the price of tacos. This elasticity measures the relative responsiveness of the consumption of hamburgers to the price of tacos. Once again, we may simplify Equation 5.8:

$$\text{cross-price elasticity of demand} = \frac{\Delta Q_H}{\Delta P_T} \times \frac{\overline{P_T}}{\overline{Q_H}} \tag{5.9}$$

where

$$\Delta Q_H = Q_{HA} - Q_{HB}; \ \Delta P_T = P_{TA} - P_{TB}; \ \overline{P_T} = \frac{P_{TA} + P_{TB}}{2};$$

$$\text{and } \overline{Q_H} = \frac{Q_{HA} + Q_{HB}}{2}$$

We can distinguish among the three different effects that a change in the price of one good can have on the demand for another good (see Table 5-4). The effects of substitutes and complements are of interest to agricultural economists. For example, Pepsi and Coca-Cola beverage products are substitutes. If the price of any Pepsi product increases, all other factors invariant, then the quantity demanded of the corresponding Coca-Cola product will rise. At major universities such as Texas A&M University, pizza and beer are complements. Complements are goods that are commonly consumed together.

Commodities with large positive (negative) cross-price elasticities are close *substitute* (*complementary*) commodities. Cross-price elasticities close to zero are indicative of commodities that are unrelated. Generally speaking, goods having cross-price elasticities close to zero are referred to as **independent goods**.

Table 5-4
CROSS-PRICE ELASTICITY CLASSIFICATIONS

If the Cross-Price Elasticity Is:	The Goods Are Classified as:
Positive	Substitutes
Negative	Complements
Zero	*Independent*

OTHER GENERAL PROPERTIES

We now focus on other properties of demand curves. The larger (smaller) the number of substitutes, the more (less) elastic the demand curve. Thus, a commodity such as salt is likely to exhibit a very inelastic demand, and a commodity such as Hunt's catsup is likely to exhibit a very elastic demand. There are several substitutes for Hunt's catsup (e.g., Heinz and Del Monte). Aggregates generally are more inelastic than their components. The demand for food is more inelastic than the demand for hamburgers. Further, the demand for Hunt's catsup is more elastic than the demand for catsup in general. The greater the number of alternative uses for a commodity, the greater its price elasticity.

Another general property deals with the budget share of the commodity. If a good or service constitutes a relatively large budget share or proportion of household income, its demand curve will be more elastic. When expenditures for a good or service are sizable, such as for automobiles, furniture, and appliances, consumers typically are more sensitive to changes in its price, *ceteris paribus*. Salt expenditures comprise a relatively small percentage of total expenditures made by a consumer. Thus, salt is not likely to exhibit a high elasticity of demand. The demand for cabbage also is inelastic by virtue of its negligible budget share. The own-price elasticity of cabbage has been estimated to be about -0.04. As such, a 50% increase in the price of cabbage will have very little effect on the quantity demanded, even though there are several substitutes for cabbage. There are relatively few substitutes for housing services, and its budget share is relatively large. Consequently, the elasticity of demand for housing services is expected to be large. To support this contention, Houthakker and Taylor (1970) estimate that the own-price elasticity of demand for housing is approximately equal to -1.

Another general property of demand curves is that short-run demand is more inelastic than long-run demand. With the passage of time, consumers find that they are better able to adjust to price changes. Suppose that the price of a product rises from P_1 to P_2 in Figure 5-3A and B. In the short run, a consumer's immediate response is to reduce his or her consumption of the product from Q_1 to Q_2 (Figure 5-3A). As consumers make adjustments to their consumption habits over a longer period of time, however, the cutback in consumption will be more magnified (Figure 5-3B). During the energy crisis of the 1970s, and most recently as well, consumers were not able to fully adjust their purchases of gasoline in the short run when the price of gasoline rose sharply. With the manufacturing of cars that got better gas mileage (e.g., hybrids), consumers were able to adjust to higher gasoline prices by lowering gasoline consumption over the long run.

A final property of demand curves is that the price elasticity of demand for farm products is greater at the retail level than at the farm level. George Brandow (1961) conducted a landmark study of selected elasticities of the demand for agricultural

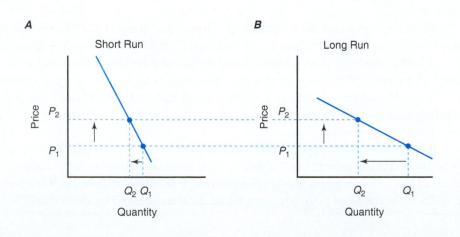

Figure 5-3

Consumer demand curves become more elastic (flatter) over time as consumers adjust to changing prices.

commodities in the United States at the farm level and at the retail level of the marketing channel. The differences in magnitude of the elasticities in these two markets in the food chain primarily are attributable to the relative level of prices in the two markets and to the value added to the product between these markets.

To summarize, the determinants of the elasticity of demand for a specific commodity include

The concept of elasticity, originated by Alfred Marshall, is a key input in making sound business decisions. Determinants of the elasticity of demand of a commodity include availability of substitutes for the commodity, the type of market, the level of the marketing channel, the percentage of the budget spent on the commodity, and time.

- availability of substitutes for the commodity,
- alternative uses for the commodity,
- type of market (e.g., farm level versus retail level or domestic market versus export market),
- the percentage of the budget spent on the commodity, and
- time.

Given this number of determinants, the elasticity of demand for a commodity is not a constant. To emphasize this point, agricultural economist Fredrick Waugh (1964) stated there is no such thing as a (i.e., single) demand elasticity.

Some Real-World Examples

Economists have estimated specific own-price, cross-price, and income elasticities of demand for various products. The U.S. Department of Agriculture houses estimates on price and income elasticities (www.ers.usda.gov/data-products/commodity-and-food-elasticities.aspx). For example, the own-price elasticity for milk is reported to be on the order of −0.25. Most of the demand elasticities are from academic and government research studies. Own-price and income elasticities for several food products at the retail level are presented in Table 5-5. As exhibited in Table 5-5, in the United States, the demand for grapes is elastic (−1.3780), the demand for bananas is inelastic (−0.4002), and the demand for oranges is unitary elastic (−0.9996).

Own-Price Elasticities   The own-price elasticity of demand for farm products in the United States generally has been in the inelastic range. Increases in the output of farm commodities because of excellent weather conditions and/or increases in productivity will depress prices rather dramatically.

Consider again the definition of own-price elasticity expressed in Equation 5.2 or the percentage change in quantity over range *AB* divided by the percentage change in price over range *AB*. Using Brandow's estimate of the own-price elasticity for farm products of −0.34, can you use this equation to defend the statement that a 1% increase in quantity coming onto the market would depress farm prices by almost 3%? (*Hint:* Given a 1% change in quantity and the elasticity of −0.34, you are left with one equation and one unknown—the percentage change in price—for which to solve.)

With respect to specific commodities, George and King (1971) found that the price elasticity of demand for beef at the retail level was −0.64. Thus, a 1% fall in the price of beef at the retail level would increase the demand for beef at the retail level by 0.64%. Tweeten (1970) suggests that the short-run own-price elasticity of demand for wheat and soybeans during the 1990s was −0.475 and −0.347, and the corresponding long-run elasticities for these commodities were actually elastic (−1.220 and −1.002, respectively).

Perhaps the most comprehensive study of retail price and income elasticities is the study by Huang reported in Table 5-5. This table suggests that a 1% increase in the retail price of sweeteners would have practically no effect on demand. However, a 1% increase in the retail price of grapes would decrease demand by more than 1%.

Table 5-5
ESTIMATED OWN-PRICE AND INCOME ELASTICITIES AT THE RETAIL LEVEL

Commodity	Own-Price Elasticity	Income Elasticity
Beef and veal	−0.6166	0.4549
Pork	−0.7297	0.4427
Chicken	−0.5308	0.3645
Turkey	−0.6797	0.3196
Eggs	−0.1452	−0.0283
Cheese	−0.3319	0.5927
Fluid milk	−0.2588	−0.2209
Wheat flour	−0.1092	−0.1333
Rice	−0.1467	−0.3664
Potatoes	−0.3688	0.1586
Butter	−0.1670	0.0227
Margarine	−0.2674	0.1112
Apples	−0.2015	−0.3514
Oranges	−0.9996	0.4866
Bananas	−0.4002	−0.0429
Grapes	−1.3780	0.4407
Grapefruits	−0.2191	0.4588
Lettuce	−0.1371	0.2344
Tomatoes	−0.5584	0.4619
Celery	−0.2516	0.1632
Onions	−0.1964	0.1603
Carrots	−0.0388	−0.1529
Cabbage	−0.0385	−0.3767
Other fresh vegetables	−0.2102	0.2837
Fruit juice	−0.5612	1.1254
Canned tomatoes	−0.3811	0.7878
Canned peas	−0.6926	0.3295
Canned fruit cocktail	−0.7323	0.7354
Sugar	−0.0521	−0.1789
Sweeteners	−0.0045	−0.0928
Coffee and tea	−0.1868	0.0937
Ice cream and other frozen dairy products	−0.1212	0.0111

Source: Huang KS, U.S. demand for food: a complete system of price and income effects, Washington, D.C., 1985, U.S. Department of Agriculture.

Income Elasticities Schultze (1971) found that the income elasticity for farm products in this country during the early 1970s was only 0.08. The magnitude of income elasticities for farm products today is still very small on the order of 0.1 to 0.2. This elasticity was shown to vary from 0.15 in Canada to 0.75 in both West Germany and France. This relatively low income elasticity in the United States suggests that a 10% increase in income would expand the demand for farm products by less than 1%. Thus, a substantial increase in consumer income would not necessarily lead to appreciable changes in the consumption of food products.

The income elasticities for major individual food items are reported in Table 5-5. This table suggests that eggs, rice, fluid milk, and other selected products are inferior goods; beef, veal, pork, chicken, and cheese are normal goods; and fruit juice is a luxury good.

Cross-Price Elasticities Consider the elasticities estimated by Capps, Seo, and Nichols (1997) for various spaghetti sauces reported in Table 5-6. The numbers along the

Table 5-6
MATRIX OF OWN-PRICE AND CROSS-PRICE ELASTICITIES OF DEMAND FOR SPAGHETTI SAUCES[a]

Item	Prego	Ragu	Classico	Hunt's	Newman's Own	Private Label
Prego	−2.5502	0.8103	0.0523	0.3918	0.1542	0.1386
Ragu	0.5100	−2.0610	0.1773	0.1381	0.0750	0.0448
Classico	0.2747	0.9938	−2.6361	0.1432	0.2496	0.4194
Hunt's	1.0293	0.5349	0.0752	−2.7541	−0.0605	−0.0316
Newman's Own	1.0829	0.9066	0.5487	−0.0861	−3.4785	0.3562
Private Label	0.6874	0.4368	0.6430	−0.0111	0.2469	−2.8038

[a]Values along the diagonal = own-price elasticities; other values = cross-price elasticities.

Source: O Capps Jr., OS Seo, Nichols JP, On the estimation of advertising effects for branded products: an application to spaghetti sauces, *Journal of Agricultural and Applied Economics*, 29. Copyright © Dec, 1997 by Southern Agricultural Economics Association. Used by permission of Southern Agricultural Economics Association.

diagonal in this table are own-price elasticities. The remaining elasticities are cross-price elasticities. The cross-price elasticity for Prego with respect to the price of Ragu is 0.8103. Thus, a 1% increase in the price of Ragu would have a notable effect on the quantity of Prego demanded (i.e., Ragu spaghetti sauce is a very close substitute for Prego). According to Table 5-6, most spaghetti sauces are substitutes for each other. As well, the own-price elasticities are in the elastic range.

APPLICABILITY OF DEMAND ELASTICITIES

Estimates of own-price, cross-price, and income elasticities of demand have a variety of applications. They are useful in policy debates, wage contract negotiations, and trade negotiations at the macroeconomic level. Firms often are interested in estimating these elasticities in order to better understand the sensitivities of their customers to changes in prices. Cross-price elasticities often are used in examining anti-trust cases, particularly mergers and acquisitions.

Applicability to Policymakers

One of the means the U.S. secretary of agriculture has historically had to support farm prices and incomes of farmers is to change the percentage of land that farmers must set aside or idle if they are to receive federal farm program benefits.[2] Tom Vilsack, the current secretary of agriculture, for example, could increase the amount of wheat land idled (i.e., increase set-aside requirements) if surplus production was expected to increase stocks and depress wheat prices and income. This policy action would lower current production and eventually lead to higher wheat prices and incomes. Importantly, if the demand curve was highly inelastic (i.e., the demand curve is very steep), a relatively small amount of land would need to be idled to achieve a specific price level. The less inelastic the demand curve, the more land the secretary would have to idle to achieve a specific price objective. Policymakers should not idle land if demand is elastic, because this action would cause revenue to fall.

---

[2]The historical features of federal government farm programs and how they have historically affected the levels of production, farm commodity prices, farm incomes, and other aspects of the nation's food and fiber industry will be discussed primarily in Chapter 11.

Table 5-7

OWN-PRICE ELASTICITY AND IMPACTS OF SUPPLY CHANGE ON FARM REVENUES

If the Own-Price Elasticity Is:	Increase in Supply Will:	Decrease in Supply Will:
Elastic	Increase revenue	Decrease revenue
Unitary elastic	No change in revenue	No change in revenue
Inelastic	Decrease revenue	Increase revenue

Applicability to Farmers

The secretary of agriculture's actions have historically had a direct impact on farmers. If the own-price elasticity of demand for wheat is less than one in absolute value (i.e., inelastic), actions taken to limit the quantity coming into the market will have the desired effect of raising wheat prices by a greater percentage than the cutback in quantity, thus raising the revenue of wheat farmers. If the own-price elasticity of demand for wheat is greater than one in absolute value (i.e., elastic), and the federal government takes actions to limit the quantity coming onto the market to support farm prices and incomes, the opposite will happen. Here we get the undesired outcome of a drop in revenue accruing to farmers (see Table 5-7).

Applicability to Consumers

Another obvious application of the own-price elasticity is predicting what a change in price will mean for consumer expenditures. In the wake of inelastic demands, increases in supply will, *ceteris paribus*, lower the cost of food and fiber products to consumers.

According to Table 5-5, apples at the retail level have a highly inelastic own-price elasticity of −0.2015. Therefore, a plentiful crop of apples should mean much cheaper apple and apple product prices for consumers. A hard freeze in apple-producing areas would mean substantially higher prices for consumers. Specifically, a 10% increase (decrease) in the quantity of apples will lead to a nearly 50% decrease (increase) in the price of apples.

Applicability to Input Manufacturers

Estimates of demand elasticities also can guide farm input manufacturer and supplier decisions by indicating the potential degree to which their market might change because of the derived nature of the demand for farm inputs. These manufacturers and suppliers depend upon a healthy farm input demand to promote the growth of their businesses. Given an inelastic demand for farm products, policies that idle land to support prices at a specific level also mean that input purchases will decline by a smaller amount than would occur if the farm level own-price elasticity of demand were more elastic.

A good example of the derived relationship between farm production and the level of farm input use is the effect the federal government's payment-in-kind (PIK) program in 1983 had upon input demand. This program made income support payments to farmers denominated in bushels of wheat, corn, and other surplus commodities rather than in dollars. This policy action dramatically reduced production and the sales of manufactured inputs to farmers in 1983.

Applicability to Food Processors and Trade Firms

We also can draw conclusions about the impacts changing market conditions have on food processing firms and wholesale and retail trade firms, based on published own-price, cross-price, and income elasticities. Thus, an increase in vegetable production will decrease the retail price of vegetables and, *ceteris paribus*, will increase the quantity of vegetables purchased. But the revenue received by retail food businesses will fall because the percentage drop in retail prices will exceed the percentage increase in vegetable consumption.

Importantly, another general property of elasticities is that the price elasticity of demand for farm products is greater at the retail level than at the farm level. Continuing with the example of the inelastic own-price elasticity of demand for vegetables at the retail level, the demand for vegetables at the wholesale level will be even more inelastic. Thus, changes in vegetable production affect the revenue received by not only retailers but also wholesalers and food processors. Most food products have an income elasticity substantially less than one, and some are negative (indicative of inferior goods) (Table 5-5). Thus, a rapid growth in consumer income nationwide will not necessarily translate into a market expansion in the demand for food products.

SUMMARY

The major points made in the chapter are summarized as follows:

1. The own-price elasticity of demand measures the percentage change in the quantity demanded for a good, given a 1% change in price. If this elasticity is greater than one, demand is said to be elastic (i.e., the percentage change in quantity exceeds the percentage change in price). If this elasticity is less than one, demand is said to be inelastic (i.e., percentage changes in quantity are smaller than percentage changes in price). If this elasticity is equal to one, demand is said to be unitary elastic (i.e., percentage changes in quantity are equal to percentage changes in price).

2. The income elasticity of demand measures the percentage change in the quantity demanded for a good, given a 1% change in income. When the income elasticity of demand is between zero and one, the good is classified as a necessity; when this elasticity exceeds one, the good is classified as a luxury good. Both luxury goods and necessities

are normal goods. When the income elasticity of demand is negative, the good is classified as an inferior good.

3. If demand is inelastic, a rise (reduction) in price will lead to increased (decreased) consumer expenditures. If demand is elastic, a rise (reduction) in price will lead to a reduction (increase) in consumer expenditures. Finally, if demand is unitary elastic, expenditures are unchanged as price changes.

4. Cross-price elasticity measures the change in the demand for one good in light of a 1% change in the price of another good. If this elasticity is positive (negative), the two goods are said to be substitutes (complements). If this elasticity is equal to zero, the two goods are independent in demand.

5. Determinants of the elasticity of demand of a commodity include availability of substitutes for the commodity, alternative uses for the commodity, type of market (e.g., farm level versus retail level or domestic market versus export market), percentage of the budget spent on the commodity, and time.

KEY TERMS

Arc elasticity	Independent goods	Perfectly elastic demand
Complements	Inelastic	Perfectly inelastic demand
Cross-price elasticity	Inelastic demand	Point elasticity
Elastic	Luxuries	Price flexibility
Elastic demand	Necessities	Substitutes
Income elasticity	Own-price elasticity of demand	Unitary elastic

TESTING YOUR ECONOMIC QUOTIENT

1. If McDonald's launches a successful ad campaign for Big Macs, what will happen to Big Mac demand? Draw a graph to show this effect. What other determinants for demand are there?

2. Mabel Cranford eats only syrup with pancakes.
   a. What is the technical name for this relationship between syrup and pancakes?
   b. Suppose the price of pancakes goes up.
      i. Represent the effect of this price increase on Mabel's pancake demand curve.
      ii. Represent the effect of this price increase on Mabel's syrup demand curve.
   c. What can we say about the cross-price elasticity between syrup and pancakes for Mabel Cranford?

3. Based on the following graph, estimate the own-price elasticity between points A and B and the own-price elasticity between points B and C. Are they elastic or inelastic? Why are the elasticities different? To increase revenue, at least in the short run, would you recommend a price increase or a price decrease?

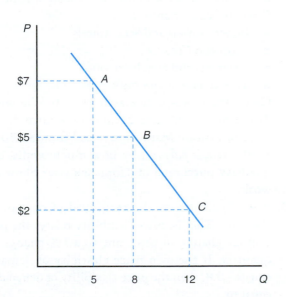

4. The Dixie Chicken currently sells 3,000 burger platters per month for $7, and the own-price elasticity for this platter has been estimated to be −1.3. If Dixie Chicken raises prices by 70 cents, how many platters will be sold?

5. Calculate the income elasticity from the following graph between points A and B and between points

B and C. Define as specifically as possible the type of good represented by each income elasticity.

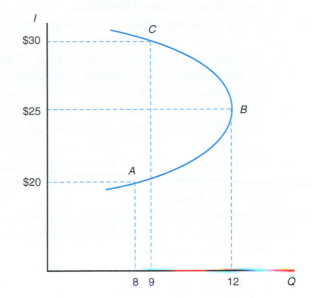

6. The cross-price elasticity for hamburger demand with respect to the price of hamburger buns is −0.6. If the price of hamburger buns rises by 5%, ceteris paribus, what change will occur for hamburger consumption? What is the relationship of these goods? Why?

7. Assume that a retailer sells 1,000 six-packs of Pepsi per day at a price of $3/six-pack. You, as an economic analyst, estimate that the cross-price elasticity between Pepsi and Coca-Cola is 0.4. If the retailer raises the price of Coca-Cola by 10%, how would sales of Pepsi be affected, ceteris paribus? Why?

8. You read in the Bryan-College Station Eagle that Texas A&M expects the price of tuition to rise by 3% for this coming fall semester. Texas A&M also expects the number of admission applications to drop by 2% because of this tuition hike. Assuming all other factors held constant, you conclude that the own-price elasticity of demand for applications to Texas A&M is equal to _____.

9. The concept of price elasticity of demand was originated by _____.

For questions 10 through 15, circle the correct answer.

10. Suppose that the own-price elasticity for Schweppes ginger ale is −1.25. In order for Cadbury Schweppes

to increase total revenue, at least in the short run, it would be advisable to

a. do nothing.

b. lower the price of the ginger ale.

c. raise the price of the ginger ale.

d. can't tell; insufficient information.

11. Generally speaking, which of the following is true?

a. The own-price elasticity at the retail level of the marketing channel is greater than the own-price elasticity at the farm level.

b. The greater the number of substitutes of a commodity, the greater the own-price elasticity.

c. The own-price elasticity is more inelastic in the short run than in the long run.

d. All of the above.

12. If the own-price elasticity is equal to –0.8, then

a. a 1% change in quantity demanded gives rise to a –0.8% change in price.

b. a 10% increase in price gives rise to an 8% decrease in quantity demanded.

c. a 1% increase in price leads to an 8% decrease in quantity demanded.

d. none of the above.

13. If the own-price elasticity for a good is –1, then the demand for the good is said to be

a. elastic.

b. inelastic.

c. unitary elastic.

d. none of the above.

14. Assume that a retailer sells 1,000 six-packs of Pepsi per day at a price of $3/six-pack. You, as an economic analyst, estimate that the cross-price elasticity between Pepsi and Coca-Cola is 0.6. If the retailer raises the price of Coca-Cola by 5%, how would sales of Pepsi be affected, *ceteris paribus?*

a. Sales of Pepsi would rise by 3 units.

b. Sales of Pepsi would rise by 30 units.

c. Sales of Pepsi would fall by 30 units.

d. None of the above.

15. From question 14, we may conclude that

a. Pepsi and Coca-Cola are complements.

b. Pepsi and Coca-Cola are substitutes.

c. Pepsi and Coca-Cola are independent.

d. can't tell; insufficient information.

16. Consider the following demand function for bananas.

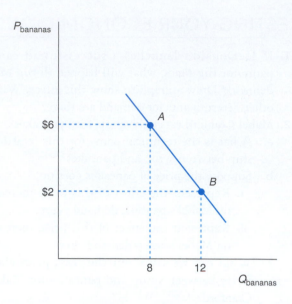

Calculate the own-price elasticity of demand.

17. a. If the cross-price elasticity between two goods is positive, then the goods are _____.

b. If the income elasticity for pork is 0.75, then what kind of good is pork? _____

18. Which of the following combination of goods most closely fits the definition of a complement? Circle the correct answer.

a. Bounty towels and Scott towels

b. Pepsi and Coca-Cola

c. Spaghetti and Spaghetti sauce

d. Tide detergent and milk

19. Given the cross-price elasticity of 0.6 between Tropicana and Minute Maid orange juice, if the price of Minute Maid were raised from $2.50 to $3.00, *ceteris paribus*, by how much would the quantity purchased of Tropicana rise? Show all work.

20. Price flexibility is the reciprocal of own-price elasticity. That is, price flexibility relates the percentage change in price due to a 1% change in quantity. If the own-price elasticity of demand equals –0.8, then the price flexibility of demand is equal to _____?

21. Using Brandow's estimate of the own-price elasticity for farm products of –0.34, defend the statement that a 1% increase in quantity coming onto the market would depress farm prices by almost 3%.

22. According to Table 5-5, apples at the retail level have a highly inelastic own-price elasticity

of −0.2015. Suppose that a hard freeze in apple-producing areas would lead to a 10% decrease in the quantity of apples. What is the repercussion of this freeze on the price of apples?

23. According to this diagram, the income elasticity of demand for salmon is

   a. less than 0.5.

   b. equal to 0.5.

   c. greater than 0.5.

   d. can't say; insufficient information.

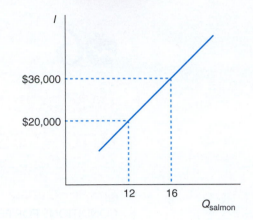

REFERENCES

Blaylock JR, DM Smallwood: U.S. demand for food: household expenditures, demographics, and projections, *ERS Technical Bulletin* 1713: February 1986.

Brandow GE: *Interrelationships among demand for farm products and implications for control of market supply*, University Park, Penn, 1961, Agricultural Experiment Station.

Capps, Jr., O, S Seo, and JP Nichols: On the estimation of advertising effects for branded products: an application to spaghetti sauces, *Journal of Agricultural and Applied Economics* 29, 2 (December 1997): 291–302.

George S, GA King: *Consumer demand for food commodities in the United States with projections for 1980*, Giannini Foundation Monograph 26, Davis, Calif, 1971, California Agricultural Experiment Station.

Houthakker HH, L Taylor: *Consumer demand in the United States: analyses and projections*, Cambridge, Mass, 1970, Harvard University Press.

Huang KS: U.S. Demand for food: a complete system of price and income effects, *USDA Technical Bulletin* 1714: December 1985.

Schultze CL: *The distribution of farm subsidies: who gets the benefits?* Washington, D.C., 1971, The Brooking Institution.

Tomek WG, KL Robinson: *Agricultural product prices*, 2nd ed., Ithaca, N.Y., 1981, Cornell University Press.

Tweeten LG: *Foundations of farm policy*, Lincoln, 1970, University of Nebraska Press.

Waugh FV: Demand and price analysis, *ERS Technical Bulletin* 1316: November 1964.

U.S. Department of Agriculture website of price elasticities, www.ers.usda.gov.

Introduction to Production and Resource Use

Businesses employ a wide variety of resources during the process of producing a particular good or service. A wheat farm, for example, uses a different combination of resources than does a cotton gin or a meat-packing firm. Some businesses use labor on a more seasonal basis than others do. And some businesses require more land when producing their product than others do.

Despite the major differences in resources used by certain types of businesses involved in the U.S. food and fiber industry, there are a number of features associated with the use of resources by these businesses that can be generalized. First, inputs can be grouped into several specific categories that facilitate their description and analysis. All these firms also share specific production and cost relationships that provide the foundation for the economic decisions made by a business, including the level of output to produce and the level of variable inputs to employ as it utilizes its current productive capacity.

In this chapter, we will assume the existence of **perfect competition**. The farm sector comes closer than any other sector of the economy to satisfying the conditions of perfect competition. We will also focus initially on understanding the effect that varying the use of a *single input* has upon the production of a *single product*. Both of these assumptions will be reviewed in later chapters when we examine multiple input choices, multiple product choices, and the nature of decisions under **imperfect competition**, which exists in several sectors of the food and fiber industry.

Forms of Competition are classified as either perfect or imperfect. Farmers and ranchers come as close as any sector in the economy of satisfying the conditions for perfect competition.

The focus of this chapter is on introducing some important physical and economic relationships, which will be broadened in later chapters. A thorough understanding of the concepts presented in this chapter is essential before proceeding with the remaining chapters in this book.

CONDITIONS FOR PERFECT COMPETITION

What does it take to have a perfectly competitive economic situation? Before we discuss how economic decisions are made, let us look at the economic environment in which they are made. An input or product market structure can be classified as perfectly competitive if the following conditions hold:

- The products sold by businesses in a sector are homogeneous. In other words, the product sold by one business is a perfect substitute for a product sold by the other businesses. This enables buyers in the market to choose from a number of sellers.
- Any business can enter or leave the sector without encountering serious barriers for entry. Resources must be free to move into the sector without encountering barriers to entry (e.g., patents, licensing). The same condition holds for resources leaving the sector.
- There must be a large number of sellers of the product. No single seller has a disproportionate influence on price; each one is a price taker.
- Perfect information must exist for all participants regarding prices, quantities, qualities, sources of supply, and so on.

When all four conditions hold, we can say a market's structure is perfectly competitive. Businesses that satisfy these conditions are also, by definition, perfectly competitive. A perfectly competitive business is a price taker, or it accepts the price of the product it receives in the product market or pays in the input as given. A corn farmer is a good example of a perfect competitor. There are thousands of corn producers who produce a homogeneous product (e.g., no. 2 yellow corn), each with equal access to extensive information when making planting and marketing decisions, and each with little or no ability to control the price of corn he or she receives or the price he or she pays for fuel.

CLASSIFICATION OF INPUTS

It is helpful to have some broad classifications in mind when discussing production relationships. These classifications not only promote efficient communication but also help to conduct economic analyses. Although not uniformly accepted, classification of inputs into land, labor, capital, and management has proven useful.

Input categories consist of land, labor, capital, and management.

Land

Land includes not only the land forms associated with the earth's crust but also resources such as minerals, forests, groundwater, and other resources given by nature. Such resources are classified as either *renewable* (e.g., forests), or *nonrenewable* resources (e.g., minerals). An example of a key land input in farming activities is productive topsoil, which has many of the attributes of a nonrenewable resource identified in Chapter 2.

Labor

Labor includes all labor services used in production with the exception of managerial activities. In crop production, labor activities include seed bed preparation, planting, irrigation, chemical applications, and harvesting. Labor activities in a canning plant

Several combines picking and shelling corn demonstrate the combination of land, capital, and labor to produce a raw agricultural product.   Credit: makaule/Fotolia.

include receiving and grading of fruit and vegetables arriving from the field, blanching, inspection, canning, and warehousing.

Capital

The term *capital* takes on different meanings in different contexts. When using the term *capital*, a banker is referring to stockholders' equity appearing on a bank's balance sheet. In a discussion of input use in the context of production, however, capital refers to manufactured goods such as fuel, chemicals, tractors, trucks, and buildings that provide productive services to their users.[1]

A key aspect of capital goods is that they do not provide consumer satisfaction directly but rather aid in the production of other goods and services. Nondurable capital inputs such as fuel and chemicals are entirely used up during the current production period. Durable capital inputs such as machinery and buildings, on the other hand, are utilized over a period of years.

Management

The final input category is management. Its functions are varied and are easier to conceptualize than to measure. Like the leader of an orchestra, farmers and agribusinesses must make decisions as to how, when, and what to produce when organizing their inputs, when and how to market the business's output, how large to grow, and how to finance business expansion.

In this chapter, we abstract from most of management's differences and instead highlight some concepts common to all inputs, with particular emphasis on technical relationships. Input and product prices will be meshed with these technical relationships

---

[1] Jargon often comes under attack by those outside a discipline of study. For example, Edwin Newman in *Strictly Speaking*, New York, 1980, Warner Books, has satirized the excessive use of such language. Economics seems to be no better or worse than other disciplines. Indeed, in most introductory classes, learning the jargon is an important portion of the course.

in this chapter when we discuss those input–output combinations that achieve a specific economic goal, such as **profit** maximization.

IMPORTANT PRODUCTION RELATIONSHIPS

Several key relationships between the level of output and the level of input use must be understood before we consider the prices of these inputs and outputs. These relationships include the concept of a production function that reflects this input–output relationship and the concepts of marginal and average product.

The Production Function

A **production function** characterizes the physical relationship between the use of inputs and the level of output. Suppose you are a salesperson for a fertilizer company and a farmer asks you to recommend the amount of fertilizer to apply per acre to maximize profit. Before you can recommend the quantity of fertilizer the farmer should apply, you must have some knowledge of the physical relationship between yields and the level of fertilizer use. If the application of more fertilizer has no effect on crop yields, the answer is simple: a profit-maximizing farmer obviously should not apply any additional fertilizer.

> **Production function** The relationship between outputs and inputs is captured in a production function.

In the general case, where there are *n* number of identifiable inputs, a production function may be expressed as

$$\text{output} = f(\text{quantity of input 1, quantity of input 2, ...,}$$
$$\text{quantity of input } n) \tag{6.1}$$

which, in words, simply states that the level of output is a function of (i.e., depends on) how much of input 1, input 2, ..., and input *n* you use. For example, in an early 1880s agricultural setting, crop output was a function of the services provided by labor, land, seed, a workhorse, a few basic implements, and management.

A production function is a rule associating an output to given levels of the inputs used. Output is measured in physical units such as bushels of wheat, gallons of milk, and cases of canned peas.[2] If one input in a production is varied and all the other inputs are held fixed at their existing level, we can rewrite Equation 6.1 as

$$\text{output} = f(\text{labor} \mid \text{capital, land, and management}) \tag{6.2}$$

in which the bar separating the first input from all other inputs indicates that only the first input is being varied and the other inputs are held fixed at existing levels.[3] This enables us to examine the relationship between labor and output as opposed to the other inputs employed in the firm's production process.

Column 2 in Table 6-1 reports the potential output levels the hypothetical firm TOP-AG can achieve per day with specific levels of labor use. One input (labor) is varied here, and all other inputs (capital, land, and management) are held fixed at their current levels. Column 1 in Table 6-1 indicates the level of daily labor use associated with the levels of output in column 2. For example, if the daily use of labor is 10 hours a day, column 2 in Table 6-1 indicates that the level of output would be one unit.

---

[2] Another way of characterizing a production function is to think of it as a cooking recipe. For example, it takes a specific combination of inputs to bake a cake. Baking a cake requires flour, eggs, water, and other ingredients. It also requires labor to blend the ingredients and capital (in the form of an oven and energy) to bake the ingredients.
[3] The terms *output* and *total physical product* can and will be used interchangeably throughout this and subsequent chapters.

Table 6-1
PRODUCTION RELATIONSHIP FOR TOP-AG, INC.

Point on Figure 6-1	(1) Daily Labor Use	(2) Daily Output Level	(3) Marginal Physical Product, $\Delta(2) \div \Delta(1)$	(4) Average Physical Product, $(2) \div (1)$
A	10.0	1.0		0.10
B	16.0	3.0	0.33	0.19
C	20.0	4.8	0.45	0.24
D	22.0	6.5	0.85	0.30
E	26.0	8.1	0.40	0.31
F	32.0	9.6	0.25	0.30
G	40.0	10.8	0.15	0.27
H	50.0	11.6	0.08	0.23
I	62.0	12.0	0.02	0.19
J	76.0	11.7	−0.02	0.15

Total Physical Product Curve

The total physical product (TPP) curve shows the relationship between output and one input, holding other inputs in the production function constant.

If we were to connect this and other combinations of output and labor use reported in columns 1 and 2 in Table 6-1, we would obtain the input–output relationship known as **the total physical product (TPP) curve** presented in Figure 6-1. By reading along the X, or horizontal axis to a particular input level, reading up to the TPP curve, and then reading over to the Y, or vertical axis, you can determine the level of output associated with this input use.

The TPP curve typically will initially increase at an increasing rate, then increase at a decreasing rate, peak at some point, and finally decrease over a full range

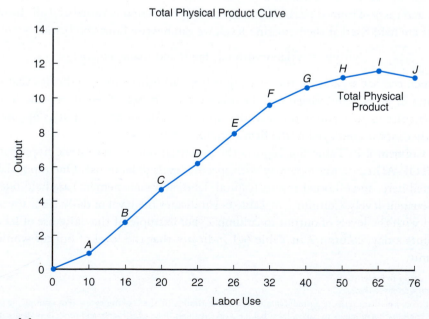

Figure 6-1

The marginal physical product curve for TOP-AG crosses the horizontal axis (i.e., negative) when the total physical product curve has reached its peak and begins to decline.

of potential input use levels. These and other properties of the total physical product curve can be better understood by calculating two additional product curves.

Marginal Physical Product Curve

If a farmer adds another pound of fertilizer per acre, will corn yields increase? If so, by how much? Or would adding more fertilizer burn out the crop and cause yields to decline? Does the addition of another employee at a grain elevator expand its output? These questions give rise to the important concept of **marginal physical product.**[4]

The marginal physical product (MPP) for an input is the change in the level of output associated with a change in the use of a particular input, where all other inputs used in the production process remain fixed at their existing levels. Stated in equation form, the marginal physical product is equal to

$$\text{marginal physical product} = \frac{\Delta \text{output}}{\Delta \text{input}} \qquad (6.3)$$

in which the Δ sign stands for marginal or "change in."

To illustrate the calculation and interpretation of the marginal physical product, consider the data for TOP-AG in Table 6-1. There may be some confusion regarding which level of output to associate with a marginal physical product as you read this table. Why are the values in column 3 on a different line than the other columns in this table? While the other columns represent levels of activity, column 3 reflects *changes in* levels of activity.

For example, when the daily labor use at TOP-AG is increased from 10 hours to 16 hours, the marginal physical product would be 0.33 units of output (i.e., $0.33 = [3.0 - 1.0] \div [16.0 - 10.0]$). Thus, if labor use is increased by 1 hour, the business can complete assembling one-third of another unit of product. This value is listed between the rows associated with 10 hours and 16 hours of labor and one and three units of output. If 20 hours of labor were used, the marginal physical product would be 0.45 units of product (i.e., $0.45 = [4.8 - 3.0] \div [20.0 - 16.0]$) and so on. The marginal physical product curves for TOP-AG are plotted in Figure 6-2.

An important relationship exists between the total and marginal physical products. The slope of the total physical product curve (with respect to the use of labor) is approximately equal to the marginal physical product. The marginal physical product is approximately equal to the slope of the total physical product curve. In other words, the marginal physical product curve measures the rate of change in output in response to a change in the use of labor.

The marginal physical product curve takes certain twists and turns as we move along the total physical product curve. For example, the marginal physical product curve cuts the average physical product from the top at approximately 30 hours of labor, where the total physical product curve in Figure 6-1 began to increase at a decreasing rate.[5] A particularly important twist is that when the total physical product curve is *decreasing*, the marginal physical product curve will be *negative*. This can be seen in Table 6-1 at the point at which TOP-AG increases its use of labor from 62 hours a day to 76 hours. Column 3 shows that the marginal physical product becomes negative at about 70 hours. The point at which the marginal physical

The MPP curve shows the change in output from the use of another unit of one input, holding other inputs in the production function constant.

---

[4] Some refer to marginal physical product as simply "marginal product." We will follow the time-honored tradition of using the word *physical* in our discussion, which makes it clear that the units of measurement are in physical units rather than dollars.

[5] The nature of the linear segments comprising the total physical product (TPP), marginal physical product (MPP), and average physical product (APP) curves precludes the MPP curve from intersecting the APP curve at its exact maximum and the MPP curve from being precisely zero at the peak of the TPP curve in Figure 6-1. A smoothing of these relationships or small linear segments would more closely approximate these conditions.

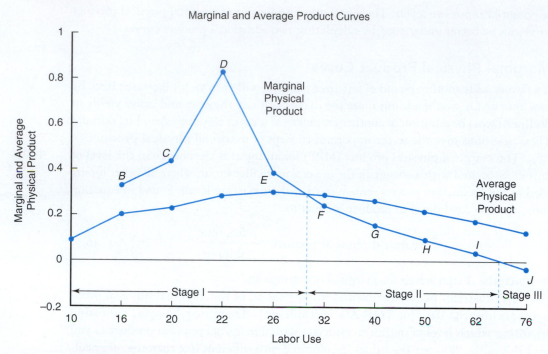

Figure 6-2

The marginal physical product curve illustrates the change in output associated with a change in labor input use by TOP-AG. The marginal physical product curve for labor falls below zero at approximately 70 hours of labor use.

product becomes negative corresponds to the point where the total physical product curve begins to decline.[6] Remember the marginal physical product approximates the slope of the total physical product curve. This characteristic will help us identify the rational range of production over the total physical product curve. This issue will be addressed shortly when we discuss the stages of production.

Average Physical Product Curve

The APP curve shows the level of output from a level of use of a specific input, holding other inputs in the production function constant.

A final input–output relationship is the **average physical product**. The average physical product (APP) is related to the level of output relative to the level of input use instead of their incremental change. Stated in equation form,

$$\text{average physical product} = \frac{\text{output}}{\text{input}} \tag{6.4}$$

In the context of our example, the average physical product represents the output per hour of *labor* with all other input levels held constant.

Column 4 in Table 6-1 presents the value of the average physical product for labor use for TOP-AG. The average physical product is shown to rise as output rises, but then falls as output expands at a decreasing rate. The average physical product curve for TOP-AG in Figure 6-2 is intercepted from above by the marginal physical product curve as total output begins to increase at a decreasing rate.

---

[6] You can find the point at which the TPP curve changes from increasing at an increasing rate to increasing at a decreasing rate by drawing a ray emanating from the origin and seeing where this ray is tangent to the TPP curve. At this point of tangency, TOP-AG's marginal physical product is equal to its average physical product.

A review of Table 6-1 and Figure 6-2 suggests the following conclusions:

- If the marginal physical product curve is above the average physical product curve, the average physical product curve must be rising.
- If the marginal physical product curve is below the average physical product curve, the average physical product curve must be falling.
- The marginal physical product curve therefore cuts the average physical product curve from above at that point where the average physical product curve reaches its maximum.

We now have the three input–output relationships for which we need to evaluate ranges of rational and irrational regions of production.

Stages of Production

To understand the production relationships illustrated in Table 6-1, you may divide them into stages of production. **Stage I of production** is found where the marginal physical product lies above the average physical product curve. **Stage II of production** begins at the end of stage I and continues until the value of the marginal physical product becomes zero. **Stage III of production** begins at the point where input use lies to the right of stage II, or at which the marginal physical product is negative.

Returning to the example in Figure 6-2, the stages of production associated with the use of labor by TOP-AG are outlined in Table 6-2. The irrational nature of stage III under normal economic conditions can be easily explained. If input is increased beyond 70 units of labor, output will fall. It is irrational to increase the use of an input if it only leads to less output. Would you recommend that TOP-AG increase its daily use of labor to 76 hours daily knowing that output would fall?

Electing to stop in stage I is also irrational, although it may be more difficult to see why. A good grade on your last examination (marginal physical product) that raises your semester average (average physical product) is an example of an outcome occurring during stage I. If you were permitted to take a make-up test for the last examination and substitute this grade for your original test score, you would likely take this make-up test only if you felt you could increase your semester average grade.

In the context of Table 6-1, look at the values in columns 3 and 4. A marginal physical product of 0.85 units per hour increases the average physical product from 0.24 units to 0.30 units. And a marginal physical product of 0.40 units increases the average physical product from 0.30 units to 0.31 units. But a lower marginal physical product of 0.25 units brings the average physical product down to 0.30 units. The first two observations are in stage I, and the third observation would occur in stage II.

Why should you stop producing if the output of your business is increasing at an increasing rate? As long as the average physical product is rising, you should expand input use. Thus, stopping production in stage I makes little sense. We will leave the presentation of the economic rationale for support of this argument for later in this chapter. At this point, we recognize that stage II appears to hold primary interest for a firm that wishes to maximize its profits.

Stage II of production represents the range of interest to economists. Why stop in stage I, and why produce in stage III?

Table 6-2 STAGES OF PRODUCTION		
Stage	Usage of Labor	Operate?
I	Between 0 and 30	Yes
II	Between 30 and 70	Yes
III	Greater than 70	No

Figure 6-2 shows that the marginal physical product of labor use is falling throughout stage II. This phenomenon is so frequently observed that it is called a law—the **law of diminishing marginal returns**. This law states that

as successive units of a variable input are added to a production process with the other inputs held constant, the marginal physical product eventually decreases.

Therefore, in the region of greatest economic interest, we would expect to observe diminishing marginal physical product for variable inputs.

A farmer obviously would not knowingly apply lime to the point where stage III occurs. If a farmer asked you to identify the application rate in stage II that makes the most economic sense, what would you advise this farmer? Do you have all the information you need to provide the farmer an answer? The answer is *no*. Although knowledge of the production relationships discussed thus far is extremely important, we must also know the costs of production.

ASSESSING SHORT-RUN BUSINESS COSTS

The total cost of production is the costs associated with the use of *all* inputs to production. A business's **total costs** in the short run can be divided into **fixed costs** and **variable costs**. We learned that fixed costs are those costs that *do not* vary with the level of input use and that variable costs are those costs that *do* vary with the level of input use. It is important to understand the measurement of these and other concepts of cost and their relationship to the level of production activity when making short-run economic decisions.

Two additional cost concepts related to the level of production that are extremely important to economic decision making are marginal costs and average costs. Each of these cost concepts, and how they are related to the total, average, and marginal product curves, is addressed in this section.

Total Costs and the TPP Curve

Figure 6-1 shows that the TPP curve typically first increases at an increasing rate, then increases at a decreasing rate, and finally decreases. In light of this curve, what do the **total variable cost** curve and **total fixed cost** curve look like?

A complete cost schedule for our hypothetical business TOP-AG is presented in Table 6-3. We have assumed that TOP-AG pays $5 per hour for labor and has fixed

Variable and fixed costs
Variable costs vary with the level of production. Fixed costs do not.

Table 6-3
SHORT-RUN COST SCHEDULE FOR TOP-AG, INC., AND SELECTED COST CONCEPTS

	(1)	(2)	(3)	(4)	(5)	(6)	(7)	(8)
Point on Figure 6-1	Total Output	Total Fixed Cost	Average Variable Cost, (2) ÷ (1)	Total Variable Cost	Average Variable Cost, (4) ÷ (1)	Total Cost, (2) + (4)	Marginal Cost, Δ(6) ÷ Δ(1)	Average Total Cost, (3) ÷ (5)
A	1.0	100.00	100.00	50.00	50.00	150.00		150.00
B	3.0	100.00	33.33	80.00	26.67	180.00	15.00	60.00
C	4.8	100.00	20.83	100.00	20.83	200.00	11.11	41.67
D	6.5	100.00	15.38	110.00	16.92	210.00	5.88	32.31
E	8.1	100.00	12.35	130.00	16.05	230.00	12.50	28.40
F	9.6	100.00	10.42	160.00	16.67	260.00	20.00	27.78
G	10.8	100.00	9.26	200.00	18.52	300.00	33.33	27.78
H	11.6	100.00	8.62	250.00	21.55	350.00	62.50	30.17
I	12.0	100.00	8.33	310.00	25.83	410.00	150.00	34.17
J	11.7	100.00	8.55	380.00	32.48	480.00	n/a	41.03

costs of $100 per hour. This includes $75 of explicit costs and $25 of implicit costs. No matter what happens to output in column 1, fixed costs in column 2 remain the same. The current property tax assessment owed by a business will be the same whether it produces at its capacity or produces nothing at all.

Total variable costs per hour for our hypothetical business TOP-AG are shown in column 4 of Table 6-3. Looking at columns 1 and 4, we see that when the level of output rises, the level of total variable costs rises. A factory will need more labor and will incur more labor costs if management decides to expand production.

Figure 6-3A graphically illustrates the nature of the total cost, total variable cost, and total fixed cost series reported in Table 6-3. The total fixed cost curve is parallel to the horizontal axis, thus illustrating its fixed nature when output rises. The total variable cost curve, on the other hand, rises when the level of output rises. Finally,

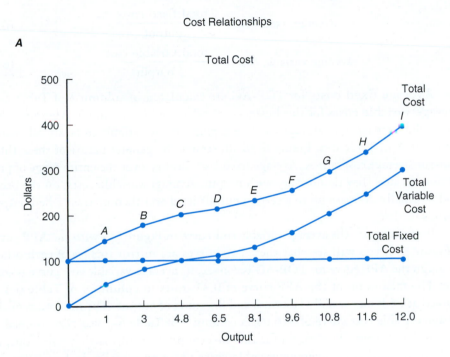

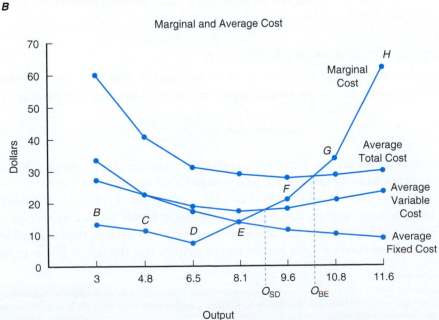

Figure 6-3

The cost relationships calculated for TOP-AG in Table 6-3 are plotted. Output levels O_{BE} and O_{SD} illustrate that the marginal cost curve intersects the average variable and average total cost curves at their minimum.

the total cost curve, which reflects both fixed and variable costs, rises when output rises. The constant gap between the total cost curve and total variable cost curve when output rises is equal to $100, or the level of fixed costs for our hypothetical business.

These total cost measures serve as the basis for average variable, average fixed, average total costs, and for marginal costs.

Average Costs and the APP Curve

The concept of **average cost** involves measuring costs per unit of output, or the level of cost associated with the level of output. The concepts of total costs, fixed costs, and variable costs discussed may be expressed in terms of average costs as

$$\text{average total cost} = \frac{\text{total costs}}{\text{output}} \tag{6.5}$$

$$\text{average fixed costs} = \frac{\text{total fixed costs}}{\text{output}} \tag{6.6}$$

$$\text{average variable costs} = \frac{\text{total variable costs}}{\text{output}} \tag{6.7}$$

Profit maximization To determine where in stage II we should produce to maximize profits, we need to study average and marginal costs associated with production.

Average fixed costs for TOP-AG are calculated in column 3 of Table 6-3. **Average variable costs** for this business are calculated in column 5. **Average total costs**, which are equal to average fixed costs plus average variable costs, are calculated in column 8 in Table 6-3. Figure 6-3B illustrates the general nature of these three short-run average cost curves. Average fixed costs decline over the entire range of production because they do not vary with output. Average variable costs, on the other hand, normally decrease up to a certain output level and then increase when output expands further.

If you compare the average variable cost curve in Figure 6-3 with the APP curve in Figure 6-2, you will see that these two curves are mirror images of each other. Although the APP curve for TOP-AG is convex, its average variable cost curve is concave. The maximum of the APP curve of 0.31 units in column 4 of Table 6-1 is attained at 26 hours of daily labor use and 8.1 units of output. The minimum of the average variable cost curve of $16.05 in column 5 of Table 6-3 was also attained at 8.1 units of output. The reason for this inverse relationship is that when output per unit of labor rises, average variable costs must necessarily decline. (Review Equation 6.7 if this is unclear.)

Marginal Costs and the MPP Curve

Marginal cost is perhaps the most important cost concept. Marginal cost is the change in the business's total cost per unit of change in output. Marginal cost is measured as

$$\text{marginal cost} = \frac{\Delta \text{total cost}}{\Delta \text{output}} \tag{6.8}$$

in which Δ represents the "change in" a particular item (e.g., costs, output). Marginal cost also represents the slope of the total cost and total variable cost curves.

The level of marginal costs for TOP-AG is calculated in column 7 of Table 6-3. The marginal cost curve for this business is plotted in Figure 6-3B. Like average variable cost, marginal cost first falls and then rises. It cuts the average total cost and average variable cost curves at their minimums (see outputs O_{BE} and O_{SD} in Figure 6-3B).[7]

---

[7] Subscripts BE and SD refer to the *break-even* and *shutdown* levels of output. These levels of output will take on special significance later in this chapter when we examine how changing market price levels affect the level of output desired by profit-maximizing businesses.

ECONOMICS OF SHORT-RUN DECISIONS

Now that we have gained an understanding of the physical aspects and cost aspects of production, the next logical issue is to determine the level of output and input use that will maximize the business's current economic profit. Before we can do this, however, we must discuss two additional revenue concepts: marginal revenue and average revenue.

Marginal and Average Revenue

In the last section, we discussed the calculation of marginal and average costs of production. They have their counterpart on the revenue side. The change in **total revenue** is called **marginal revenue**, and it represents the change in revenue from producing more output, or

$$\text{marginal revenue} = \Delta \text{total revenue} \div \Delta \text{output} \qquad (6.9)$$

Marginal revenue The marginal revenue under conditions of perfect competition is the price the producer receives from the market.

which, under perfect competition, will also be equal to the per-unit sales price of (price per bushel, per ton, per pound, etc.) the business's product.[8]

If TOP-AG increased its production from 11.6 units per hour to 12 units per hour, its total revenue would increase from $522 an hour (i.e., 11.6 units of output multiplied by a product price of $45 per unit) to $540.

If TOP-AG expands its output from 11.6 units to 12 units per hour, its

$$\begin{aligned}
\text{marginal revenue} &= (\$540 - \$522) \div (12.0 - 11.6) \\
&= \$18 \div 0.40 \text{ units of output} \\
&= \$45 \qquad (6.10)
\end{aligned}$$

which is identical to the $45 market price assumed for TOP-AG's product.

The concept of **average revenue** reflects the revenue per unit of output the business receives for its product, or

$$\text{average revenue} = \text{total revenue} \div \text{output} \qquad (6.11)$$

which simply represents another way of looking at the price of the product. If TOP-AG produced 12 units of output an hour and received $45 for each unit it produced, its total revenue would be $540 per hour. TOP-AG's average revenue under these circumstances would be

$$\begin{aligned}
\text{average revenue} &= \$540 \div 12 \text{ units of output} \\
&= \$45 \qquad (6.12)
\end{aligned}$$

which is identical both to the market price this business receives when it sells its product in the marketplace and to the marginal revenue calculated in Equation 6.10. The marginal revenue and average revenue curves under conditions of perfect competition assumed in this chapter will be perfectly flat, which we will illustrate shortly. This reflects the notion that a business is a price taker; nothing it does will change the price in the market. In the example presented earlier, the intercept of these two curves on the vertical, or Y, axis would be $45.

Level of Output: MC = MR

Expansion of the business's variable input use in the current period is profitable at the margin or as long as the marginal revenue exceeds the marginal cost. A business should not increase the use of an input if the marginal cost exceeds the marginal revenue, or if the change in cost of purchasing additional inputs is greater than the

---

[8] Under imperfect market structures, marginal revenue will differ from market price, which we will discuss in Chapter 9.

Table 6-4
DETERMINATION OF TOP-AG's PROFIT-MAXIMIZING LEVEL OF OUTPUT

Point on Figure 6-1	(1) Total Output	(2) Market Product Price $	(3) Total Revenue, (1) × (2) $	(4) Total Costs $	(5) Economic Profit, (3) − (4) $	(6) Marginal Cost, Δ(4) ÷ Δ(1) $	(7) Marginal Revenue, Δ(3) ÷ Δ(1) $
A	1.0	45.00	45.00	150.00	−105.00		
B	3.0	45.00	135.00	180.00	−45.00	15.00	45.00
C	4.8	45.00	216.00	200.00	16.00	11.11	45.00
D	6.5	45.00	292.50	210.00	82.50	5.88	45.00
E	8.1	45.00	364.50	230.00	134.50	12.50	45.00
F	9.6	45.00	432.00	260.00	172.00	20.00	45.00
G	10.8	45.00	486.00	300.00	186.00	33.33	45.00
H	11.6	45.00	522.00	350.00	172.00	62.50	45.00
I	12.0	45.00	540.00	410.00	130.00	150.00	45.00
J	11.7	45.00	526.50	480.00	46.50	n/a	n/a

revenue the business would receive from their use. Furthermore, as long as a higher profit is preferred over a smaller profit, a business should not stop expanding production if marginal revenue exceeds marginal cost.

This logic leads to the following economic strategy, under conditions of perfect competition in the short run, in which you produce at the point at which

$$\text{marginal revenue} = \text{marginal cost} \qquad (6.13)$$

or to the point at which the marginal revenue from the sale of another unit of output equals the marginal cost of producing this unit.

MC = MR criterion
Profit from production is maximized when the firm operates where marginal cost is equal to marginal revenue.

Let us expand our discussion of TOP-AG developed in Tables 6-1 and 6-3 to determine its profit-maximizing level of output. Rows *F* and *G* of Table 6-4 suggest that when TOP-AG expands its production from 9.6 units of output to 10.8 units of output, the business will achieve an economic profit of $186 per hour. TOP-AG's total revenue would be $486 per hour, while its total costs would be $300 per hour. The entry in column 5 of row *G* of $186 is the largest entry in this column, which suggests that profit is maximized at 10.8 units of output.

It is very important that you understand the economic rationale underlying the profit-maximizing level of output. If TOP-AG expanded its output to 11.6 units of output, the business's economic profit would fall from $186 to $172 an hour (see row *H* in Table 6-4). Obviously, TOP-AG's management would not wish to further expand its operations in the current period if its goal is to maximize profits.

Measuring profit The level of profit is equal to the level of output multiplied by average profit, or the difference between price and average total cost.

As TOP-AG expanded its output from 9.6 units of output to 10.8 units, its marginal cost was $33.33 as shown in column 6, and its marginal revenue was $45 as shown in column 7. The net benefits from this expansion are positive. But if TOP-AG further expanded its operations from 10.8 units per hour to 11.6 units, the marginal cost of doing so would be $62.50 as compared with a marginal revenue of only $45. The level of output that maximizes profit, or the point at which the marginal revenue associated with the expansion just equals the marginal cost associated with the expansion, occurs when marginal cost equals $45. This will occur somewhere between rows *G* and *H* or between 10.8 and 11.6 units of output as shown in Figure 6-4.

Reviewing Figure 6-4, we see that the marginal revenue curve is perfectly flat, reflecting the fact that this business is a price taker. The business thinks that the level of its production is small enough not to have a perceptible impact on the market

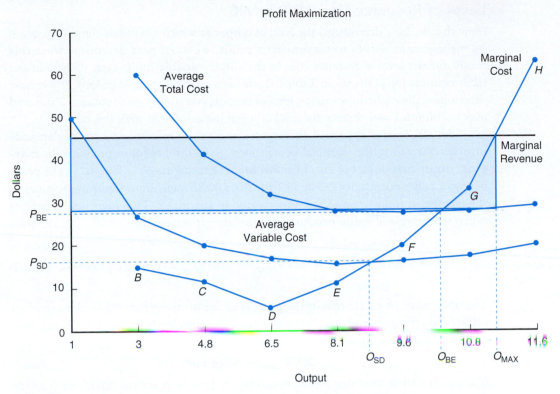

Figure 6-4

We can determine graphically what Table 6-4 could only hint at given the levels of production studied in that table. Given the MC = MR criterion expressed in Equation 6.13 for profit maximization, we see that profits would be maximized at 0_{MAX}, or slightly more than 11 units of output per hour.

price. Because we have assumed the presence of perfect competition in this chapter, the marginal revenue curve also reflects the average revenue. The intersection of this marginal revenue curve and the marginal cost curve indicates that output 0_{MAX}, or slightly more than 11 units of output, would maximize TOP-AG's profit.

Two additional output levels deserve special mention here—the *breakeven* level of production and the *shutdown* level of production. If the marginal revenue curve falls in a parallel fashion to the point where it is just tangent with the minimum point on the average total cost curve, the business's average total costs will be exactly equal to its average revenue. The business would just be able to meet both its fixed and variable costs of production with the revenue it received during the current period, or **break even**.

If the marginal revenue curve were to fall further (again in a parallel fashion) to the point at which it is just tangent to the minimum point on TOP-AG's average variable cost curve, TOP-AG would be just able to cover its variable costs but none of its fixed costs. Further declines in marginal revenue would cause the business to cease operations in the current period. TOP-AG could no longer pay its fuel bill, meet its hired labor payroll, or pay other expenses that vary with the level of production. The fuel supplier will stop making deliveries, hired workers will leave, or other factors critical to production will be curtailed. Thus, output level 0_{SD} represents the **shutdown** level of production, and the minimum point on TOP-AG's effective marginal cost curve, or its supply curve.

Finally, the level of economic profits in Figure 6-4 is equal to the shaded rectangle formed by the difference between TOP-AG's average revenue and average total cost per unit at output 0_{MAX}, or $45 minus approximately $28, multiplied by the quantity of output at 0_{MAX}, or approximately 11.1 units. The level of economic profit here would be approximately $189 (i.e., [$45 − $28] × 11.1).

Level of Resource Use: MVP = MIC

Now that we have determined the level of output at which a business should operate if its management wishes to maximize its profit, we must next determine what this means for the level of resource use. In the single-variable input case, such as in our labor example for TOP-AG in Table 6-1, this is a relatively simple process. If you have determined the profit-maximizing level of output, you simply go to column 2 and read over to column 1 and observe the level of input use associated with this output.

An alternative approach to determining profit-maximizing input demands involves comparing the marginal benefit for a given level of input use with the **marginal input cost** or cost of an additional input. Because revenue is equal to the product price times output, it is clear that the marginal benefit from input use is equal to the change in total revenue per unit change in the input. This marginal benefit is called the **marginal value product**, and for labor it is equal to

$$\text{marginal value product for labor} = \text{MVP}_{labor} = \text{MPP}_{labor} \times \text{product price} \qquad (6.14)$$

> **Marginal value product (MVP)** represents the additional revenue earned from employing another unit of input. Marginal input cost (MIC) is the cost of using that additional unit of input.

The optimum, or profit-maximizing, level of input use occurs when the marginal value product equals the marginal input cost. For the case of labor, the optimal level of labor use is given by

$$\text{MVP}_{labor} = \text{wage rate} \qquad (6.15)$$

If additional labor were employed beyond this point, the marginal input cost (i.e., the wage rate) would exceed the marginal benefits (i.e., the marginal value product for labor).

Let us illustrate the determination of the profit-maximizing level of input use by applying Equation 6.15 to the information presented in Table 6-5. This table illustrates that the price of TOP-AG's product is $45 per unit and the cost of labor is equal to $5 per hour. Multiplying the marginal physical product in column 2 by $45 gives us the marginal value product reported in column 3, or the marginal benefit from adding another hour of labor.

Column 4 in Table 6-5 indicates that TOP-AG is a price taker in the labor market because its increased use of labor had no effect on hourly wage rates in the labor market. The contribution to profit (marginal net benefit) is reported in column 5. This value is found by subtracting the marginal cost in column 4 from the marginal benefit reported in column 3.

Table 6-5
DETERMINATION OF TOP-AG'S PROFIT-MAXIMIZING LEVEL OF LABOR USE

	(1)	(2)	(3)	(4)	(5)	(6)
Point on Figure 6-1	Use of Labor	Marginal Physical Product[a]	Marginal Value Product, (2) × $45	Wage Rate	Marginal Net Benefit, (3) − (4)	Cumulative Net Benefit
A	10					
B	16	0.33	14.85	5.00	$9.85	9.85
C	20	0.45	20.25	5.00	15.25	25.10
D	22	0.85	38.25	5.00	33.25	58.35
E	26	0.40	18.00	5.00	13.00	71.35
F	32	0.25	11.25	5.00	6.25	77.60
G	40	0.15	6.75	5.00	1.75	79.35
H	50	0.08	3.60	5.00	−1.40	77.95
I	62	0.03	1.35	5.00	−3.65	74.30
J	76	−0.02	−0.90	—	—	—

[a]Column 3 in Table 6-1.

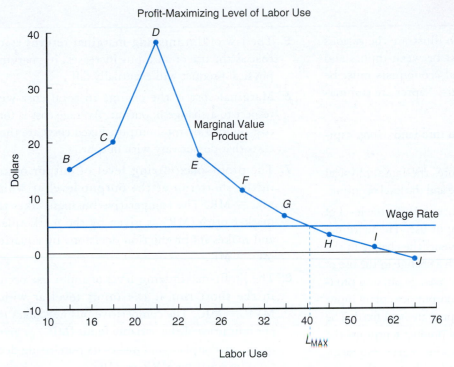

Figure 6-5
Equation 6.15 indicates that the profit-maximizing level of input use in the short run occurs at the point at which the marginal value product of labor equals the marginal input cost. In the case of TOP-AG, we see that this occurs at approximately 41 hours of labor per day if the wage rate is $5 an hour, or at quantity L_{MAX}.

The information in Table 6-5 illustrates that TOP-AG should use slightly more than 40 hours of labor. The wage rate in column 4 would equal the marginal value product of using additional labor in column 3. If labor were expanded to 50 hours, the cumulative net benefit in column 6 would be declining.

The marginal value product and marginal input cost relationships calculated in Table 6-5 are plotted in Figure 6-5. Because the marginal value product curve is nothing more than the marginal physical product curve multiplied by a fixed product price, the marginal value product curve looks very much like a marginal physical product curve. As Table 6-5 suggests, the labor use that maximizes profits is about 41 hours, the point at which the marginal value product curve intersects the marginal input cost curve.

The analysis presented in Table 6-5 and Figure 6-5 can be extended to other inputs. It represents a general way of characterizing the profit-maximizing level of input use. Profit maximization requires that the marginal value product (marginal benefit) of each variable input equal its marginal input cost simultaneously.

WHAT LIES AHEAD?

The focus of this chapter was on the economic decisions faced by a business in the short-run or current period. We looked at the effects that varying the use of one input (hired labor) would have on output and the business's costs of production. This chapter determined the profit-maximizing level of output in the short run and the profit-maximizing use of a variable input.

Chapter 7 will broaden the focus of the business's decisions by examining the determination of the least-cost combination of variable inputs in the short run and the optimal expansion path for labor and capital over the long run. This chapter will also address the profit-maximizing combination of products to produce.

SUMMARY

The purpose of this chapter is to illustrate the various physical relationships that exist between inputs and outputs with which agricultural economists must be familiar. The major points of this chapter are summarized as follows:

1. Farm inputs can be classified into land, labor, capital, and management.

2. A production function captures the causal physical relationship between input use and the level of output.

3. The total physical product curve reflects the level of output of a given level of input use. The marginal physical product curve represents the change in the level of output associated with a change in the use of a particular input. Finally, the average physical product curve reflects the level of output per unit of input use. In each case, all other inputs are held fixed. The value of the marginal physical product represents the slope of the total physical product curve. No rational farmer would want to produce beyond the point at which the marginal physical product equals zero, because further input use would cause the level of output to fall.

4. There are three stages of production:

   - Stage I is the point at which the marginal physical product curve for a particular input is rising but still lies above the average physical product curve.
   - Stage II is the point at which the marginal physical product equals the average physical product and continues until the marginal physical product for the input in question reaches zero.
   - Stage III is the point at which stage II left off or where the total physical product curve begins to decline and the marginal physical product curve becomes negative.

5. The law of diminishing marginal returns states that as the use of an input increases, its marginal physical product will eventually fall.

6. Marginal cost is the change in total cost with respect to a change in output. Average cost is total cost divided by total output. Fixed costs are those costs that do not vary with output.

7. The profit-maximizing level of output occurs in the short run at the output level at which MC = MR. The competitive business takes the market price (MR) as given by the marketplace and makes its production decisions by equating MC = MR.

8. The profit-maximizing level of input use occurs in the short run at the input level at which MVP = MIC. The competitive business takes the per unit price of the variable input (MIC) as given by the marketplace and makes its purchasing decisions by equating MVP = MIC.

9. The business will break even (TR = TC) in the short run at the output level where the price the business receives for its product falls to the point at which AR = ATC or where average profit per unit of output is zero. The business may continue to operate in the short run if AR < ATC, because it can minimize its losses (i.e., cover at least some of its fixed costs).

10. The business will cease operations, or shut down, in the short run if the price the business receives for its product falls to the point at which AR < AVC. When this occurs, the business will no longer be able to cover its variable costs of production (e.g., pay its fuel bill) and will be unable to acquire additional inputs.

KEY TERMS

Average cost	Law of diminishing marginal	Shutdown
Average fixed costs	returns	Stage I of production
Average physical product	Marginal cost	Stage II of production
Average revenue	Marginal input cost	Stage III of production
Average total costs	Marginal physical product	Total costs
Average variable costs	Marginal revenue	Total fixed cost
Breakeven	Marginal value product	Total physical product curve
Fixed costs	Perfect competition	Total revenue
Imperfect competition	Production function	Total variable cost
Input categories	Profit	Variable costs

TESTING YOUR ECONOMIC QUOTIENT

1. Please insert the appropriate labels in the blanks in the following graph. Examine the graph carefully to note all labels. Then clearly indicate below the graph the particular significance of point *A*, point *B*, and point *C*.

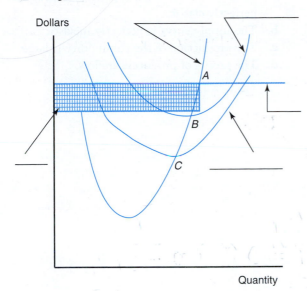

2. Define the supply curve of a perfectly competitive firm.

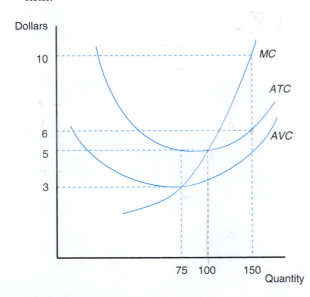

Use the preceding graph to answer questions 3 through 5.

3. Find the shutdown point. What is the quantity produced, average total cost, average variable cost, total cost, total variable cost, and profit (loss) at this point?

4. Find the break-even point. What is the quantity produced, average total cost, average variable cost, total cost, total variable cost, and profit (loss) at this point?

5. If MR = 10, then what is the quantity produced, average total cost, average variable cost, total cost, total variable cost, and profit (loss) at this point?

6. Define in words and write the formula for TFC, TC, TVC, MC, AVC, ATC, and AFC. There may be more than one formula for each one.

7. Fill in the missing cells. Assume the firm operates in a perfectly competitive environment in both the input and output markets. Calculate the profit (loss) when the firm receives $0.40 for the product.

L	Q	P(L)	TFC	TVC	TC	MC	ATC	AVC	AFC
2	40	5	110						
	65	5				.4			
	80	5							.375
	90	5			150				

8. List four conditions for perfect competition.
   a.
   b.
   c.
   d.

9. The following information pertains to a production schedule for sorghum from a West Texas farm.

		Sorghum			
Land (Acres)	Fertilizer (Pounds)	Yield (Tons)	MPP	APP	Stage of Production
4	40	68	—		—
4		75		1.25	

   a. Which input is the variable input?
   b. Which input is the fixed input?
   c. Fill in the blanks in the table.

10. Complete the following table:

Input	Output	TFC	TVC	TC	MC	AFC	AVC	ATC
2	20			100	125	—		
	40		—				10	

11. A profit-maximizing firm will use an input up to the point where the cost of the input equals the marginal revenue received by the firm.   T   F

12. Marginal cost is the additional cost created by the next or marginal unit of the variable input.   T   F

13. _____ are those costs that do not vary with input changes.

14. The firm's supply curve is represented by the firm's _____ that lies above the _____.

15. The marginal physical product of labor is
   a. The output which labor could produce without other factors of production.
   b. The additional revenue received by the firm by selling the output of one additional worker.
   c. The amount of extra output that is produced when one extra worker is added and other factors of production are held constant.

   d. The amount of extra output that is produced when one worker is added and other factors of production are increased proportionally.

16. Which of the following would provide the best evidence that a commodity is being produced under conditions of perfect competition?
   a. The demand curve facing any one producer is perfectly elastic.
   b. The supply curve is perfectly inelastic.
   c. The production of the commodity is large.
   d. The profits of producers are low.

Economics of Input and Product Substitution

The example of labor use by TOP-AG in Chapter 6 focused on varying use of a single input to produce a single product. This allowed us to introduce a number of important production concepts, their relationship to the cost of production, and the profit-maximizing level of output and input use in the short run. Let us now expand this discussion to include two variable inputs and input substitution. This requires shifting the bar appearing after the first input (labor) in Equation 6.2 so that it appears after the second input (capital).

In virtually every setting, a business can alter the combination of capital and labor used in production. For example, weeds can be pulled or hoed (a labor-intensive practice) as they were at the turn of the century, or they can be killed with herbicides (a capital-intensive practice).[1] The choice between capital-intensive and labor-intensive operations becomes an issue in the long run and is influenced by such things as the relative cost of capital and labor and changes in technology.

As illustrated in Chapter 2, farming operations have become much more capital intensive during the post–World War II period. This trend not only has implications for farm input manufacturers and farm laborers but also has environmental consequences, which will be discussed in Chapter 10.

---

[1] Remember, the term *capital* can include both variable inputs, such as fuel, fertilizer, and rented land or machinery, and fixed inputs such as owned machinery, buildings, and land.

The purpose of this chapter is to explain the economics of input substitution in the short and long run. In the short run, we determine the least-cost combination of labor and variable capital inputs, given the business's existing fixed resources and technology. Because all inputs are variable in the long run, the business will also have an interest in the optimal expansion path of labor and all capital over time.

CONCEPT AND MEASUREMENT OF ISOQUANTS

If we attempted to graph a total physical product curve for two inputs, it would take three dimensions: two dimensions for the two inputs and one dimension for output. However, three-dimensional figures are difficult to draw and understand; therefore, in this chapter, two-dimensional figures will be used. This can be done by focusing on the combinations of two inputs that, when used together, result in a specific level of output.

A curve that reflects the combinations of two inputs that result in a particular level of output is called an **isoquant** curve. The term *iso* here has the same meaning (i.e., equal) as it did in Chapter 3 when we were discussing isoutility, indifference curves for two goods faced by consumers. An isoquant consists of a locus of points that correspond to an equal or identical level of output. Along any isoquant, an infinite number of combinations of labor and capital that result in the same level of output are depicted. As the quantity of labor (capital) increases, less capital (labor) is necessary to produce a given level of output.

To illustrate this point, think of quantities of capital as being divisible units of fuel and machinery (e.g., hours of tractor use). When the tractor and its complementary equipment and fuel are increased in quantity, fewer hours of labor are required to produce a given level of wheat production, for example. Similarly, with less capital available, more hours of labor are required to produce the same amount of wheat. We can conclude from this discussion that capital and labor are *technical substitutes*.

An isoquant captures unique combinations of two inputs that result in the same level of output. The prefix *iso* is the Greek word meaning equal.

Rate of Technical Substitution

To determine the rate of substitution between two inputs, which represents (the negative of) the slope of an isoquant, we must measure the **marginal rate of technical substitution**. This concept is illustrated in Figure 7-1. As we move from range *A* to *B* on the isoquant corresponding to 10 units of output, we see that less capital and more labor are required.

Consider the three separate 1-unit changes in labor illustrated: ranges *A*, *B*, and *C* each represent different reductions in the use of capital for three separate 1-hour increases in labor use on the isoquant associated with 10 units of output. Figure 7-1 implies that the marginal rate of technical substitution of capital for labor falls from approximately 4 over range *A* to 1 over range *B*, and to 0.25 over range *C*. The rate of substituting capital for labor can be expressed mathematically as

Slope of isoquant The slope of an isoquant will play a key role later in this chapter when we determine the optimal or least-cost combination of two inputs when producing a product.

$$\frac{\Delta \text{capital}}{\Delta \text{labor}} = \frac{\text{MPP}_{\text{labor}}}{\text{MPP}_{\text{capital}}} \tag{7.1}$$

in which $\text{MPP}_{\text{capital}}$ and $\text{MPP}_{\text{labor}}$ represent the marginal physical products for capital and labor, and Δ represents the change in a variable.

The expression in Equation 7.1 indicates that changes in labor must be compensated by changes in capital, if the level of output is to remain unchanged.[2] For

---

[2] If output is to remain unchanged (i.e., remain on the same isoquant), the loss in output from the decrease in labor must equal the gain in output from the increase in capital: $-\Delta \text{labor} \times \text{MPP}_{\text{labor}} = +\Delta \text{capital} \times \text{MPP}_{\text{capital}}$. Equation 7.1 simply represents a rearrangement of this statement.

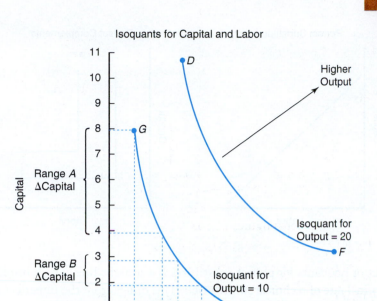

Figure 7-1
The slope of an isoquant for a particular level of output typically changes over the full range of the curve.

example, if output is to remain unchanged and the marginal rate of technical substitution of capital for labor is equal to 3, capital use must be reduced by 3 hours if labor is increased by 1 hour.[3]

These observations illustrate that when labor is substituted for capital along an isoquant (output remaining unchanged), the marginal rate of technical substitution of capital for labor falls. A declining marginal rate of technical substitution is a consequence of the law of diminishing returns (discussed in Chapter 6). When labor increases, its marginal physical product falls. Reductions in capital imply an increase in its marginal physical product.

How do the isoquants in Figure 7-1 relate to the stages of production discussed in Chapter 6? Focusing on the isoquant for 10 units of output, the marginal physical product of capital is negative above point *G* and to the right of point *H*. You will recall that the marginal physical product was negative in stage III. Because stage III is not of economic interest, the economic region of production is bounded by points *G* and *H* for the isoquant corresponding to an output of 10 units and by points *D* and *F* for the isoquant associated with an output of 20 units. Thus, only certain regions of input–output relationships are of interest to businesses seeking to maximize their profit.

Increases in output are reflected in Figure 7-1 by isoquants that lie farther away from the origin. In this figure, the isoquant for an output of 20 units lies farther from the origin than the isoquant associated with an output of 10 units.

Finally, isoquants at the extreme can be either perfect substitutes or perfect complements. Each case is illustrated in Figure 7-2.

Choice of isoquant
Higher isoquants represent higher levels of output. Does this mean that the firm can be or desires to be on a higher isoquant? The answer so far is that we cannot tell with the information we have.

---

[3] Because the marginal rate of technical substitution is negative in all rational areas of production (i.e., stage II), most economists do not bother to include the minus sign.

Figure 7-2
A graphical illustration of extreme perfect substitutes and complements.

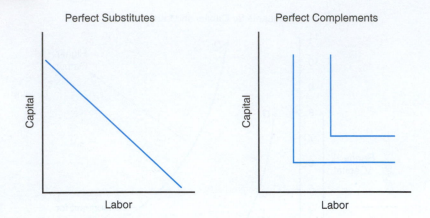

A set of isoquants for perfect substitutes is a straight line, which implies a constant marginal rate of technical substitution. This differs from the imperfect substitution implied by the isoquants in Figure 7-1, which have a decreasing marginal rate of technical substitution as one moves down the isoquants. A set of isoquants for perfect complements forms 90-degree angles, indicating that both capital and labor are required to produce a specific level of output. That is, it takes a certain proportion of labor and capital to produce a product.

THE ISO-COST LINE

Assume that a business uses two inputs (labor and capital) to produce a particular product. The total cost of production in this case would be equal to the wage rate times the hours of labor used plus the cost of capital times the amount of capital used. The concept of wage rates paid to labor is familiar, but the cost of capital will require further explanation.

Input costs   All inputs to production (land, labor, capital, and management) have a cost. The cost of two inputs can be captured in something called an iso-cost line.

We have learned that capital includes both variable and fixed inputs. We cited fuel as an example of a variable input and land as an example of a fixed input. The cost of capital therefore equals the price of fuel and other variable inputs purchased times the amount purchased and the **rental rate of capital** for using fixed inputs such as tractors and other machinery, buildings, and land. In the short run, the business can rent an additional tractor or land. The annual rental payment for leased fixed inputs is a variable cost of production. Owned capital has its costs, too. The owner of a building, for example, has the option of leasing or selling the building to someone else and using those monies in his or her next best alternative. The revenue forgone from not selling or leasing the building is a cost. Economists call this an implicit or opportunity cost. Thus, our cost of capital is a composite variable that reflects in the short run the cost of both variable inputs: labor and rented capital. The prices for these two inputs, or the wage rate for labor and the rental rate for capital, are treated as fixed in the short run; they will not vary with the level of input use by a single firm.

Suppose Frank Farmer has $1,000 available daily to finance a business's production costs. The wage rate for labor is $10 per hour, and the rental rate for capital is $100 per day. The business's daily budget constraint therefore is

$$(\$10 \times \text{use of labor}) + (\$100 \times \text{use of capital}) = \$1,000 \qquad (7.2)$$

Frank's choice of how much capital and labor to employ must be no more than $1,000. The combination of labor and capital Frank can afford for a given level of total cost is illustrated by line *AB* in Figure 7-3. This relationship is referred to as an **iso-cost line**.

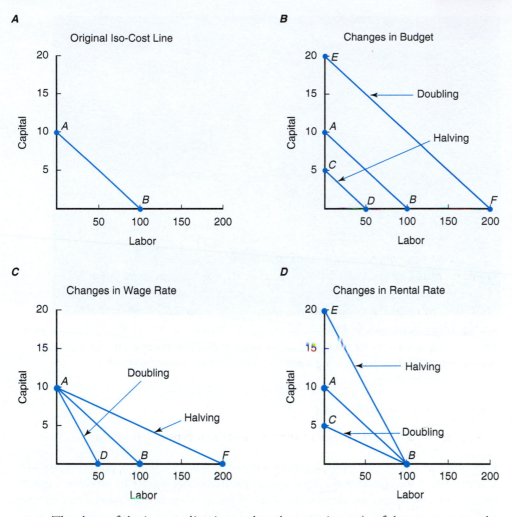

Figure 7-3
The iso-cost line plays a key role in determining the least-cost combination of input use. (A) The slope of curve AB is given by the ratio of the wage rate for labor to the rental rate for capital. (B) A doubling of the production budget changes both intercepts but not the slope of the iso-cost line. (C) A doubling of the wage rate (holding the rental rate constant) would make the iso-cost line steeper as shown by line AD. (D) A doubling of the rental rate (holding the wage rate constant) would make the iso-cost line flatter as shown by line CB. Declines in these input prices would have the opposite affect

The slope of the iso-cost line is equal to the negative ratio of the wage rate to the rental rate of capital,[4] or

$$\text{slope of iso-cost line} = -\frac{\text{wage rate}}{\text{rental rate}} \qquad (7.3)$$

If these two input prices change by a constant proportion, the total cost will change, but the slope of the iso-cost line will remain constant.

To illustrate the nature of the iso-cost line, suppose the budget the firm allocated to these two inputs was doubled. Total costs may double, but the iso-cost line *EF* would still have the same slope as line *AB* (Figure 7-3B). Only changes in the relative price of inputs (or input price ratio) will alter the slope of the iso-cost line.

For a given total cost, a rise (fall) in the price of capital relative to that of labor will cause the iso-cost line to become flatter (steeper) (Figure 7-3D). If labor's wage rises (falls), the iso-cost line would become steeper (flatter) (Figure 7-3C).

Suppose that the wage rate was $20 an hour instead of $10. The new iso-cost line *AD* would be steeper than line *AB* (Figure 7-3C). The new iso-cost line would still intersect the capital axis at point *A*, because the rental price of capital remained fixed and a maximum of 10 units of capital can still be purchased, if the producer's total budget is limited to $1,000. If the capital price rose to $200 per unit, the iso-cost line *CB* would be flatter than the original iso-cost line *AB* (Figure 7-3D).

The slope of an iso-cost line is represented by the ratio of two inputs. This line allows us to use economics to determine the least-cost combination of two inputs.

---

[4] Equation 7.3 can be rearranged algebraically to give the iso-cost line and its slope as follows:

$$\text{hours of capital} = \frac{\$1,000}{\text{rental rate}} - \frac{\text{wage rate}}{\text{rental rate}} \times \text{hours of labor}$$

The use of herbicides and other chemicals has replaced hand-weeding and use of harrows to control weeds as well as insects and diseases in crops today. This illustrates the substitution relationship between a number of inputs to produce a raw agricultural product.
Credit: ASP Inc./Fotolia.

LEAST-COST USE OF INPUTS FOR A GIVEN OUTPUT

There are essentially two input decisions a business faces in the short run that pertain to input use. One concerns the least-cost combination of inputs to produce a given level of output. Here, we assume the level of production is not constrained by the business's budget. The other is the least-cost combination of inputs and output constrained by a given budget. This section focuses on the first of these two concerns.

Short-Run Least-Cost Input Use

The first of these two perspectives on the least-cost use of inputs requires that we find the lowest possible cost of producing a given level of output with a business's existing plant and equipment. Technology and input prices are assumed to be known and constant. We know from Figure 7-1 that the alternative combinations of capital and labor produce a given level of output that forms an isoquant and that the relative prices of inputs help shape the iso-cost line in Figure 7-3.

Least-cost criteria The point of tangency of an isoquant and an iso-cost line, and not where they might cross, represents the least-cost combination of two inputs to produce a particular level of output.

We need to find the least-cost combination of inputs that will allow the business to produce a given level of output in the current period. Any additional capital is rented (variable cost) through a short-term leasing arrangement rather than owned (fixed cost), or represents nonlabor variable inputs (e.g., fuel and chemicals). Graphically, the least-cost combination of inputs is found by shifting the iso-cost line in a parallel fashion until it is tangent to (i.e., just touches) the desired isoquant. This point of tangency represents the least-cost capital/labor combination of producing a given level of output and the total cost of production.

Figure 7-4 can be used to determine the least-cost combination of labor and capital to produce 100 units of output using the business's current productive capacity. Assume that iso-cost line *AB* reflects the existing input prices for labor and capital and current total costs of production. The least-cost combination of labor and capital to produce 100 units of output is found graphically by shifting line *AB* out in a parallel fashion to the point where it is just tangent to the desired isoquant.

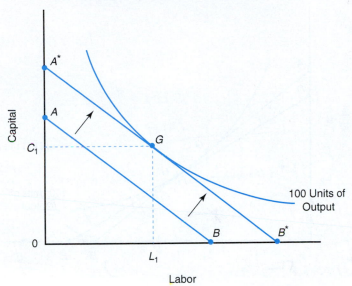

Least-Cost Input Choice for Given Output

100 Units of Output

Figure 7-4
The least-cost choice of input use is given by the point where the iso-cost curve is tangent to the isoquant for the desired level of output. If the iso-cost line is line AB, the least-cost combination would occur at point G (where 100 units of output are produced).

Figure 7-4 shows that line A^*B^* is tangent to the isoquant associated with 100 units of output at point G. The new total cost at point G in Figure 7-4 can be determined by multiplying the quantity of labor (L_1) times the wage rate and adding that to the product of the quantity of capital (C_1) times the rental rate for capital.

A fundamental interpretation to the conditions underlying the least-cost combination of input use is illustrated in Figure 7-4. The slope of the isoquant is equal to the slope of the iso-cost line at point G. At this point, the marginal rate of technical substitution of capital (fertilizer, fuel, feed, rental payments, etc.) for labor, or the negative of the slope of the isoquant, is equal to the input price ratio, or the negative of the slope of the iso-cost line. Thus, the least-cost combination of inputs requires that the market rate of exchange of capital for labor (i.e., the ratio of input prices) equal their rate of exchange in production (i.e., their marginal rate of technical substitution).

We can express the foregoing conditions for the least-cost combination of labor and capital in mathematical terms as

$$\frac{MPP_{labor}}{MPP_{capital}} = \frac{\text{wage rate}}{\text{rental rate}} \qquad (7.4)$$

We can rearrange Equation 7.4 as

$$\frac{MPP_{labor}}{\text{wage rate}} = \frac{MPP_{capital}}{\text{rental rate}} \qquad (7.5)$$

Equation 7.5 suggests that the marginal physical product per dollar spent on labor must equal the marginal physical product per dollar spent on capital. This is analogous to the condition for consumer equilibrium described in Equation 4.2, and it represents a recurring theme in economics. In the present context, a firm should allocate its expenditures on inputs so the marginal benefits per dollar are spent on competing equally.[5]

---

[5] Another way to think of this equilibrium is that marginal benefit equals marginal cost. Suppose that the marginal value product of labor usage (marginal physical product times the price of output) is $5 and the corresponding marginal benefit is $7 for capital. The opportunity cost of expending a dollar on increased labor usage is the $7 gain if this expenditure were instead used to purchase another unit of capital services. Therefore, the marginal benefit ($5) is less than the marginal cost ($7), and labor usage should be reduced. If output is to remain constant when labor is reduced, capital must be expanded until marginal benefit equals marginal cost.

Figure 7-5

The shift in the iso-cost line from AB to AB* caused by a lower wage rate suggests that labor should be increased to L_2 and capital use should be reduced to C_2. Line DE represents a parallel shift of line AB* to a point of tangency with the desired isoquant.

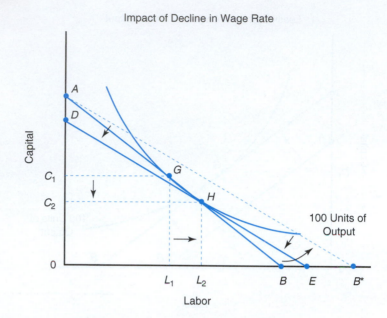

Change in input price   A change in one or both of the prices of two inputs will cause the iso-cost line to shift in one fashion or another. This will affect the desired use of inputs and even the level of production.

Change in budget   An increase in the budget available to purchase two inputs will mean that the firm can reach a higher isoquant or produce more output. The reverse will occur if the firm's budget is cut.

The discussion presented earlier can be summarized as follows: input use depends on input prices, desired output, and technology. Such cost-minimizing input use is often referred to as conditional demand because it is conditioned by the desired level of output.

Effects of Input Price Changes

Now let us see what would happen to these input demands if we allowed the price of an input to change. Because total production costs equal the sum of expenditures on each input, total cost will also change. A fundamental principle of economic behavior is that a firm will use less of an input as its per unit cost rises (Figure 7-5).

Figure 7-5 shows that as the relative price of labor (wage rate divided by price of capital) falls, the iso-cost line becomes flatter, as illustrated by the shift of iso-cost line AB to line AB*. We know from the previous discussion that our next step must be to find the point of tangency with the desired isoquant. Moving line AB* inward in a parallel fashion to the point where it (let's use a new label, line DE) is tangent to the isoquant, we see that the least-cost combination of inputs for 100 units of output shifts from point G to point H.

Therefore, when the price of labor falls (rises) relative to the price of capital, labor is substituted for capital, causing the capital/labor ratio to fall (rise). Because of diminishing marginal products, equilibrium is attained by reducing capital (from C_1 to C_2) use and using more labor (L_2 instead of L_1).

LEAST-COST INPUT USE FOR A GIVEN BUDGET

The previous section illustrated how to determine the least-cost combination of inputs in the current period to produce a given level of output. A somewhat different twist to this analysis is to determine the least-cost combination of inputs and output in the current period *for a given production budget*. We will continue to use the concept of the iso-cost line and the isoquant for specific levels of output.

Assume that a firm has a specific amount of money to spend on current production activities and wants to know the least-cost combination of capital (currently owned plus rented) and labor to employ. Figure 7-6 contains four isoquants that

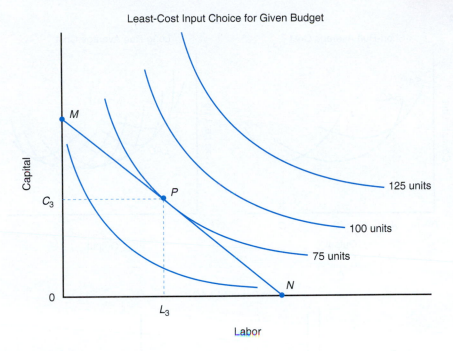

Least-Cost Input Choice for Given Budget

Figure 7-6
The least-cost choice of input use for a given budget is found by plotting the iso-cost line associated with this budget and observing the point of tangency with the highest possible isoquant.

present all the information we need to answer this firm's question. The isoquant for 50 units of output shows all the combinations of labor and capital that are needed to produce this level of output. Similar isoquants are shown for 75, 100, and 125 units of output.

Line *MN* in Figure 7-6 represents a total cost of production that completely exhausts the amount of money in the firm's budget. The point of tangency between this iso-cost line and the highest possible isoquant will indicate the least-cost combination of inputs associated with the firm's current budget constraint. This occurs at point *P* in Figure 7-6, which suggests that the firm would utilize C_3 units of capital and L_3 units of labor to produce 75 units of output. The economic conditions set forth in Equation 7.4 are satisfied by this combination of inputs (remember, the left-hand side of this equation represents the slope of the isoquant, and the right-hand side represents the slope of the iso-cost line).

The firm simply could not afford to operate on a higher isoquant in the current period. An output of 100 units is beyond the firm's current budget. The only way the firm could move out to the isoquant associated with 100 units of output is if it were able to attract additional funds or if both input prices declined to the point where the iso-cost line became tangent with this higher isoquant.

LONG-RUN EXPANSION OF RESOURCE USE

In the previous section, we learned that some costs are variable in the short run, and other costs are fixed. In the long run, however, a business has the time to expand the size of its operations, and thus all costs can be considered variable. The purpose of this section is to discuss the long-run average cost curve and the factors that influence its shape.

Long-Run Average Costs

Figure 7-7 depicts three sets of cost curves. The presence of fixed inputs in the short run ensures that these short-run average cost curves depicted here are U-shaped. Each short-run average cost curve in Figure 7-7A reflects the full average cost of the

Figure 7-7

The long-run average cost (LAC) curve plays a key role in determining the minimum costs of operation in the long run. Often referred to as the planning curve, the LAC curve represents an envelope of a series of short-run average cost curves (A and B).

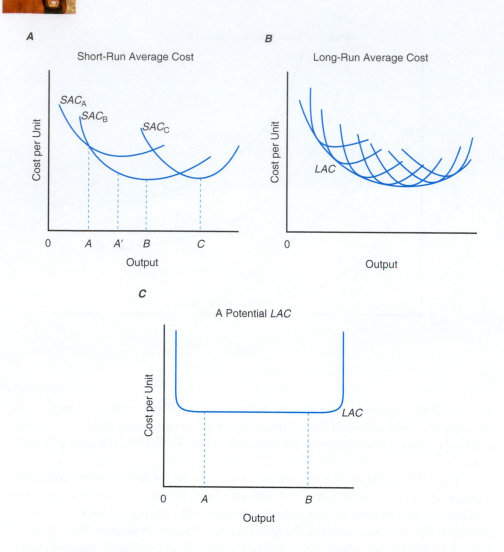

business for three separate sizes. Size *A* is the smallest, with costs represented by SAC_A. This curve might correspond to TOP-AG using a specific amount of capital to produce 12 units of output in the example discussed previously.

Size *B* is larger and can operate at much lower costs. Curve SAC_B is much lower, except at its extreme left end. This curve might reflect the capital needed by TOP-AG to minimize its cost of producing more than 12 units of output. Size *C* is still larger, but the curve SAC_C represents a higher cost structure than size *B*.

A business desiring to minimize its production costs will wish to operate at size *A* in Figure 7-7 if an output equal to *OA* or less is desired. The average cost of production will be substantially lower than size *B* at outputs less than *OA*. If output *OB* or *OC* is desired, however, the business will prefer size *B* or size *C*, respectively.

If the business desires to produce an output larger than *OA*, size *B* would be better than size *A* because its average costs are lower. Output *OA'* has special significance for a business of size *A*, because it represents its minimum cost of operation. Obviously, it would be better, when producing more than *OA*, to operate a business of size *B* at less than capacity than it would be to continue operating a business of size *A*, because the average costs of production would be lower.

The Long-Run Planning Curve

When deciding how big the business should be in the long run, management must consider a relevant range of size options. Management may be aware of this range, either from its own experience or from economic feasibility studies conducted for other businesses of a similar nature. The short-run average cost curves associated with different sizes over this range enable the business to determine its long-run average cost (*LAC*) curve as illustrated in Figure 7-7B.

Often referred to as the *long-run planning curve*, the **long-run average cost curve** is an envelope of a series of short-run average cost curves. The LAC illustrates to the business how varying its size will affect the business's economic efficiency. It also indicates the minimum product price at which any output can be produced in the long run as given by the minimum point on the LAC.

What causes the long-run average cost curve to decline, become relatively flat, and then increase? The answer lies in what economists call *returns to scale*. If an increase in output is exactly proportional to an increase in inputs, constant returns to scale are said to exist. This means that a doubling or a tripling of inputs used by the firm will cause a doubling or tripling of its output.[6] If the increase in output is more (less) than proportional to the increase in input use, we say the returns to scale are increasing (decreasing). Decreasing (increasing) returns to scale will exist if the firm's long-run average costs are increasing (decreasing) when the firm is expanded.

Increasing Returns to Scale   Some of the physical causes of increasing returns to scale are purely dimensional in nature. If the diameter of a pipe is doubled, the flow through it is more than doubled. The carrying capacity of a truck also increases faster than its weight. After some point, such increases in dimensional efficiency stop. As the size of the pipe is increased, it has to be made out of thicker and stronger materials. The size of the truck will also be limited by the width of streets, the height of overpasses, and the capacity of bridges.

A closely related technical factor that helps to explain the existence of increasing returns to size is the indivisibility of inputs. In general, indivisibility means that equipment is available only in minimum sizes or in a specific range of sizes. As the scale of the firm's operations increases, the firm's management can switch from using the minimum-sized piece of equipment to larger, more efficient equipment. Thus, the larger the scale of the operation, the more the firm will be able to take advantage of large-size equipment that cannot be used profitably in smaller operations.

Another technical factor contributing to increasing returns to scale comes from the potential benefits from specialization of effort. For example, as the firm hires more labor, it can subdivide tasks and become more efficient.[7] When the firm expands the size of its operations, it can buy specialized pieces of equipment and assign special jobs to standardized types of machinery.

A frequently noted factor that also helps us explain the existence of increasing returns to scale is volume discounts on large purchases of production inputs. Lower input prices paid by larger farming operations can be yet another major reason why average costs decline as farm size is increased.

Constant Returns to Scale   Increasing returns to scale cannot go on indefinitely. Eventually, the firm will enter the phase of constant returns to scale, where doubling of all inputs doubles output. The phase of constant returns to scale is virtually nonexistent in the U-shaped LAC curves when discussing economies of scale before the phase of decreasing returns to scale sets in. Empirical evidence suggests, however, that the phase of constant or nearly constant returns to scale can cover a large range of output levels.

The long-run average cost curve is comprised of points on a series of short-run average cost curves. This curve helps economists determine the profitability of different sizes of operations.

There are three measures of returns to scale: increasing, constant, and decreasing. Each of these sizes has much to say about the "staying power" of the firm should prices fall.

---

[6] A word of caution: the phrase *the economies of mass production* carries several meanings, some of which are irrelevant here and therefore are potential sources of confusion. For example, the greater efficiency frequently observed for larger production units (in contrast to smaller ones) is often caused because larger units are newer and use better production techniques than the older and smaller units. However important this may be, improvements in technology are not part of the concept of returns to scale, which assumes a given technology.
[7] The benefits gained from specialization are well known. Adam Smith—in his book *The Wealth of Nations*, published in 1776—addressed the gains from the division of labor.

Decreasing Returns to Scale   Can a business keep on doubling its inputs indefinitely and expect its output to double? Most likely, the answer is *no*. Eventually, there must be a decreasing return to scale. The farmer may actually be part of the reason for decreasing returns to scale. While all other inputs can be increased, his or her ability to manage larger operations may not. The managerial skills needed to coordinate efforts and resources usually do not increase proportionately with the scale of operations.

In Figure 7-7B, the long-run average cost curve is the *envelope* of the set of short-run average cost curves; that is, the long-run average cost curve is tangent to the short-run average cost curve when it is declining. When the long-run average cost curve is rising, it touches the short-run average cost curves to the right of their minimum points. The minimum point on the long-run average cost curve is the only point that touches the minimum point on the short-run average cost curve. The declining portion of the long-run average cost curve suggests the existence of increasing returns to scale. Beyond the minimum point on this curve, decreasing returns to scale exist.

Economists are concerned with the shape of the long-run average cost curve for several reasons. The minimum point on the long-run average cost curve represents the most efficient use of resources in the long run in the sense that the business's average costs of operation are minimized. Another concern is the risk that the price of the product falls below one or more of the short-run average cost curves. Firms to the right or the left of the minimum point on the LAC must either increase the size of their operations (left side) or downsize their operations (right side), or go out of business.

The long-run cost curve depicted in Figure 7-7B reflects the conventional shape illustrated in most textbooks. Although the long-run average cost curve no doubt decreases over some range of output before eventually turning up, its shape is not likely to be perfectly U-shaped.

Studies by agricultural economists suggest that there may be some range of output where the long-run average cost curve is relatively flat. In California, Hall and LaVeen (1978) found that the long-run average cost curve becomes relatively flat after initially declining rapidly. They reported that the costs of producing highly mechanized crops generally continued to decline slowly over the entire range of surveyed farm sizes. For vegetables and fruit crops, however, Hall and LaVeen found little or no decline after the initial benefits from expansion were achieved.

Figure 7-7C illustrates the general nature of these findings. Between outputs *OA* and *OB,* the long-run average cost curve is relatively flat. Over this range, all business sizes will have approximately the same average costs. Thus, the long-run average cost curve in practice is more L-shaped than U-shaped.

ECONOMICS OF BUSINESS EXPANSION

In the long run, businesses have time to expand (or contract) the scale of their operations. Suppose that the short-run marginal cost and average cost curves for an existing business are represented in Figure 7-8 by SMC_1 and SAC_1, respectively. If the market price for the product was equal to P and the business was produced at the point where $P=SMC_1$, the business would sustain a loss on each unit of output produced. At this point, the business would have two options: (1) it could go out of business, or (2) it could expand the scale of its existing operations, if it could convince its banker of the benefits from this expansion.

If the business expanded to the scale represented by SAC_2 and SMC_2, it would produce quantity Q_2 and earn an average profit per unit equal to P minus the short-run average cost at quantity Q_2. A profit-maximizing business may want to expand to the size represented by SAC_4 and SMC_4. By producing at Q_4, the business would be operating at the point where P is equal to SMC_4.

While this long-run adjustment for an existing business is taking place, the number of businesses may also increase because of attractive economic profits. Some of

Long-run average cost curve   Long-run average cost curves are rarely if ever perfectly U-shaped. They are more likely to fall relatively sharply and then decline very slowly before turning upward when decreasing returns to scale appear.

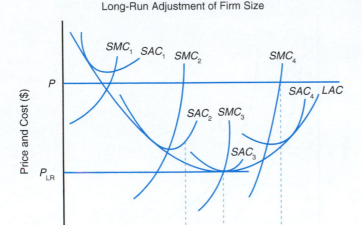

Figure 7-8
A profit-maximizing business may desire to expand the scale of its operations to the level corresponding to the SAC$_4$ and SMC$_4$ cost curves, given the level of product price P. When others respond to the existence of economic profits, total market supply will expand and the market price will fall. In a free market setting, businesses will cut back their output and size as best they can. Some may cease producing altogether. The market will be in long-run equilibrium at the point where P = MC = AC, which would result in the business producing Q$_{LR}$ units of output at a market price of P$_{LR}$.

these entrants will be newly created businesses. Others may be firms that have shifted out of less profitable enterprises. As these new and modified operations begin to produce output, the market supply of the product will increase. This, in turn, will cause the price of the product to fall.[8] When each business responds to the new lower market price, the output of each will generally become smaller than before. Those businesses that were just preparing to expand the scale of their operations in response to the earlier price will be able to adjust rapidly. Other businesses that have just completed expansion of their firm will obviously respond more slowly. Those businesses that cannot contract their operations by either idling existing capacity or selling off a portion of their business will lose more money than those that can. This process conceivably will continue until all remaining firms are breaking even (zero profits) and the incentive for additional firms to enter the market has been eliminated. Those existing businesses that are losing money will eventually cease producing this product.

The market will be in long-run equilibrium at price P_{LR}. At this price, the firm will be operating at the point at which product price is just equal to the minimum point on the *LAC* curve in Figure 7-8. This figure shows that the optimal scale of operations in the long run would see it producing an output equal to Q_{LR}. Businesses expanding beyond this output run the risk of eventually scaling back or "downsizing" their operations. In a downturn in the economy, for example, airlines might furlough flight crews and park jet airplanes until economic conditions improve.

Market equilibrium
A market achieves equilibrium at the minimum point on the industry's long-run average cost curve.

Capital Variable in the Long Run

Thus far, the firm has been limited to expanding its use of variable inputs (including rented capital like farmland). The ownership of capital, held constant in the short run, is allowed to vary in the long run. This input can be thought of as plant size

---

[8] Once we develop the market supply curve for all businesses producing a particular product in Chapter 8, this sequence of events will become more clear.

Figure 7-9
In the long run, the size of the firm's operations can be expanded if the economic incentive to do so is there. This requires increasing the capital input to its least-cost level (i.e., the point at which the marginal rate of technical substitution of capital for labor is equal to their price ratio). You will recall that this is the same stipulation made for the short-run case. Unlike the short run, however, the firm can now increase its use of capital beyond 1 unit. Producing 20 units of output instead of 10 units can be most efficiently done by operating at point B, not C.

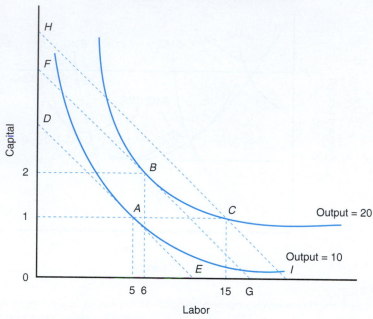

Input substitution   The firm may face the choice of increasing labor and/or capital when expanding the scale of its operations. The choice will be influenced not only by the productivity of these inputs but also by their relative cost.

(e.g., the number of manufacturing lines, the capacity of a feedlot, or the number of grain silos owned by an elevator). In the long run, the capital structure of the firm can be adjusted to its least-cost level that occurs when the marginal rate of technical substitution of capital for labor is equal to the input price ratio.

Figure 7-9 shows that at 1 unit of capital, labor is at its least-cost level of use only if output is equal to 10 units (point *A*). At this output level, the least-cost combination of inputs would be 5 units of labor and 1 unit of capital (see point *A*).

If an output of 20 units is desired in the long run, the firm has two options: (1) expand labor use to 15 units and operate at point *C*, or (2) expand capital to 2 units and operate at point *B* at which you would employ 6 units of labor. The least-cost combination of inputs to produce 20 units would occur at point *B*. If capital is held constant at 1 unit, the only way this firm can produce 20 units of output would be to employ 15 units of labor (point *C*). The total cost of producing 20 units would be given by the iso-cost line *HI*. Because iso-cost line *FG* lies to the left of iso-cost line *HI*, the total cost of producing 20 units of output at point *B* would be less than the total cost of producing 20 units of output at point *C*. Suppose the wage rate for labor was $10 per hour and the rental rate for capital was $50 per hour. The total production costs associated with the three iso-cost lines in Figure 7-9 appear in Table 7-1.

Thus, it would cost $200 an hour to operate at point *C* or to produce 20 units of output with 1 unit of capital and 15 units of labor. It would cost only $160 an hour to produce the same quantity of output with 2 units of capital and 6 units of labor.

Table 7-1
TOTAL PRODUCTION COSTS

Iso-Cost Line	Calculation of Total Cost	Total Cost
DE	(1 × $50) + (5 × $10)	$100
FG	(2 × $50) + (6 × $10)	$160
HI	(1 × $50) + (15 × $10)	$200

Therefore, there is economic incentive for the business desiring to produce 20 units of output to expand its capital to the level indicated by point *B*. Using 2 units of capital and 6 units of labor will minimize the cost of producing 20 units of output.

CONCEPT AND MEASUREMENT OF THE PRODUCTION POSSIBILITIES FRONTIER

So far we have examined issues associated with the combination of inputs used by a business. Part of our interest focused on the degree to which one input could be substituted for another in producing a given level of output. It is also important to understand the substitution among the different products the business can produce. To do this, we shall examine the production possibilities frontier (PPF) a business faces, both in the short run, when some inputs are fixed, and in the long run, when all inputs are variable.

Production Possibilities Frontier

Chapter 6 introduced the concept of technical efficiency by indicating the minimal number of hours required to produce given levels of output. In the multiproduct case, we can think of technical efficiency in terms of the maximum outputs possible from the given levels of inputs.

Suppose SunSpot Canning Company has the option of canning all fruit, all vegetables, or some combination of these two products as shown in Table 7-2. As fruit (vegetable) canning is increased, vegetable (fruit) canning must be decreased because of the plant's fixed current canning capacity in the same sense that inputs were substituted for one another in Chapter 6.

If SunSpot specialized in fruit canning, it could can 135,000 cases of canned fruit a week. If it specialized in vegetable canning, SunSpot could can 90,000 cases of canned vegetables a week.[9] Column 3 in Table 7-2 reflects the physical trade-off this canning plant faces for these two products, or the marginal rate of product transformation. It represents the slope of the **production possibilities frontier**.

The production possibilities frontier illustrates the maximum output for different combinations of two products a firm can produce, given its existing resources.

Table 7-2
PRODUCTION POSSIBILITIES FOR SUNSPOT CANNING

	(1) Cases of Canned Fruit	(2) Cases of Canned Vegetables	(3) Marginal Rate of Product Transformation, $\Delta(1) \div \Delta(2)$
A	135,000	0	
B	128,000	10,000	−0.7
C	119,000	20,000	−0.9
D	108,000	30,000	−1.1
E	95,000	40,000	−1.3
F	80,000	50,000	−1.5
G	63,000	60,000	−1.7
H	44,000	70,000	−1.9
I	23,000	80,000	−2.1
J	0	90,000	−2.3

[9] We will assume that the fruit pack or canning season and the vegetable canning season overlap and thus compete for use of SunSpot's existing resources.

Product Substitution

The **marginal rate of product transformation** represents the rate at which the canning of fruit must contract (expand) for a one-case increase (decrease) in vegetable canning. The marginal product rate of transformation in absolute terms is given by

$$\text{marginal rate of product transformation} = \frac{\Delta \text{canned fruit}}{\Delta \text{canned vegetables}} \quad (7.6)$$

In Table 7-2, the marginal rate of product transformation of vegetables for fruit is initially very small (i.e., Δ canned fruit relative to Δ canned vegetables is quite small). In column 3, however, the marginal rate of product transformation becomes much higher (i.e., Δ canned fruit relative to Δ canned vegetables becomes quite large). This increasing marginal rate of product transformation is a widely observed and measured phenomenon and has the same general lawlike acceptance as the declining marginal rate of technical substitution discussed for two inputs.

The substitution relationship between two products can be illustrated further by plotting the combinations of fruit and vegetables shown in Table 7-2. Points A through J in Figure 7-10 represent production levels of fruit and vegetables for a canning plant with a given technically efficient use of capital and labor. Point A, for example, represents specialization in the canning of fruit. Point J, on the other hand, represents specialization in vegetable canning. Point C would result in the canning of some of both commodities with the *same* inputs. Point K can be ruled to be technically *inefficient* because a smaller amount of output is being produced with the same quantity of inputs. It is not utilizing its current production capacity.

A curve drawn through these points is called a production possibilities frontier, which gives the product combinations that can be *efficiently* produced using the business's existing resources. Finally, point L is impossible to attain with SunSpot's existing resources because it lies outside the production possibilities frontier.

Figure 7-10 suggests that vegetable and fruit canning operations at SunSpot are close—but not perfect—substitutes in competing for the firm's scarce resources in production (i.e., the PPF is neither linear nor has a constant slope of -1.0).

Figure 7-10

The downward-sloping production possibilities frontier illustrates the physical trade-offs this business faces in choosing between canning fruit and vegetables as documented in Table 7-2. The concave shape of this curve reflects the less-than-perfect substitutability of input use in switching from canning fruit to canning vegetables.

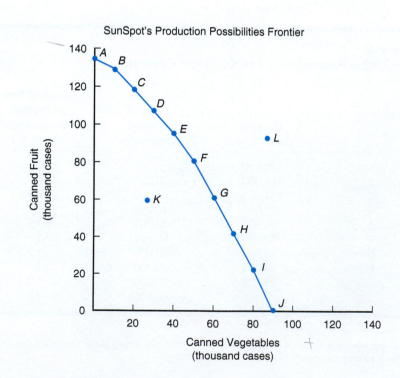

CONCEPT AND MEASUREMENT
OF THE ISO-REVENUE LINE

We need to account for the price received by the canning firm for these two products before we may determine what combination maximizes SunSpot's profits. The **iso-revenue line** represents the rate at which the market is willing to exchange one product for another. We may begin to define an iso-revenue line for SunSpot by defining its total revenue, which, for fruit and vegetables, is given by

$$\text{total revenue} = (\text{price of canned fruit} \times \text{cases of canned fruit})$$
$$+ (\text{price of canned vegetables} \times \text{cases of canned vegetables}) \qquad (7.7)$$

If no canned vegetables are produced by this canning plant, then the number of cases of canned fruit produced for a specific level of revenue is given by the level of revenue divided by the price of canned fruit. Similarly, if no canned fruit is produced, then the number of cases of canned vegetables produced for a specific level of revenue is given by the level of revenue divided by the price of canned vegetables.

The slope of the iso-revenue line is the ratio of the price of the two products, or the price of canned vegetables this business receives by selling a case of canned vegetables in the market divided by the price of canned fruit. All points on the iso-revenue line result in the same level of revenue. Stated mathematically, the slope is given by

$$\text{slope of iso-revenue line} = -\frac{\text{price of vegetables}}{\text{price of fruit}} \qquad (7.8)$$

For example, if the wholesale price of a case of canned fruit that SunSpot receives is $33.33, and the price of a case of canned vegetables it receives is $25.00, the slope of the iso-revenue line would be −0.75 (i.e., the negative of $25.00 divided by $33.33). One case of canned vegetables is worth three-fourths of a case of canned fruit.

Figure 7-11 illustrates the general nature of the iso-revenue line for these two products. You may wonder why the slope of the line plotted in Figure 7-11 is the negative of the ratio of the price of canned vegetables to the price of canned fruit, when the vertical axis is labeled "cases of fruit" and the horizontal axis is labeled "cases of vegetables."[10] This is entirely consistent with the discussion of the budget constraint in Chapter 3, in which we determined the slope of the budget line.

The original iso-revenue line associated with a revenue of $1 million, the price of a case of canned fruit of $33.33, and the price of a case of canned vegetables of $25.00 are the basis for iso-revenue line *AB* plotted in Figure 7-11A. The maximum number of cases of canned fruit associated with this level of revenue is 30,000 cases (i.e., $1 million ÷ $33.33), and the maximum number of cases of canned vegetables is 40,000 cases (i.e., $1 million ÷ $25.00). Thus, SunSpot would achieve a revenue of $1 million if it could process 30,000 cases of fruit, 40,000 cases of vegetables, or the specific combinations of these two products that appear on the iso-revenue line.

Figure 7-11B shows that if consumer expenditures for SunSpot's products fell by half, the iso-revenue line would shift in a parallel fashion from line *AB* to line *CD*. Only 15,000 cases of fruit, 20,000 cases of vegetables, or specific combinations of the two products could be sold. A doubling of consumer expenditures for these products would shift the iso-revenue line from line *AB* to line *EF*. Under these conditions, SunSpot would sell 60,000 cases of canned fruit, 80,000 cases of canned vegetables, or specific combinations of these two products. Figure 7-11C shows what would happen to the

Role of input price   We cannot determine the optimal combination of two products to produce without knowing the price of these two products.

Shifting iso-revenue curve   The iso-revenue curve will shift or change slope as the price of the two products changes.

---

[10] Equation 7.8 can be rearranged algebraically to form the iso-revenue line and its slope:

$$\text{case of canned fruit} = \frac{\text{total revenue}}{\text{price of fruit}} - \frac{\text{price of vegetables}}{\text{price of fruit}} \times \text{cases of canned vegetables}$$

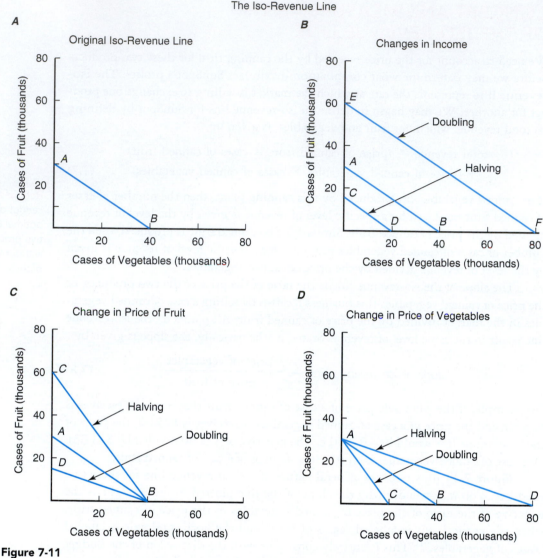

Figure 7-11

The iso-revenue line plays an important role in the determination of the profit-maximizing combination of two products. The slope of this line is equal to the negative of the price ratio for the two products.

iso-revenue line if the price of fruit were either doubled (line *BD*) or cut in half (line *BC*). Figure 7-11*D* shows what would happen if the price of vegetables doubled (line *AC*) or were cut in half (line *AD*). In Figure 7-11*C* and *D*, the slope of the iso-revenue line became either flatter or steeper as the price of one of the commodities changed.

PROFIT-MAXIMIZING COMBINATION OF PRODUCTS

We can determine the profit-maximizing combination of products under conditions of perfect competition by considering both the physical and economic trade-offs from the alternatives currently available. This requires uniting the concepts of the production possibilities frontier and the iso-revenue line.

Choice of Products in the Short Run

The technical rate of exchange between canned fruit and canned vegetables for SunSpot in the current period is captured by the production possibilities frontier in

Condition for profit maximization   The profit-maximizing firm will operate where the slope of the iso-revenue curve is tangent to the production possibilities curve. The firm has maximized its revenue, and, by definition, it has maximized its efficiency.

Table 7-3
PROFIT-MAXIMIZING COMBINATION OF PRODUCTS FOR SUNSPOT

(1) Cases of Canned Fruit	(2) Cases of Canned Vegetables	(3) Revenue, $33.33 × (1) + $25.00 × (2)	(4) Marginal Rate of Product Transformation, Δ(1) ÷ Δ(2)	(5) Ratio of Price of Vegetables to the Price of Fruit, $25.00 ÷ $33.33
135,000	0	$4,499,550		0.75
128,000	10,000	4,516,240	−0.70	0.75
119,000	20,000	4,466,270	−0.90	0.75
108,000	30,000	4,349,640	−1.10	0.75
95,000	40,000	4,166,350	−1.30	0.75
80,000	50,000	3,916,400	−1.50	0.75
63,000	60,000	3,599,790	−1.70	0.75
44,000	70,000	3,216,520	−1.90	0.75
23,000	80,000	2,766,590	−2.10	0.75
0	90,000	2,250,000	−2.30	0.75

Figure 7-10. We know from Equation 7.6 that the slope of this curve, called the marginal rate of product transformation, is equal to the ratio of the change in the production of these two products. The slope of this curve is negative, indicating an increasing opportunity cost of product substitution.

The profit-maximizing business seeks to maximize the revenue for the least-cost combination of inputs. In the present context, the business will want to determine the point at which the marginal rate of product transformation is equal to the relative prices of the products being produced. Table 7-3 suggests that the profit-maximizing combination of canned fruit and vegetables for SunSpot in the current period, given existing input prices, would be between 119,000 and 128,000 cases of canned fruit and between 10,000 and 20,000 cases of canned vegetables.

The absolute value of the marginal rate of product transformation in column 4 of Table 7-3 will equal the absolute value of the price ratio in column 5 of 0.75 in this range. At this point, the marginal rate of product transformation (slope of the production possibilities curve) for fruit and vegetables equals the ratio of the price of vegetables to the price of fruit (slope of the iso-revenue line). We can, therefore, state the conditions for the profit-maximizing combination of these two products in mathematical terms as

$$\frac{\Delta \text{canned fruit}}{\Delta \text{canned vegetables}} = -\frac{\text{price of vegetables}}{\text{price of fruit}} \qquad (7.9)$$

in which both sides of the equation will have a negative value. (Recall the negative values for the marginal rate of product transformation in Tables 7-2 and 7-3.)

Figure 7-12 suggests that the profit-maximizing combination of canned fruit and vegetables for SunSpot would be approximately 126,000 cases of fruit and 13,000 cases of vegetables, which lie in between the ranges discussed in the context of Table 7-3. Total revenue would reach approximately $4,524,580. This combination also represents maximum profits because the business is on the production possibilities curve, which assures maximum technical efficiency.

Effects of Change in Product Prices

Let's assume that we are at point *M* in Figure 7-12, and the wholesale price of fruit suddenly falls to $25. This will alter the slope of the iso-revenue line in Figure 7-13.

Figure 7-12

The profit-maximizing choice of how much fruit and vegetables to can is given by the point at which the iso-revenue line is tangent to this business's current production possibilities frontier (point M).

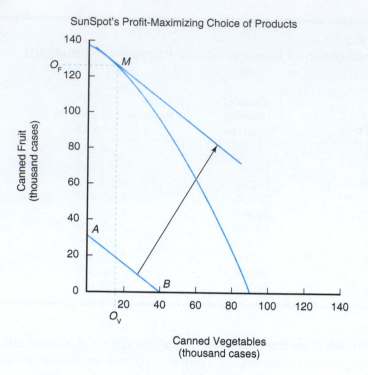

Figure 7-13

A decrease in the price of canned fruit would cause SunSpot to alter the combination of the products it cans.

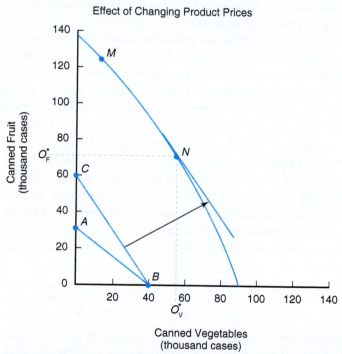

The new iso-revenue line *CB* is extended out in a parallel fashion until it is tangent to the production possibilities curve at point *N*.

In Figure 7-13, SunSpot's profit-maximizing objective is to *decrease* its fruit canning operations from O_F (Figure 7-12) to O^*_F and *increase* its vegetable canning operations from O_V (Figure 7-12) to O^*_V. This suggests that a business, with a given amount of resources, will alter the allocation of resources between the production of alternative products as their price ratio changes. The quantity of vegetables and fruit SunSpot chooses to can, therefore, will depend on the prices of all its inputs, the stock of its existing fixed inputs, the technology embodied in its labor and capital, and the relative price of fruit and the price of vegetables.

SUMMARY

The purpose of this chapter is twofold: (1) consider several input demand issues facing the business, including the size of the business in the long run and the factors that can influence this size, and (2) illustrate the physical and economic relationships associated with the issue of product choice by a business under conditions of perfect competition. The major points made in this chapter are summarized as follows:

1. The least-cost combination of inputs and level of output possible with a given budget is found graphically by looking for the point of tangency between a specific iso-cost line and the highest possible isoquant curve.

2. The cost of production associated with the least-cost combination to produce a given level of output is found graphically by looking for the point of tangency between a specific isoquant and an iso-cost line.

3. The least-cost combination of two inputs can be found numerically by searching for the equality between the ratio of the two input prices (slope of the iso-cost line) and marginal rate of technical substitution or ratio of the two input MPPs (slope of an isoquant).

4. Firms in the long run can expand the size of their operations by using more of all inputs, including forms of capital that were fixed in the short run.

5. The long-run average cost curve, often referred to as the planning curve, illustrates how varying the scale of the firm will affect its efficiency or its average cost associated with a given level of output.

6. The long-run equilibrium of the firm under conditions of perfect competition will occur at that output level where the product price is equal to both the firm's marginal and average total costs.

7. When businesses expand the size of their operations, they will incur returns to scale.

8. If the increase in output is exactly proportional to the increase in input use, the returns to scale are constant. If this increase is more (less) than proportional to the increase in input use, returns to scale are increasing (decreasing). The business will normally pass through a phase of increasing returns to scale or **economies of scale** before encountering constant and then decreasing returns to scale, or diseconomies of scale, as it expands beyond the minimum point on the long-run average cost curve. The risk in doing this is that it may have to downsize its operations eventually if prices fall.

9. The production possibilities frontier in the current period represents the different combinations of two products a business can produce, given efficient use of its existing resources. When the business expands its operations in the long run, it will be on higher production possibilities frontiers that reflect the changing nature of its resources.

10. The slope of the production possibilities curve is called the marginal rate of product transformation. This slope reflects the rate at which the business can substitute between the production of two products in the current period. If two products are perfect substitutes, the marginal rate of product transformation will be constant at all points along the production possibilities frontier.

11. The iso-revenue line reflects the rate at which the market is willing to substitute between two products as their prices change. The slope of this line therefore reflects the ratio of the prices of the two products. The intercept of this line on both axes reflects the maximum quantity of these two products that could be purchased if bought alone and also reflects a given amount of revenue and the prices of the products. Changes in the prices of these products will alter the slope of the iso-revenue line.

12. The profit-maximizing combination of two products to produce is determined by the point of tangency with the business's current production possibilities frontier and the iso-revenue line. At this point of tangency, the marginal rate of product transformation, or slope of the production possibilities frontier, will be equal to the ratio of the two product prices, or the slope of the iso-revenue line.

KEY TERMS

Economies of scale
Iso-cost line
Isoquant
Iso-revenue line
Long-run average cost curve

Marginal rate of product transformation
Marginal rate of technical substitution

Production possibilities frontier
Rental rate of capital

TESTING YOUR ECONOMIC QUOTIENT

1. A firm uses corn and protein supplement to mix a particular type of hog feed. Corn costs $0.08 per pound and protein supplement costs $0.12 per pound. Let's assume the firm has $3,000 on these two inputs. Plot the iso-cost line suggested by this information in the following graph. What is the value of this curve's slope?

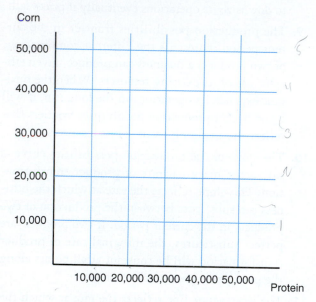

2. Suppose the wage rate for labor is $20 an hour and the rental rate for capital is $50 per hour. Based on this information, please answer the questions appearing below the graph.

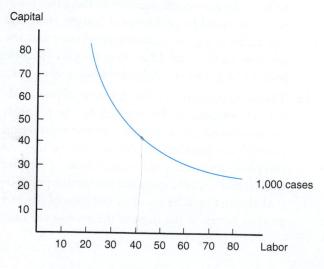

   a. Draw the iso-cost line associated with an hourly budget of $1,000.
   b. What is the least-cost level of capital and labor this business should utilize when packaging 1,000 cases of fruit juice? How did you arrive at this answer?

c. How much does it cost this business to package 1,000 cases of fruit juice? If the firm can sell the juice for $50 per case, what is its accounting profit?

3. In *each* of the following four graphs, describe what caused the iso-cost line to change *in each box* and the nature of the change *on each line*.

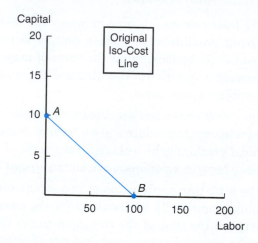

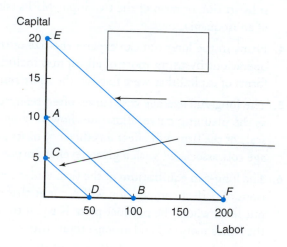

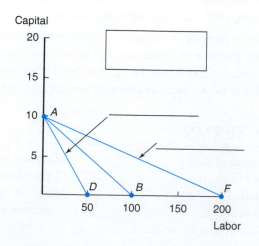

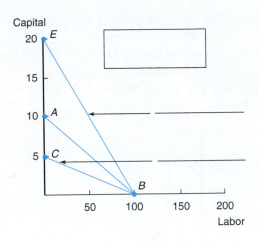

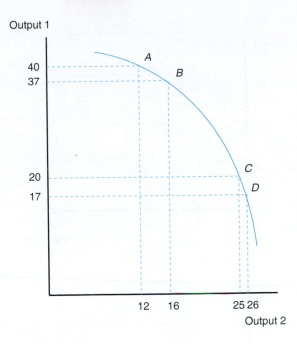

4. Given the following graph, briefly respond to the questions appearing directly below this graph.

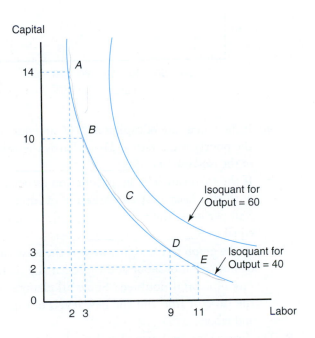

6. In *each* of the following four graphs, describe what caused the iso-revenue line to change *in each box* and the nature of the change *on each line*.

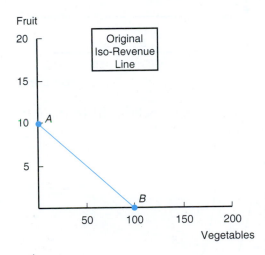

a. For an output level of 40 units, calculate the marginal rate of technical substitution between points *A* and *B*.

b. Also for an output level of 40 units, calculate the marginal rate of technical substitution between points *D* and *E*.

5. Given the following graph for two products, respond to the questions below:

a. Calculate the marginal rate of product transformation between points *A* and *B*.

b. Calculate the marginal rate of product transformation between points *C* and *D*.

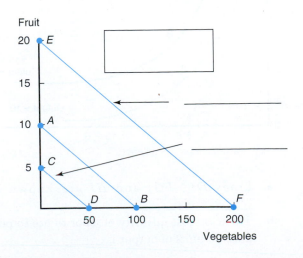

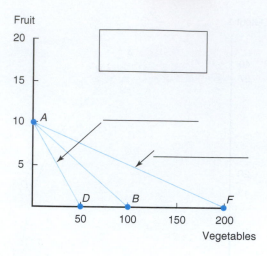

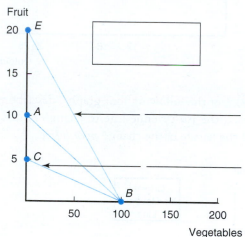

7. Given the graph below, fill in the following blanks:

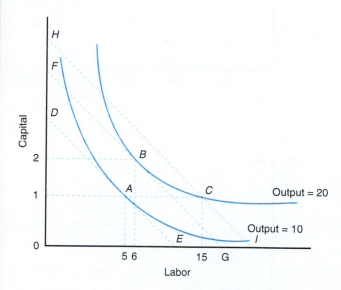

a. Lines *DE*, *FG*, and *HI* are known as _____.

b. The least-cost quantity of labor to use to produce 10 units of output is _____.

c. The least-cost quantity of capital to use to produce 20 units of output is _____.

d. The curves labeled Output = 10 and Output = 20 are known as _____.

8. Given the graph below, fill in the following blanks:

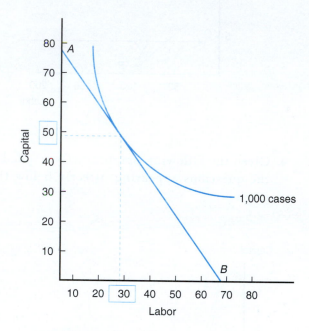

a. If the rental rate of capital is $100 an hour and the hourly wage rate is $10, what is the value of the iso-cost line? $ _____.

b. If the firm can sell the production associated with this least-cost combination of inputs at $30 per unit, the firm's total profit would be equal to $ _____.

c. In the graph to the right, what is the least-cost combination of labor and capital if the rental rate of capital doubled? Show all changes to the iso-cost line and the new levels of capital and labor.

9. The firm's marginal rate of technical substitution shows
a. the slope of the indifference curve.
b. the slope of the isoquant curve.
c. the slope of the iso-revenue curve.
d. the slope of the budget constraint.

10. The point of tangency between the iso-cost curve and the isoquant indicates
a. how much of two inputs a firm should use.
b. how much profit a firm can make using these two inputs.
c. the minimum cost of using these two inputs for a given level of output.
d. none of the above.

11. In the short run, all the resources that may be used by the firm are considered variable. T F

12. The doubling of a firm's feedlot capacity which results in a doubling of production is characteristic of increasing returns to scale. T F

13. The declining portion of the firm's short-run average cost curve is characteristic of increasing returns to scale. T F

14. A line connecting the points of tangency between successive isoquants and iso-cost lines is known as the firm's _____.

15. The horizontal segment of the long-run average cost curve is characteristic of an industry experiencing _____.

16. The slope of the iso-revenue line is determined by _____.

REFERENCE

Hall FF, EP LaVeen: Farm size and economic efficiency: the case of California, *American Journal of Agricultural Economics* 6(4): 589–600, 1978.

Market Equilibrium and Product Price: Perfect Competition

Chapter 4 discussed the derivation of the market demand curve, based on the demands of individual consumers, and its elasticity. This represented exactly one-half of the relationships needed to understand changing market conditions, including the market equilibrium price. The other part of the puzzle is the market supply curve.

   The purpose of this chapter is to explain how we can derive the market supply curve for a particular product under conditions of perfect competition and interpret what market equilibrium means for consumers and producers. Attention will also be given to the forces that cause changes in the market equilibrium price and to the nature of the adjustment to a new market equilibrium.

DERIVATION OF THE MARKET SUPPLY CURVE

The market supply curve for a particular product is based on the decisions of what and how much to produce made by individual businesses in an industry.

Firm Supply Curve

The marginal cost curve and the average variable cost curve help determine the minimum price at which a business can justify operating from an economic perspective. For our hypothetical business TOP-AG, whose costs of production were presented in Table 6-3, the minimum acceptable product price would be approximately $16, which is far below the $45 TOP-AG is currently receiving for its product. If the price of TOP-AG's product fell to $10, should the business continue to operate? Would TOP-AG be covering all of its costs of production at this product price? Would the business even be covering its variable costs of production? In the discussion to follow,

Holstein cows in a milking power on a dairy farm. The number of dairy farmers producing fluid milk is just one example of the type of enterprises in the farm sector that approximate the conditions of perfect competition. Credit: Inzyx/Fotolia.

we will see that the marginal cost curve lying above the minimum average variable cost represents the **firm supply curve** in the current period.

Output O_{BE} in Figure 8-1 represents the *break-even* level of production for TOP-AG, or the point at which the marginal cost curve in Figure 6-4 intersects the average total cost curve. At this level of output, average revenue is just equal to average total cost. This means that at P_{BE}, economic profits are equal to zero. Output O_{SD} in Figure 8-1 is identical to the point in Figure 6-4 at which the marginal cost curve intersects the average variable cost curve. This means that if the price fell to P_{SD}, average revenue would just equal average variable cost. The business could minimize its losses in the current period by continuing to produce if prices were below price P_{BE}. If the product price corresponding to the segment of the marginal cost curve lies between prices P_{BE} and P_{SD}, the firm could cover all of its variable costs and some, though not all, of its fixed costs by continuing to produce.

A rational competitive firm will cease producing in the short run only when the product price falls below the average variable costs of production, which occurs at price P_{SD} in Figure 8-1. Operating when the price is below point P_{SD} on the marginal cost curve will only add to the firm's losses because TOP-AG is no longer covering all variable costs. Furthermore, the suppliers of variable inputs (such as fuel and hired labor) will likely cease supplying these services to the business when the checks start to bounce. This is why the level of output associated with price P_{SD} on the marginal cost curve (output O_{SD}) is known as the *shutdown* point. This explains why we present only the portion of TOP-AG's marginal cost curve that appears above its average variable cost curve when illustrating this business's supply curve in Figure 8-1.

Market Supply Curve

Figure 4-6 illustrates the fact that the market demand curve represents the summation of the quantities desired by all consumers at specific market prices. The market demand curve for the two-consumer example depicted in that figure was found by horizontally summing the individual demands of both consumers. The firm's supply

The firm's supply curve in the short run is that portion of its marginal cost curve that lies above its average variable cost curve.

The market supply curve can be seen as a summation of all the firm supply curves, or the quantity each firm would be willing to supply for specific prices.

Figure 8-1

The portion of this business's marginal cost curve that lies above the average variable cost curve represents a business's current supply curve under conditions of perfect competition.

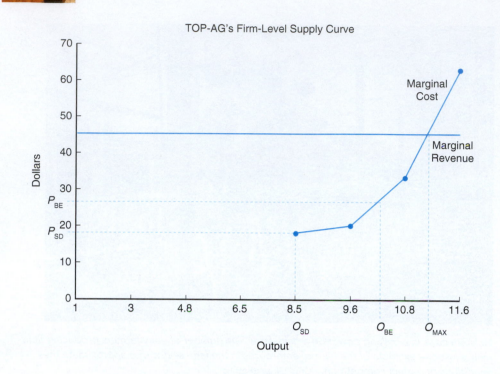

curve is the portion of its average variable cost curve as illustrated in Figure 8-1. The **market supply curve** under conditions of perfect competition is determined in a similar manner. The market supply curve represents the summation of the quantities all firms are *willing to supply* at specific market prices.

Figure 8-2A suggests that Gary Grower would be willing to supply 1 ton of fresh broccoli if the market price were $1.00 per pound, 2 tons if the price were $1.50 per pound, and so on. Figure 8-2B shows that Ima Gardner would decline to produce at a market price of $1 per pound, but would supply 1 ton of broccoli at a market price of $1.50 per pound, and so on. Figure 8-2C shows that if the market supply were limited to these two producers, the total supply of broccoli forthcoming at a market price of $1 per pound would be 1 ton, 3 tons at a market price of $1.50 per pound, and so on.

Like the demand curve, we can also characterize the properties of the market supply curve by examining the elasticity of this curve.

Own-Price Elasticity of Supply

The market supply curve for a particular product generally has a positive slope because the quantity supplied by businesses increases when the price it receives goes up. It is helpful to think of the behavioral response of producers in the context of their own-price **elasticity of supply**. This elasticity is expressed as

$$\frac{\text{own-price}}{\text{elasticity of supply}} = \frac{(Q_{SA} - Q_{SB})/[(Q_{SA} + Q_{SB})/2]}{(P_A - P_B)/[(P_A + P_B)/2]} \tag{8.1}$$

in which Q_{SA} is the quantity supplied *after* the change in price from P_B to P_A and Q_{SB} is the quantity *before* the change in price. An own-price elasticity of supply exceeding one indicates an elastic supply, and an own-price elasticity of less than one suggests an inelastic supply.

For example, if the own-price elasticity of supply for a product is 1.5, a 1% increase in product price would cause businesses producing this product to increase their production by 1.5%. Because the percentage change in revenue is equal to the

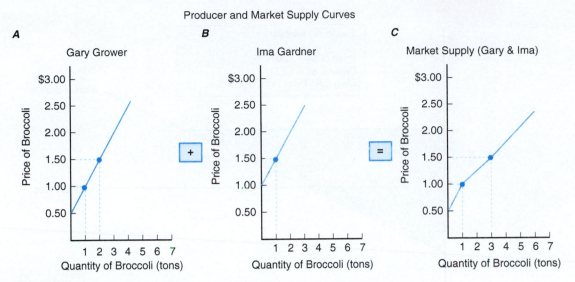

Figure 8-2

The market supply curve is found by horizontally summing the quantities supplied by all producers for given levels of market price.

percentage change in price plus the percentage change in the quantity supplied, the total revenue of producers would increase by 2.5%.

Finally, the more (less) elastic or flatter (steeper) the market's supply curve is, the greater (lower) the impact a price change will have on total revenue, all other things constant.[1] What would the impact of a 1% increase in product price be on quantity supplied if the market supply curve were perfectly inelastic? What would the change in total revenue be under these conditions?

Producer Surplus

Economic profit, or **producer surplus**, is the economic return above the firm's cost of production.[2] When economic profit exists, surpluses are accruing to businesses. This surplus may be measured for an individual business by examining the business's returns above costs of production.

A business will supply the first unit of output at a price equal to the marginal cost of producing the first unit. If this marginal cost were $1 and the price of the product were $4, the business would receive a $3 surplus from producing and exchanging the commodity. If the marginal cost of producing the one-hundredth unit were $3, the surplus would be $1 for producing the unit.

By similar reasoning, the area above the market supply curve and below market price represents the producer surplus accruing to businesses participating in the market. This surplus is represented by area *ABC* in Figure 8-3 when the product price is $4. If the product price rises to $6, producer surplus increases to area *CED*. These areas represent the total economic profit received by firms in their market. Area *AEDB* represents *the gain* in producer surplus resulting from the rise in the product price from $4 to $6. Hence, producer surplus represents a measure of the gain in economic welfare that businesses receive from supplying a particular product in the current period.

> **Producer surplus** represents the profit realized by firms in the market for specific quantities supplied.

---

[1] If firms produce more than one product, and these products are independent of one another, the discussion presented earlier applies to each product considered separately.

[2] This reflects both variable and fixed costs as the supply curve reflects the marginal cost curves of firms in the industry.

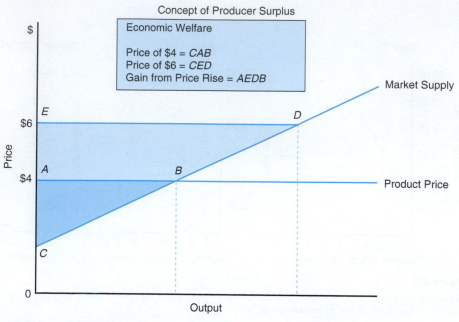

Figure 8-3

The change in the economic welfare of businesses in this market can be approximated through the concept of economic rent or producer surplus. The value of this surplus at a price of $4 is given by the darkly shaded area above the market supply curve and below the $4 equilibrium market price for the product. This area reflects the revenue received by a business above the minimum price at which it would have been willing to supply its product.

MARKET EQUILIBRIUM UNDER PERFECT COMPETITION

One of the conditions for perfect competition presented at the beginning of Chapter 6 is that the business's product is homogeneous, or a perfect substitute for the product sold by the other businesses in this market. Perfect competition enables buyers in the market to choose among a large number of sellers. Another condition enables any business that desires to enter or leave the sector to do so without encountering serious barriers. There must be a large number of sellers and buyers in the market to have perfect competition. No single buyer or seller should have a disproportionate influence on price. Finally, sufficient information must be available to all participants regarding prices, quantities, qualities, sources of supply, and so on.

When all four conditions hold, we can say the market's structure is perfectly competitive. Businesses supplying goods in this market are also, by definition, perfectly competitive. Each is a **price taker**, or accepts the price of the product as given. Agriculture probably comes as close as any sector in the economy to satisfying these conditions. There are thousands of corn growers in the United States producing No. 2 yellow corn, each having similar access to daily market information, and none confront legal barriers to enter or leave the sector.[3]

Market Equilibrium

The equilibrium price in a perfectly competitive market is established by the point of intersection of the market's demand and supply curves. Let D_M represent the market demand schedule for the sector's product and S_M represent the market supply

Area of producer surplus
Producer surplus is found graphically by calculating the area lying above the market supply curve and below the market equilibrium price.

---

[3] The corresponding profit of an individual firm operating at MR = MC is equal to the average profit, or average revenue (P) minus average total cost, times the quantity produced.

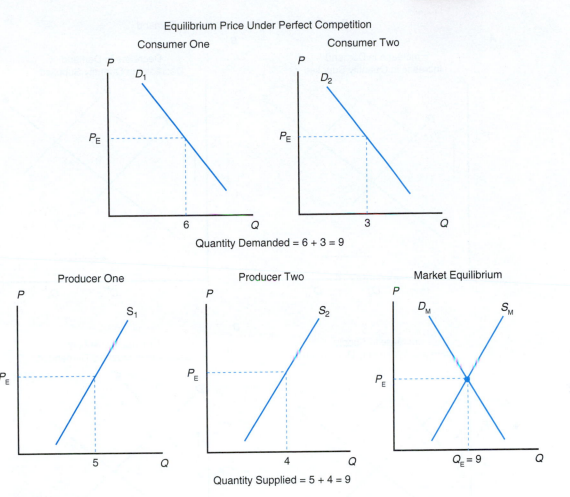

Figure 8-4

The equilibrium price in a competitive market is given by the intersection of the market demand and supply curves. As shown in this figure, this would result in a price P_E and quantity Q_E. At this price, consumers collectively would demand 9 units, and producers collectively would supply 9 units.

schedule for all businesses in this market. We have assumed, for ease of presentation, that there are only two buyers (consumers) and two sellers (producers) in this market. Figure 8-4 shows that the equilibrium price in this market would be equal to P_E. At this price, businesses would be willing to supply quantity Q_E (or 9 units) and the buyers of this product would desire to purchase quantity Q_E (also 9 units). Thus, P_E and only P_E is the price per unit that will "clear the market."

Shifts in either the demand curve or the supply curve will result in a new equilibrium market price. Four possible events can occur that will affect the market equilibrium price and quantity:

1. Demand increases, shifting the demand curve to the right.
2. Demand decreases, shifting the demand curve to the left.
3. Supply increases, shifting the supply curve to the right.
4. Supply decreases, shifting the supply curve to the left.

The effects of each situation on the equilibrium price that clears the market are illustrated in Figure 8-5. In Figure 8-5A, for example, we see that an increase in demand (perhaps consumer disposable income increased) will result in a higher market price (P^*_e). Buyers will now demand, and businesses will supply, a quantity equal to Q^*_e

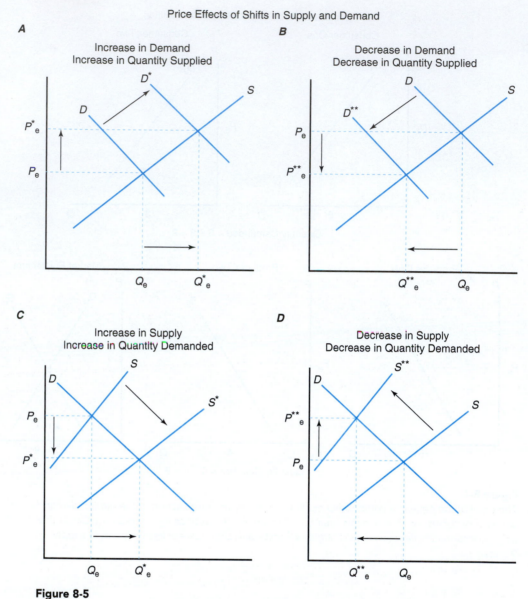

Figure 8-5

It is important to distinguish between changes in supply or demand and changes in the quantity supplied or demanded.

instead of Q_e. The opposite effect occurs when demand decreases (see Figure 8-5B). In both cases, there is a *change in demand* and a *change in the quantity supplied*.

Turning to supply, let us assume that the supply of the product increases or that the supply curve shifts to the right. Figure 8-5C illustrates that this shift will lead to a decline in the market clearing price from P_e to P^*_e. At this new price, firms will supply, and buyers will demand, a quantity equal to Q^*_e. Figure 8-5D shows that the opposite outcome will occur if there is a decrease in supply. In both cases, there is a change in *supply* and a *change in the quantity demanded*.

The elasticity of the demand and supply curves plays an important role in determining how much the equilibrium price will change if either demand or supply changes. For example, the more inelastic or steeper the demand curve, the greater the rise (fall) in the market price will be for a given decrease (increase) in supply. The relatively inelastic nature of the demand for farm products, coupled with a volatile supply curve that can shift to the right or to the left depending on the vagaries of weather, helps explain the high variability of farm income that we often see from one period to the next. More will be said about this when we discuss the traditional farm problem in subsequent chapters.

Total Economic Surplus

In Figure 4-9, we learned that consumer surplus is given by the area below the demand curve and above the equilibrium price. We also learned that producer surplus is given by the area above the supply curve and below the equilibrium price (see Figure 8-3). These areas represent the economic well-being achieved by consumers and producers at the equilibrium market price. If we add these two triangular areas together, the newly formed triangle represents the economic well-being achieved by all market participants in this particular market. In Figure 8-6, the summation of consumer surplus (area 1) plus producer surplus (area 2) represents the total area above the supply curve and below the demand curve, and hence the **total economic surplus** received by all market participants.

Now suppose that because of low yields, the supply curve for this market shifts inward to the left from S to S^* (Figure 8-7). Producer surplus would now be equal to

Total economic surplus is equal to consumer plus producer surplus.

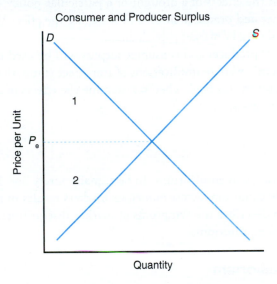

Consumer and Producer Surplus

Figure 8-6

Area 1 represents consumer surplus and area 2 represents producer surplus. The sum of both areas represents the economic well-being of society from participating in this market.

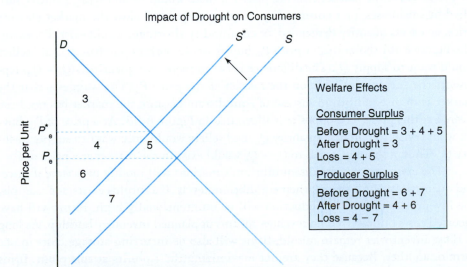

Impact of Drought on Consumers

Welfare Effects
Consumer Surplus
Before Drought = 3 + 4 + 5
After Drought = 3
Loss = 4 + 5
Producer Surplus
Before Drought = 6 + 7
After Drought = 4 + 6
Loss = 4 − 7

Figure 8-7

This figure shows what would happen to the economic welfare of consumers and producers if a drought caused the aggregate supply curve to shift from S to S*.

area 4 plus area 6. The numbers here merely refer to an area in the graph, and do not represent the value of that area. Thus, if areas 4 plus 6 sum to less than areas 6 plus 7, we may conclude that the economic well-being of producers would have declined. We can also conclude that consumer surplus declined because consumers received the equivalent of areas 3, 4, and 5 before but now receive only area 3 and thus are worse off than before. Finally, we can conclude that this decrease in supply means that the economic well-being of market participants in general would have fallen by an amount equal to area 5 plus area 7.

Applicability to Policy Analysis

The concept of producer and consumer surplus is an important analytical tool to economists. We may be examining the economic welfare implications of a major drought, such as the drought of 2012, which results in a shift in the current supply curve as discussed previously. Or we may be examining a change in agricultural or macroeconomic policy that causes a shift in either the demand or the supply curve in the current period. The concept of producer and consumer surplus helps in assessing the relative effects of these externalities. It is not unusual for economists, when testifying before Congress on the effects of a drought or a particular policy change, to use the concept of consumer and producer surplus to illustrate the effect this would have on the economic well-being of market participants.

The concept of producer and consumer surplus will be used in Chapter 9 when we assess the economic welfare implications of imperfect competition. It will also be used extensively later in this book when we examine the effects of agricultural policy on consumers and farmers.

ADJUSTMENTS TO MARKET EQUILIBRIUM

Markets are not always in equilibrium. In fact, many rarely are. Instead, changing demand and supply conditions across numerous markets results in market disequilibrium. This section describes the symptoms of market disequilibrium and how markets adjust to a new equilibrium.

Market Disequilibrium

A commodity surplus or shortage will occur when the market is in disequilibrium.

At prices above the market clearing price P_e, there would be an excess quantity supplied by businesses, or a **commodity surplus**. At prices below the market clearing price, an excess quantity demanded, or **commodity shortage**, would exist. For example, Figure 8-8A shows that at price P_s, buyers would wish to purchase Q_d, and sellers would want to supply Q_s. The difference between these two quantities ($Q_s - Q_d$) represents the surplus available on the market at the price P_s. This suggests that the market is in disequilibrium instead of equilibrium because the market has not been cleared at this price. The opposite is illustrated in Figure 8-8B. At a price of P_d, buyers would wish to purchase quantity Q_d, and sellers would only want to supply quantity Q_s. Thus, a shortage equal to $Q_d - Q_s$ would exist in the market at a price of P_d.

The existence of these disequilibrium situations will modify over time if prices and quantities are free to seek their equilibrium levels. If a surplus exists, for example, the inventories of unsold production will be unintentionally high. Firms will have incurred costs but received no revenues for this unplanned inventory buildup. As long as these inventories remain unsold, firms will also be incurring storage costs in one form or another. Because they are not maximizing their profits at this point, firms will find it profitable to decrease their level of production and accept a lower price for their inventories.

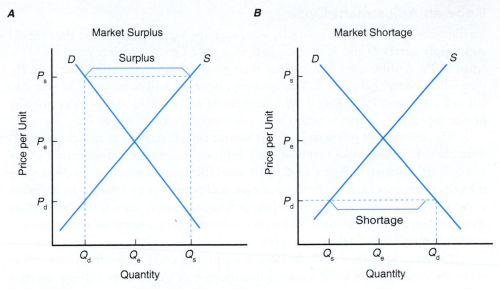

Figure 8-8

The equilibrium price in a competitive market is given by the intersection of the market demand and supply curves. If, instead, the price were equal to P_s (A), producers would be willing to supply more than consumers would demand. This phenomenon is referred to as surplus. If the price were instead equal to P_d, the quantity demanded by consumers would be greater than the quantity producers would be willing to supply. This excess demand situation is commonly referred to as a shortage.

This adjustment process will stop after prices have fallen from P_s to P_e. If a shortage exists, buyers would compete for available supplies by offering to pay higher prices. This will encourage firms to raise and market more of this commodity. This adjustment process will stop after prices have risen from P_d to P_e. At this point, the quantity demanded will be exactly equal to the quantity supplied, and market equilibrium will be restored.

Length of Adjustment Period

The adjustment processes discussed earlier may suggest that the quantities demanded and supplied are both determined by current prices. In some sectors like agriculture, however, adjustment to market equilibrium takes time. One reason is the biological nature of the production process itself. Once the crop has been planted, for example, little can be done to adjust the supply response of producers until the next production season. Furthermore, when farmers plant their crop, they do not know what the market price will eventually be when they sell their crop several months later.

Let us assume for the moment that farmers base their production plans for this year on last year's price. Price and quantity are now *sequentially determined* rather than simultaneously determined. Last year's price determines this year's production response. This year's quantity marketed, however, will affect this year's price, which will affect next year's production, and so on.[4] If prices were high last year, for example, farmers under free market conditions would respond by expanding their production activities with the anticipation of eventually marketing more output. The increased level of production will lead to lower prices, all other things constant. This pattern of price and quantity responses forms a pattern like a spider's cobweb over time.

---

[4] The demand and supply functions in this instance would be given by $P_t = f(Q_t)$ and $Q_t = f(P_{t-1})$, respectively. The response to last period's price in the supply function is thus different from the response to current price assumed thus far.

Cobweb Adjustment Cycle

Cobweb adjustment One way to think of how markets adjust to a new equilibrium is the cobweb theorem, which leaves a pattern much like a spider web.

To illustrate cobweb market behavior, let us examine Figure 8-9A. Given the demand and supply curves D and S, let's suppose that the price of corn last year (year 1) is equal to P_1. Assume corn farmers base their production intention for year 2 on P_1, they will produce Q_2 in year 2. This quantity, however, will cause prices in year 2 to fall to P_2. As shown in Figure 8-9B, corn farmers will respond to this lower price by producing only Q_3 in year 3, which will cause market prices to rise to P_3.

This behavior of prices and quantities over time is referred to as a cobweb pattern, after the cobweb-like nature of the solid lines tracing the movements of prices and quantities shown in Figure 8-9C. This panel illustrates the nature of a converging cobweb. Here, prices and quantities will eventually converge to a market equilibrium at price P_E. This cobweb pattern will occur when the *supply curve* is steeper or more inelastic than the demand curve. A diverging cobweb occurs when the *demand curve* is steeper or more inelastic than the supply curve.[5]

Events causing changes in demand or supply can cause an interruption to these cycles and lead to a new set of market adjustments over time. As we will discuss later in Chapter 11, federal programs that are designed to modify the booms and busts associated with fluctuating prices and quantities exist for some farm commodities.

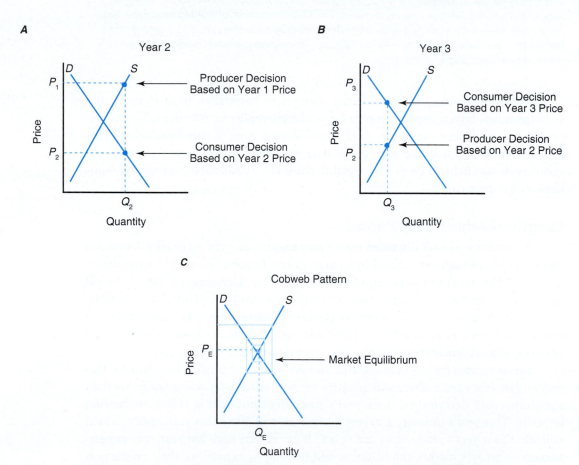

Figure 8-9

When producers respond to the previous period's price, markets will adjust to market equilibrium in a cobweb pattern.

---

[5] A persistent cobweb will occur if the demand and supply curves have identical slopes, which means that the market will continue to oscillate around the market's equilibrium, never converging or diverging.

SUMMARY

The purpose of this chapter is to explore the determination of the market equilibrium price under conditions of perfect competition and how the market adjusts to a new market equilibrium when market conditions change. The major points made in this chapter are summarized as follows:

1. The supply curve for an individual business under conditions of perfect competition is represented by that portion of its marginal cost curve that lies above its average variable cost curve. This suggests that the business will supply this product in the current period as long as it is able to cover its variable costs of production.

2. The market supply curve for a particular commodity under conditions of perfect competition represents a *horizontal summation* of the supply responses of the businesses selling this commodity.

3. The break-even level of output occurs for a business when the price of the product is sufficient to just cover its *average total costs*. Economic profit at this output level would be equal to zero.

4. The shutdown level of output occurs when the price of the product is just sufficient to cover the business's *average variable costs*. If the price of the product were to fall below this level, the firm would cease operations.

5. Producer surplus, the supply-side counterpart to consumer surplus discussed in Chapter 4, represents the gain in economic well-being that businesses will achieve by participating in a particular market during the period. Product surplus is equal to the area above the market supply curve and below the market equilibrium price.

6. Market equilibrium under conditions of perfect competition occurs when the market demand curve intersects the market supply curve. At this price, consumers will purchase the market equilibrium quantity at the market equilibrium price, and businesses will supply the market equilibrium quantity at the market equilibrium price.

7. A shift in the market supply curve will result in a *change in supply* and a *change in the quantity demanded*. This is the opposite of a shift in the market demand curve, which results in a *change in demand* and a *change in the quantity supplied*.

8. A market disequilibrium occurs when producers respond to an expected market equilibrium price when making their production decisions that turns out to differ from the eventual market equilibrium price. This results in commodity surpluses when expectations are too high and commodity shortages when price expectations are too low. The cyclical pattern of adjustments to a new equilibrium price under these conditions can take on a pattern much like a spider's cobweb.

KEY TERMS

Commodity shortage	Elasticity of supply	Price taker
Commodity surplus	Firm supply curve	Producer surplus
Consumer surplus	Market supply curve	Total economic surplus

TESTING YOUR ECONOMIC QUOTIENT

1. Given the following graph, please label the curves where asked and answer the questions appearing below the graph.
   a. What is the profit-maximizing level of output?
   b. What is the significance of 8.5 units of output?

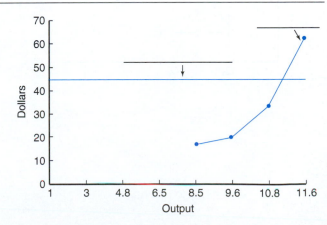

2. Describe what has happened to demand, supply, and market equilibrium in each panel.

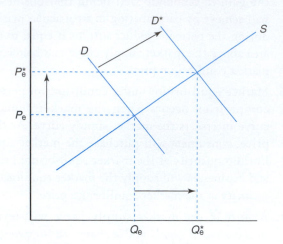

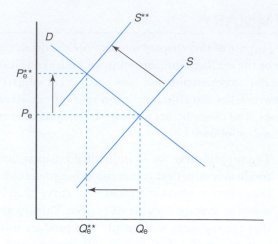

3. Given the following price and quantity-supplied data, graph this firm's supply curve on the axis provided. Also, for each price and quantity-supplied combination, calculate the amount of producer surplus.

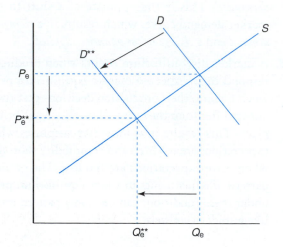

Price	Quantity Supplied
0	0
1	1
2	2
3	3
4	4
5	5
6	6
7	7
8	8
9	9
10	10

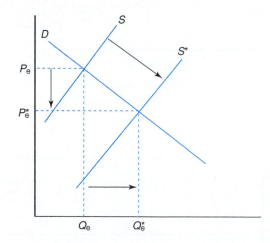

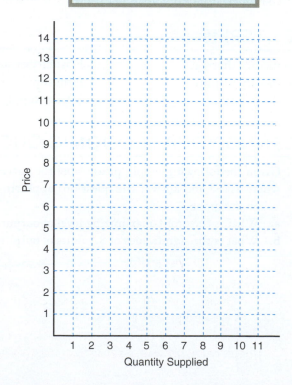

4. Assume that the following graph depicts a shift of the worldwide demand curve for beef. Starting in 2004, demonstrate a cobweb adjustment to the market equilibrium in 2005 on the following graph. Assume that the market was in equilibrium in 2004 and will eventually converge to the equilibrium depicted by the 2005 demand curve. Assume producers respond to last year's price (the expected price is last year's observed price).

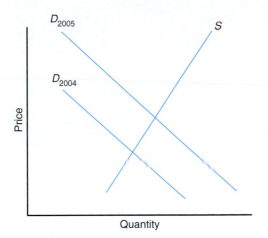

5. Given the following graph, fill in the following blanks:

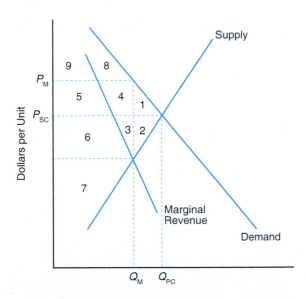

a. The level of consumer surplus under perfect competition is equal to areas _____.
b. The level of producer surplus under a monopoly is equal to areas _____.
c. The economic loss to society associated with a monopoly is equal to areas _____.
d. The difference in producer surplus between perfect competition and a monopoly is equal to areas _____.

6. Assume the graph below depicts the shift in the worldwide demand for chicken. As an economist, you are asked to explain this change in demand to Tysons Foods.

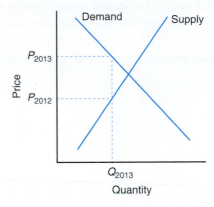

a. Based on the cobweb theory, on what price did chicken processors base their production decision in 2013?
b. What price were consumers willing to pay for the quantity produced by chicken processors?
c. What factors will determine the equilibrium price for chicken over time?
d. Given your answer to part c, what factors could change this equilibrium price?

7. In a perfectly competitive industry, the profit-maximizing level of output is that output level where the product's price is equal to _____.
8. In the long run, the economic profit for a perfectly competitive firm will be _____.
9. An outward shift in the demand curve for a perfectly competitive market will result in an increase in supply.   T   F
10. A perfectly competitive firm will cease production in the short run if marginal revenue falls below total average cost.   T   F
11. Assume the market demand and supply curves for a product in a perfectly competitive market are
$P = 100 - 0.50(Q_D)$ demand curve
$P = 2 + 4.0(Q_S)$ supply curve
If this market converges to equilibrium, the equilibrium price and quantity would be
a. $P = 108.89$ and $Q = 45.56$
b. $P = 45.56$ and $Q = 108.89$
c. $P = 110$ and $Q = 43$
d. $P = 43$ and $Q = 110$
12. Which of the following markets is likely to most nearly reflect a perfectly competitive market?
a. The farm implement market
b. The wheat market
c. The automobile market
d. The apparel market

13. Assume that a firm is operating in a perfectly competitive market at its break-even level of output. Which of the following statements is false?

a. Marginal cost and average revenue are equal.

b. Marginal cost and marginal revenue are equal.

c. Marginal cost and average variable cost are equal.

d. Marginal cost and average total cost are equal.

14. The demand curve for a firm operating under conditions of perfect competition is

a. elastic but not perfectly elastic.

b. perfectly elastic.

c. inelastic but not perfectly inelastic.

d. perfectly inelastic.

Market Equilibrium and Product Price: Imperfect Competition

Food processors, fiber manufacturers, farm input manufacturers, and other nonfarm businesses provide an important bridge between farmers and consumers of food and fiber products. These markets send important signals to farmers that help them decide *what* commodities they should produce, *how much* of each commodity they should produce, *how* these commodities should be produced, and *where* these commodities should be distributed.

Up to this point, we have assumed that the conditions required for perfect competition exist in the marketplace. In reality, the economy does not consist entirely of perfectly competitive firms. Although farms and ranches generally satisfy the conditions of perfect competition, most of the firms with which they do business in the food and fiber industry do not. In this chapter, we will examine several forms of **imperfect competition** and cite examples of each form in the U.S. food and fiber industry.

Two classic studies of imperfect competition exist in the economic literature—*The Theory of Monopolistic Competition*, by Edward H. Chamberlin, published in 1933,

Feedlots are indicative of the relatively few buyers of cattle versus the relatively large number of beef producers.   Credit: Pete McBride/Getty Images.

and *The Economics of Imperfect Competition*, by Joan Robinson, also published in 1933. Understanding the theory of imperfect competition is paramount when analyzing the economics of the farm or agribusiness sector. The economic structure and characteristics of any market define the limits within which market forces are allowed to establish prices. To gain insight into the determination of prices in an industry, we must first understand its structural characteristics and competitive behavior.

MARKET STRUCTURE CHARACTERISTICS

There are four key interrelated structural characteristics used when discussing competitive behavior of a market. These include

1. the number and size distribution of sellers and buyers,
2. the degree of product differentiation,
3. the extent of barriers to entry, and
4. the economic environment within which the industry operates (i.e., the conditions of supply and demand).

These four characteristics, combined, determine whether an industry or various segments of the industry exhibit behavior conducive to perfect competition or imperfect competition.

Number of Firms and Size Distribution

The number of firms buying and selling in a market is important in determining competitive conduct and price determination processes, because competitive conditions tend to break down when the number of firms involved declines. When competition wanes, prices are less likely to be set by the forces of supply and demand. The distribution of sales or purchases by firms in the industry helps indicate the degree of control that one or a few firms may have over prices. The fewer the number of firms and/or the greater the percentage of product sales accounted for by a few firms in the market, the greater the concentration of the market and the greater the likelihood that prices are determined by forces other than supply and demand.

Product Differentiation

The degree of product differentiation refers to the extent buyers at each level in an industry perceive a difference in the products offered to them by alternative suppliers. If buyers consider the products to be virtually identical, the products are said to be *undifferentiated* or *homogeneous*. A seller in an undifferentiated product market has difficulty in setting a higher price than what the forces of supply and demand would dictate because buyers simply would shift to other suppliers who offer lower prices to obtain the quantities of the product needed. Because the products offered by all suppliers are virtually identical, buyers have little incentive to pay one supplier a higher price than they would another supplier. Thus, sellers of undifferentiated products are price takers, not price makers.

As stated by Pindyck and Rubinfeld (2009), "Thousands of farmers produce wheat, which thousands of buyers purchase to produce flour and other products. As a result, no single farmer and no single buyer can significantly affect the price of wheat."[1]

The more differentiated a product becomes in the minds of buyers, the more opportunity the supplier has to raise the price of the product, taking advantage of buyers' perceptions concerning its unique characteristics. It is not necessary that these perceived differences be real. Product differences perceived by consumers may be due entirely to well-planned and well-executed advertising campaigns.

Finally, products can be either positively or negatively differentiated in the minds of buyers; however, only a positive attitude toward the uniqueness of a product allows its seller to set prices. A negative attitude simply drives buyers to other products.

Barriers to Entry

Barriers to entry are those forces that make it difficult for firms to enter an industry. Some barriers are created by existing firms or by governments; others are simply economic facts of life. If barriers to entry are low, even firms in highly concentrated industries (i.e., small number of firms) have little opportunity to set prices above competitive levels. A firm in an industry with low barriers to entry that is attempting to increase its profits by raising prices would likely encourage new firms to enter the industry eager to share in the profit opportunities. Four common barriers to entry are

1. **absolute unit-cost advantages,**
2. **economies of scale,**
3. **capital access and cost**, and
4. **preferential government policies.**

An absolute cost barrier exists if the unit cost of production for an established firm is lower at all levels of output than would be the case for a new entrant. In this case, there is no level of output at which a new firm could operate competitively.

An economies-of-scale barrier generally exists when the smallest efficient-sized plant is large relative to the size of the market, and smaller firms face significantly higher costs of production. New firms entering this industry would have to be large and generate a substantial amount of output, causing industry prices to drop drastically and some firms to be driven out of the market.

A capital cost barrier exists when the capital investment required for efficient operation is large. Capital cost barriers tend to accompany economies-of-size barriers, but this situation is not always true. A potential industry entrant may face a notable capital investment cost to begin efficient operation and still only hope to capture a small share of the market.

---

[1] From *Microeconomics* by Robert S. Pindyck and Daniel L. Rubinfeld. Published by Pearson Education Inc., © 2009.

Finally, the government can administer preferential policies such as patents, copyrights, and import controls that protect one or more sellers from all or some of the potential entrants to a domestic industry.

Economic Environment

Of the many aspects affecting the economic environment in which an industry operates, the one most closely related to the behavior and determination of prices is existing supply and demand conditions. Three aspects of supply and demand particularly important to determining price in a market are

1. the level of output in the industry,
2. the responsiveness of supply and demand to price changes and the responsiveness of demand to income changes, and
3. the proportion of consumer food expenditures accounted for by the industry's product.

If an industry's sales volume is relatively large, more profit opportunities will be available for firms to be involved at all levels. This situation makes it difficult for one or a few buyers or sellers to control prices. The opposite is the case if volume is low.

If demand is highly responsive to changes in price (elastic demand) and income, and consumers spend a large portion of their income on the product, the industry will grow when income rises and prices fall. Opportunities for new entrants in this industry will be present, increasing the competitiveness at each level. If demand is instead largely unresponsive to price (inelastic demand), if demand also is unresponsive to income change, and consumers, on average, spend only a small proportion of their income on the product, demand will be stable and growth will stagnate.

To summarize, we can classify forms of market structure, in part, by

- the number and size distribution of sellers and buyers,
- the degree of market concentration by the firms,
- the degree of product differentiation,
- the barriers to entry, and
- the economic environment in which firms operate.

Classification of Firms

Four major conditions, combined, define perfect competition and guarantee a market in which the forces of supply and demand determine prices. These conditions are (1) a large number of buyers and sellers; (2) a homogeneous product; (3) freely mobile resources; and (4) perfect knowledge of market conditions.

Imperfectly competitive market structures
Imperfectly competitive forms of market structure can be classified according to the number of firms and size distribution in the market, the degree of product differentiation, the extent of barriers to entry, and the economic environment within which the industry operates. The profit-maximizing level of output for any imperfectly competitive firm is determined where marginal revenue equals marginal cost.

The conditions for perfect competition are extreme, but examples of nearly perfect competition abound in agriculture. In selling activities, numerous small producers of livestock and crops are producing a nearly identical product. In buying activities, numerous consumers are purchasing food and fiber products that are not highly distinguishable. The important result of the conditions of perfect competition, or even near-perfect competition in a market, is that buyers and sellers are price takers, not price makers. The actions of any individual buyer or seller do not affect the market. The interaction of market supply and market demand determines the market price in a perfectly competitive market environment.

Market structure from the seller's perspective can be classified into four types, based on the extent to which market prices are set by the forces of supply and demand. These four types of competition in selling are

1. perfect competition,
2. monopolistic competition (imperfect competition),
3. oligopoly (imperfect competition), and
4. monopoly (imperfect competition).

Although farmers and ranchers largely satisfy the conditions of perfect competition, much of the rest of the food and fiber industry is characterized by one of the three forms of imperfect competition in selling.

IMPERFECT COMPETITION IN SELLING

To fully appreciate the complex nature of the U.S. food and fiber system, we must consider various market structures that represent imperfect forms of competition. Three forms of imperfect competition in the selling activities of firms are monopolistic competition, oligopoly, and monopoly.[2] A very important characteristic of imperfect competition in selling is that the demand curve faced by the firm is downward sloping, not perfectly elastic as in the case of perfect competition.

Monopolistic Competition

The conditions of monopolistic competition are identical to those of perfect competition with one important difference—*the products sold are no longer perfectly identical or homogeneous*. If many businesses are selling a **differentiated product** (which is different from the standardized or homogeneous product in perfect competition), then **monopolistic competition** exists. The differentiation of the product gives the seller some flexibility in pricing. If an individual seller can convince buyers that its product is superior to those offered by other sellers, it can charge a higher price for its product. Farm input manufacturers and suppliers differentiating their products with the use of advertising to promote branded feed supplements or branded hybrid seeds exemplify this type of imperfect competition. Like perfect competition, there are no barriers to entry or exit. *Thus, product differentiation is the key difference between monopolistic competition and perfect competition.*

A monopolistic competitor becomes a price maker if it can effectively differentiate its product in the market when other sellers in the market are offering similar products. The opportunity to influence prices may be limited because the action of each individual firm may have imperceptible effects on the market and because the product may not be unique. Consequently, attempts by a seller to raise the price too much may drive buyers to other sellers of similar products. How much is "too much" is determined by the degree of differentiation of the seller's product. The marketing strategy of a firm in a monopolistically competitive market must focus on creating differences in the products it sells, relative to other firms in the market. Price competition, including price specials and discounts, is an important aspect of marketing efforts.

Monopolistic competitors face the same average total cost (ATC) and marginal cost (MC) curves as perfect competitors. By differentiating their product, however, individual monopolistic competitors face a downward-sloping demand curve (and a downward-sloping marginal revenue curve that reflects the change in total revenue at each output level) instead of the perfectly elastic (i.e., flat or horizontal) demand curve faced by perfect competitors. The relationship among demand (average revenue), total revenue, and marginal revenue is exhibited in Table 9-1 and Figure 9-1.

The relationship between the demand curve and marginal revenue curve is tied to the elasticity of demand along the demand curve. For example, marginal revenue (MR) is zero at the point of unit elasticity of demand (i.e., -1.0). Up to that point,

---

[2] Imperfect competition among buyers also takes three general forms: monopsonistic competition, oligopsony, and monopsony. The suffix *poly* refers to sellers, and the suffix *sony* refers to buyers. The prefix *mono* means one, and the prefix *oligo* means few. Therefore, imperfect competition may pertain to either buyers or sellers. Imperfect competition prevails whenever individual sellers and buyers exert a measure of control over price.

Table 9-1
RELATIONSHIP BETWEEN DEMAND (AVERAGE REVENUE), TOTAL REVENUE, AND MARGINAL REVENUE

Average Revenue or Demand Curve		Total Revenue	Marginal Revenue	
Price	Quantity	(Price × Quantity)	Change in Total Revenue/Change in Quantity	Elasticity of Demand
15	0	0	—	—
14	2	28	14	−29
13	4	52	12	−9
12	6	72	10	−5
11	8	88	8	−3.29
10	10	100	6	−2.33
9	12	108	4	−1.73
8	14	112	2	−1.31
7	16	112	0	−1
6	18	108	−2	−0.76
5	20	100	−4	−0.50
4	22	88	−6	−0.43
3	24	72	−8	−0.30
2	26	52	−10	−0.20
1	28	28	−12	−0.11

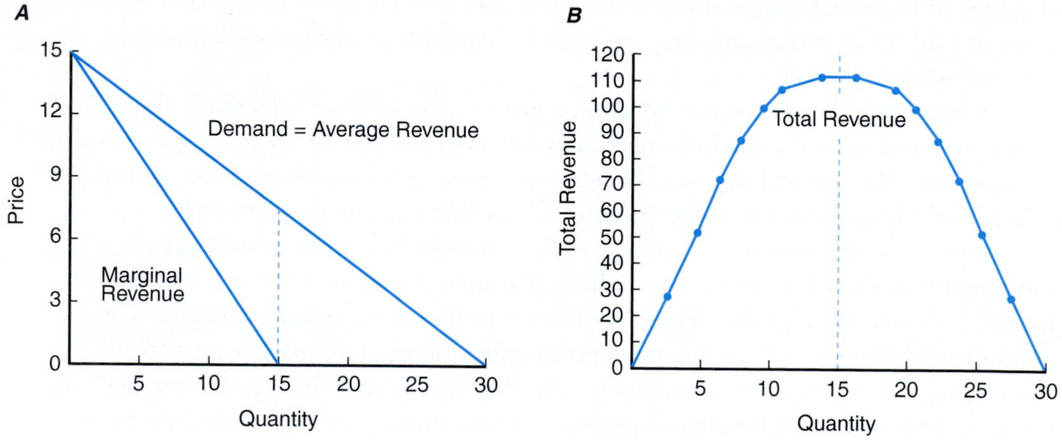

A

B

Figure 9-1

The relationships among average, marginal, and total revenue in these figures are based on the data in Table 9-1. These concepts determine equilibrium prices and quantities in the case of imperfect competition in selling.

the imperfectly competitive firm's total revenue is increasing as output increases. In Chapter 6, we learned that the demand curve for a perfect competitor is perfectly elastic and that $P = \text{MR}$. Because monopolistic competitors face a downward-sloping demand curve, it is no longer true that $P = \text{MR}$. The monopolistic competitor must be aware of the decline in its MR curve and the possibility of a negative MR at lower points on its demand curve.

Product differentiation conducted by the monopolistic competitor can be accomplished by modifying the particular product or by advertising and sales promotion activities. The object is to intensify the demand for the product by distinguishing it from other products in the mind of the buyer. Such activities can be found in

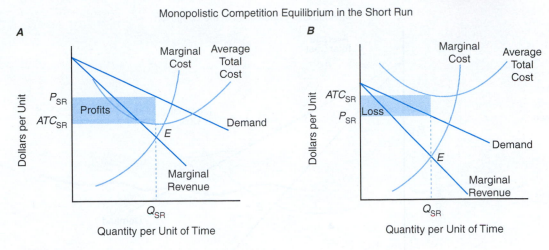

Figure 9-2

Monopolistic competitors also equate marginal cost and marginal revenue. The difference between the demand curve, which also represents average revenue, and the average total cost curve represents the average profit or loss per unit. Multiplying this difference by the quantity supplied (Q_{SR}) gives us the economic profit (A) or loss (B) for the business in the short run.

most markets in this country. The better the business is at product differentiation, the greater its influence on product price.

Short-Run Equilibrium   The equilibrium values for price and quantity under monopolistic competition in the short run are determined by the intersection of the marginal cost curve and the marginal revenue curve. In Figure 9-2, this intersection occurs at point E, suggesting a level of output in the short run of quantity Q_{SR}. The business then sets the price it charges for its differentiated product by reading up to the demand curve and over to the price axis. From Figure 9-2, we see the monopolistic competitor will charge price P_{SR} in the current period.

The gap between the demand curve (D) and the average total cost curve at output Q_{SR} indicates the business achieved either an economic profit or a loss in the current period. If price exceeds average total costs, as it does in Figure 9-2A, an economic profit exists in the current period. If price is less than average total costs (i.e., the demand curve lies below ATC), as it does in Figure 9-2B, an economic loss is incurred. The existence of profits (losses) would result in the entry (exit) of additional monopolistic competitors over time.

Long-Run Equilibrium   Additional entrants to the market when economic profits exist would shift the demand curve downward (decrease average revenue) and lower profits and possibly even create losses. Monopolistic competitors exiting the market when economic losses exist will shift the demand curve upward (increase average revenue) and reduce losses and perhaps create economic profits. Figure 9-3 suggests that the monopolistic competitor has reached its long-run equilibrium. How can we tell? There is no gap between the demand curve and the average total cost curve at Q_{LR} in Figure 9-3, which means that there are no economic profits that would attract other monopolistic competitors into the market and there are no losses that would cause some competitors to leave the market. Note that the firm is still equating marginal cost to marginal revenue at point E.

Is the equilibrium quantity here less than what a business would have supplied under perfect competition? The answer is *yes*. Note in Figures 9-2 and 9-3 that the monopolistic competitor chooses to operate to the *left* of the minimum point on the

Figure 9-3
At the point where the monopolistic competitor is operating at Q_{LR}, average revenue will equal average total cost, indicating that both average profits and average losses have been eliminated.

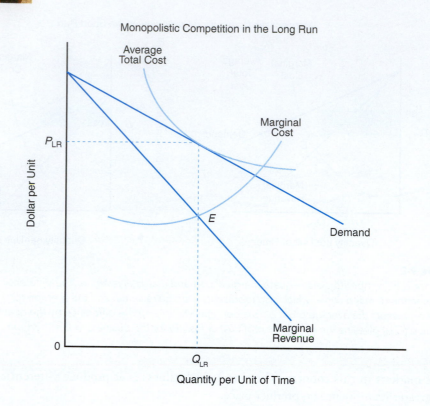

Monopolistic Competition in the Long Run

average total cost curve. You will recall that the minimum point on the *ATC* curve represents the long-run equilibrium output level for a perfectly competitive firm.

Monopolistic competition therefore is less efficient than perfect competition from the viewpoint of consumers. The market price is higher and output is lower than that occurring under perfect competition. Deciding whether this trait is undesirable for society, however, depends on whether we want a marketplace with an undifferentiated or a standardized product. Do we all want to wear white shirts and brown shoes? The fashion industry certainly hopes not. Also, do we all want to eat a certain brand of vanilla ice cream and a specific brand of white bread? The marketplace has shown a preference for differentiated products, even if it means paying a higher price. In fact, most of the food and fiber products we purchase are supplied by monopolistic competitors.

Many businesses in the United States spend large sums of money annually in an effort to differentiate their product in the eyes of the consumer. In Table 9-2, the top 10 advertising categories of 2014 are reported. Automotive was the top category with $8.4 billion of spending followed by the quick-service restaurant industry, the pharmaceuticals industry, automotive dealerships, and the motion picture industry. Wireless telephone services, department stores, auto insurance, and conventional restaurants round out this top 10 list. These aforementioned sectors are quite familiar to most consumers. Given the magnitude of expenditure on advertising and promotion, indeed the cost associated with product differentiation is nontrivial.

Conditions of monopolistic competition The conditions of monopolistic competition are identical to those of perfect competition except that the products sold are no longer homogeneous. Product differentiation is the key difference between monopolistic competition and perfect competition.

Oligopoly

Further removed from the characteristics of perfect competition in selling is **oligopoly**. The economic conditions that define oligopoly are the same as those of monopolistic competition with one major exception: there are only *a few sellers*, each of which is large enough to have an influence on market volume and price.

Oligopolists also are interdependent in their decision-making. The actions of an individual oligopolist are seen as a competitive threat to the other oligopolists in the

Table 9-2
TOP 10 ADVERTISING CATEGORIES OF 2014

Rank	Category	2014 ($ Billion)
1	Automotive	8.4
2	Restaurant—quick service	3.4
3	Pharmaceutical	3.2
4	Automotive dealerships	2.8
5	Motion picture	2.3
6	Wireless telephone services	2.2
7	Department store	2.1
8	Auto insurance	1.4
9	Restaurants	1.3
10	Direct response	1.2
	Total[a]	28.3

[a]The sum of the individual categories can differ from the total shown due to rounding.
Source: Nielsen, http://www.nielsen.com/us/en/insights/news/2015/tops-of-2014-advertising.html.

market and as such may invoke retaliation. It is this interdependence that is the key component in the marketing strategies and pricing behavior of the industry. **Nonprice competition** is the main competitive strategy, including any and all efforts to uniquely differentiate products in the market.

If an oligopolist attempts to raise its price, the other oligopolistic firms will not necessarily follow suit. The size of the subsequent drop in the firm's sales will depend on how successful it has been in differentiating its product from those offered by its competitors. Typically, the drop in sales will more than offset the price increase, leading to reduced revenues for the price-raising oligopolistic firm.

On the other hand, if an oligopolist attempts to lower its price, competing firms will immediately retaliate by lowering their prices to keep from losing their market shares. The lower price, combined with little or no change in sales, again means that the firm initiating the price decline experiences a loss in sales revenues. Further attempts by the firm to reduce price to gain market share will simply result in further rounds of price cutting with relatively small gains in sales. As a consequence, prices of products sold by firms in an oligopolistic industry tend to be stable. This situation exists today in the automobile industry and in the airline industry.

Once prices in an oligopolistic industry are established, they tend to stay at that level (Purcell, 1979). If there are differences in the prices of the products offered by competing firms in an oligopoly, they are generally the result of successful differentiation of the products by the respective firms. This stable price behavior is not the result of **collusion** but rather of rational economic decision-making by each firm in the oligopoly. Also, because oligopolistic firms refrain from competing on a price basis so that prices remain relatively stable, changes in the costs of production, processing, marketing, and so on are not easily passed on and must be absorbed to a large extent.

An oligopolistic market structure in selling often develops because of barriers to entry. No matter how freely firms can enter or exit a market, if the market is thin (has a low and stagnant volume), profit opportunities will not exist for more than just a few firms. For any firm to enter a thin market industry, the firm might need to make a sizeable capital investment to capture a large enough share of the market and attain the necessary economies of size. Other barriers to entry also can create the environment in which an oligopolistic market structure will flourish.

Although prices are set in an oligopolistic market in many ways, the most common method is price leadership. In this situation, a particular firm dominates the market either because it controls the largest share of the market or because other firms in the industry view it as more efficient in operation, more proficient in analyzing the market, more experienced, and so on. The dominant firm first sets its price so as to maximize its profits. The other firms, with no collusive behavior, simply set their prices at the same level after making any adjustments they feel are justified by the differentiating characteristics of their products. As a consequence of these actions, the price set by the oligopolistic seller is higher than that under perfect competition, making the volume produced and sold lower than would exist under perfect competition. However, the dominant firm may be cost-efficient enough to set a price low enough so as to start a price war that would eventually drive all other firms out of the market. Such a move would establish a monopoly, however, with all the difficult legal problems that would have to be faced.

The automobile industry and aircraft manufacturing industry are two nonagricultural examples of oligopolies. In agriculture, the farm machinery and equipment industry can be classified as an oligopoly. John Deere makes combines that are different from those made by J.I. Case or New Holland. The top four brands sold to farmers and ranchers account for the majority of all combine sales in the United States.

Although concentration among the industries that procure slaughter livestock has increased in the last 25 years, it has remained relatively stable in recent years. Four-firm concentration in steer and heifer procurement rose from 36% in 1980 to 81% in 1993, but since 1993 has remained fairly constant. Four-firm concentration in hog procurement rose from 34% in 1980 to 55% in 1996, remaining at about that level until moving to 64% in 2003 and 2004. Four-firm concentration in sheep and lamb procurement rose from 56% in 1980 to 77% in 1988, but decreased to 57% in 2004.

The pesticide industry and the fertilizer industry also can be classified as oligopolies. In the pesticide industry, three or four firms account for the majority of the total sales to farmers. The products developed in this industry are identified or differentiated by company brand or name.

The interdependence of pricing policies among firms and product differentiation makes it difficult to analyze oligopolistic situations. Figure 9-4 illustrates the "kinked" demand curve that economists use to explain the rigid price behavior of

Characteristics of oligopoly The economic conditions that define oligopoly are the same as those of monopolistic competition except that there are only a few sellers, each of which is large enough to have an influence on market volume and price. Oligopolists are interdependent in their decision-making. Nonprice competition is the main competitive strategy. The fact that oligopolists match all decreases but not all increases in price leads to a kinked demand curve and discontinuous marginal revenue curve.

Figure 9-4

This figure illustrates the equilibrium price and quantity associated with an oligopoly market structure. The fact that oligopolists match all price decreases but not price increases leads to a "kinked" demand curve, given by d1D. The corresponding marginal revenue curve, given by d23456, is discontinuous.

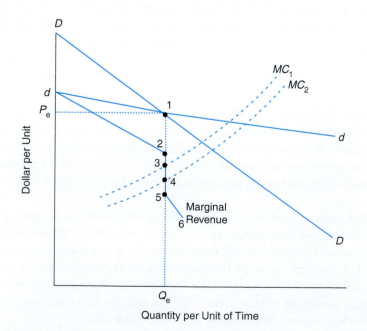

oligopolists. Let the demand curve *DD*, which passes through point 1, represent the demand curve when all sellers move prices together and share the total market. Further, let the demand curve *dd* represent the demand curve when a single oligopolist changes its price. Note that the demand curve *DD* is more inelastic than the *dd* curve. Because oligopolists take account of the reaction of other oligopolists, there is no single demand curve facing a particular oligopolist. Oligopolies typically match all price decreases by their fellow oligopolists (they do not want to be undersold), but they do not match all price increases (they want to capture a greater market share).

Below point 1, where other oligopolists match the firm's price cut, the demand curve *DD* prevails. Above point 1, the demand curve *dd* will prevail because rival oligopolists will not match the firm's price increase. The kinked demand curve, given by *d1D*, leads to a break, or vertical discontinuity, in the marginal revenue curve at output Q_e. The segment between points 2 and 5 represents the magnitude of this vertical discontinuity. Shifting marginal cost (*MC*) curves because of technological advances (downward shift from MC_1 to MC_2) intersecting the marginal revenue curve at point 4 rather than point 3 will not change the oligopoly price and quantity.

In meeting demand along the lower segment of the kinked demand curve (i.e., to the right of point 1), the oligopolist will be maintaining its market share, which explains why there is a tendency for prices to remain at P_e. Each oligopolist will earn an economic profit per unit of P_e minus its average total cost.

For a variety of reasons, including the inherent uncertainty of knowing how others will respond, oligopolists facing similar demand and cost conditions may behave in a collusive fashion (i.e., arrange to charge the same price for their output) instead of in the manner depicted in Figure 9-4. Their objective may be to maximize their joint profits. The prices and quantities observed when this situation occurs will be much the same as those charged by a single seller or monopolist. Each oligopolist would charge price P_e, produce its predetermined share of Q_E, and share in the higher level of profits.[3]

Monopoly

At the opposite end of the spectrum from perfect competition in selling is the **monopoly**. Instead of many firms or even a few firms, there is only one seller in the market. A monopoly exists for the same reason that oligopolies exist: barriers to entry. In the case of a monopolist, however, the barriers are sufficiently high to discourage all potential competitors from attempting to enter the industry. The monopoly is similar to an oligopoly, except that the monopolist has no concern for retaliation by competitors in response to changes in pricing. The monopolist sets a price that is higher than would exist under perfect competition to maximize profits. The volume sold also is below that observed under perfect competition. Because the monopolist has no competitors, the price set is usually even higher than would occur under oligopoly.

In practice, however, monopolists tend to hold the price below their profit-maximizing level to discourage both the entry of competitive firms and anti-trust litigation. Also, the lack of competitors means that a monopolist can pass on changes in cost to consumers more easily than an oligopolist. Consequently, the prices set by monopolists tend to move closely with movements in input costs that they face in production.

---

[3] Firms attempting to collude to reap profits are referred to as **cartels**. The cartel members jointly establish monopoly prices and quantities and each member's share of total sales. Perhaps the most famous cartel in recent years has been OPEC (Organization of Petroleum Exporting Countries). OPEC had a dramatic impact on the world price of oil in the 1970s by restricting the amount of crude oil coming into the world market. Explicit collusion among domestic sellers, however, is in violation of anti-trust laws, which we will discuss shortly.

Figure 9-5

A monopolist also equates marginal cost and revenue. The price charged to consumers at this equilibrium quantity is read off the demand curve. The monopolist differs from other forms of imperfect competitors in that it strives to block others from entering the marketplace. The demand curve faced by the individual firm is the market demand curve. $ONAQ_E$ represents total variable costs, area $NMBA$ represents total fixed costs, area $OMBQ_E$ represents total costs, area OP_ECQ_E corresponds to total revenue, and area MP_ECB depicts economic profits, or total revenue minus total costs.

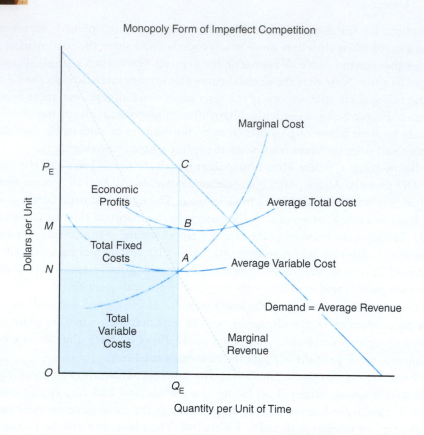

The long-run equilibrium price and quantity under a monopoly are depicted in Figure 9-5. The monopolist will maximize profits by operating where marginal cost equals marginal revenue and then determine the highest price for its product. Figure 9-5 shows that this quantity and price combination occurs at Q_E and price per unit P_E. If the monopolistic firm produced at a lower level, the firm's profits would not be maximized (i.e., it could achieve additional profits by expanding its production to Q_E). If the monopolist produced at a higher level, the costs of producing the additional units would exceed the revenue received from their sale.

Economic profits will exist under a monopoly in the long run because there are likely both economic and legal barriers preventing other firms from entering the sector. The economic barriers might be prohibitive production costs of product prices or outright restraints of trade. The legal barriers might be patents on the design of specific products. Other laws may permit a group of firms to act as a monopoly. Electric power companies and telephone companies are two examples of monopolies. In agriculture, **marketing orders** for specific products such as fluid milk and oranges permit the setting of a higher price by controlling the flow of products into the market. This arrangement enables farmers to receive a higher price than they would have otherwise received.

Monopoly At the opposite end of the spectrum from perfect competition in selling is monopoly. As the prefix indicates, there exists a single seller in this market. Electric power companies and marketing orders in agriculture are examples of this market structure.

As with the oligopolist and the monopolistic competitor, the monopolist will choose to sell a smaller quantity than would occur under perfect competition. Each of these imperfectly competitive firms prefers to operate to the left of the minimum point on their average total cost curve, where long-run equilibrium occurs under perfect competition. Thus, there is a cost to society associated with all forms of imperfect competition. The quantity supplied is smaller, the unemployment of resources is higher, and prices are higher than under perfect competition.

Table 9-3
ALTERNATIVE FORMS OF MARKET STRUCTURE IN SELLING

	Market Structure in Selling			
Item	Perfect Competition	Monopolistic Competition	Oligopolies	Monopolies
Number of sellers	Numerous	Many	Few	One
Ease of entry or exit	Unrestricted	Unrestricted	Partially restricted	Restricted
Ability to set market price	None	Some	Some	Absolute
Long-run economic profits possible	No	No	Yes	Yes
Product differentiation	None	Yes	Yes	Product unique
Examples	Corn producers, beef producers, soybean producers, and wheat producers	Soft drink bottlers, restaurants, hairdressers, and grocery stores	Manufacturers of farm tractors, cell phone service providers, and airline industry	Marketing orders

Comparison of Alternative Market Structures

Table 9-3 summarizes the basic features of both the perfectly competitive and imperfectly competitive forms of market structure from a selling perspective. For example, Table 9-3 describes the number of sellers, ease of entry and exit, the ability to set market prices, the existence of economic profits in long-run equilibrium, and the extent of product differentiation.

There are numerous corn producers, for example, who sell an undifferentiated product and who individually have no ability to influence market prices. Soft drink bottlers, primarily through advertising, try to differentiate their products. Because the four largest manufacturers of farm tractors account for 80% of total sales, these manufacturers try to differentiate their product (e.g., different colors, model styles) and have some ability to influence the price of their product. Marketing orders, such as that enforced for oranges in the United States, are one of the more visible monopolies in agriculture.

Welfare Effects of Imperfect Competition

We can utilize the concept of producer and consumer surplus introduced in earlier chapters to evaluate the economic welfare implications of imperfect competition. We have already learned that imperfect competitors operate to the left of the minimum point on their average total cost curve. This situation, of course, differs from the behavior of the perfectly competitive firm, which operates where the product price is equal to its marginal cost as long as it exceeds the minimum point on its average variable cost curve. This difference occurs because of the downward-sloping nature of the imperfect competitor's demand curve and the desire of these firms to operate at the point at which their marginal costs are equal to marginal revenue and to price their product according to the demand curve that lies above the marginal revenue curve. What does this difference in market behavior imply about the economic well-being of producers and consumers?

We can see the maximum impact of imperfect competition on producers and consumers in a particular market if we compare the two extreme forms of market structure: perfect competition and a monopoly. Figure 9-6 indicates that the market equilibrium price under perfect competition would be P_{PC} and the quantity marketed would be Q_{PC}. Total consumer surplus under these conditions would be equal to the sum of areas 1, 4, 5, 8, and 9. Total producer surplus would be equal to the sum of areas 2, 3, 6, and 7.

If this market exhibited the characteristics of a monopoly instead, the equilibrium price would rise to P_M and the equilibrium quantity market would fall to Q_M (remember that the supply curve in this case is the monopolist's marginal cost curve).

Figure 9-6

Consumer surplus would be lower under a monopoly (areas 8 and 9) than it would be under perfect competition (areas 1, 4, 5, 8, and 9). Producers would have been willing to supply Q_{PC} at a price of P_{PC} under perfect competition. A monopoly, however, would be willing to supply a quantity of Q_M at price P_M. Producer surplus would be higher under a monopoly (areas 3, 4, 5, 6, and 7) than it would be under perfect competition (areas 2, 3, 6, and 7). The economic welfare of society as a whole would decline by the sum of areas 1 and 2.

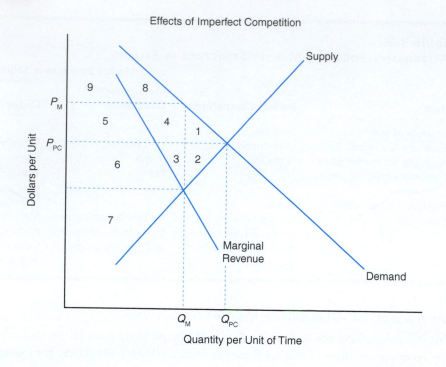

The monopolist would earn economic profits equal to the sum of areas 3, 4, 5, and 6. The total producer surplus would be equal to this economic profit plus area 7. Thus, the producer gained areas 4 and 5 and lost only area 2. Because areas 4 and 5 exceed area 2, we can say that the monopolist would be in a better position economically than all producers were collectively under perfect competition.

Total consumer surplus would fall by the sum of areas 1, 4, and 5 under conditions of pure monopoly, totaling only area 8 and area 9. So, whereas producers gained areas 4 and 5 and lost only area 2, consumers gained nothing and lost areas 1, 4, and 5. Consumers, therefore, would be in a considerably worse position under a monopoly than they would be under conditions of perfect competition.

Society as a whole would be a net loser if a market's structure were to switch from perfect competition to a monopoly. Society would gain nothing in terms of economic welfare and would lose areas 1 and 2. The sum of areas 1 and 2 is frequently referred to as a **dead-weight loss**. These results support our earlier statements about the relative efficiency of perfect competition from the perspective of consumers and society as a whole.

IMPERFECT COMPETITION IN BUYING

Up to this point, we have considered imperfect competition in selling activities. Imperfect competition in buying activities can influence the market price for resources used in production. Here the supply curve faced by the firm is upward sloping rather than perfectly flat or elastic, as was the case in perfect competition. A meat-packing firm can influence market price, for example, by its decisions concerning how many and what kind of livestock of each class and grade to buy. In a given location, this meat packer may be the *only* buyer of beef cattle. In fact, a typical agricultural situation has been one in which a large number of farmers face a single buyer for their product. A textile plant in a small rural "company town" that owns other businesses in town and the apartments that workers live in may have a buyer's monopoly in the local labor market. These situations are termed a **monopsony**.[4]

---

[4] Where there are several buyers in a similar situation, *oligopsony* is the proper term.

Table 9-4
MONOPSONY AND MARGINAL INPUT COST

(1) Units of Variable Input	(2) Price per Unit ($)	(3) Total Cost of Input ($)	(4) Marginal Input Cost
1	3.00	3	—
2	3.50	7	4
3	4.00	12	5
4	4.50	18	6
5	5.00	25	7
6	5.50	33	8
7	6.00	42	9
8	6.50	52	10
9	7.00	63	11
10	7.50	75	12

Monopsony

A buyer in a perfectly competitive input market views the input supply curve as a horizontal line. The perfectly competitive firm's purchases are relatively small and perfectly elastic or do not perceptibly affect market price. A monopsonist, however, is the only buyer in the market and faces an upward-sloping market input supply curve. As a consequence, its buying decisions affect input prices. To increase its input usage, it is necessary for the monopsonist to pay a higher input price. The monopsonist must therefore consider the **marginal input cost** (MIC) of purchasing an additional unit of a resource. *Marginal input cost* is defined as the change in the cost of a resource used in production as more of this resource is employed.

A numerical example of a set of supply and marginal input cost curves facing a monopsonist is given in Table 9-4. Columns 1 and 2 represent the input supply curve, and column 4 represents the marginal input cost curve. These curves are depicted graphically in Figure 9-7.

The supply curve for a resource facing the monopsonist, which necessarily represents the market supply curve, may also be thought of as the average cost

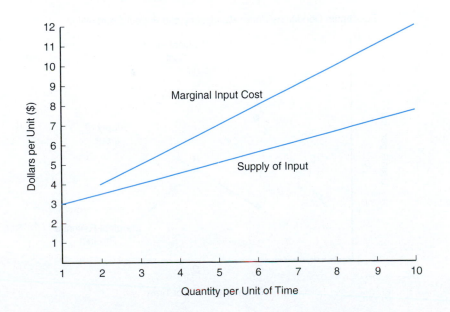

Figure 9-7

These curves reflect the cost and input use data presented in Table 9-4.

curve for the resource. The marginal input cost curve, which lies above the supply curve, illustrates that the monopsonist must pay higher prices per unit if it wishes to employ greater amounts of the resource. This situation is the mirror image of the monopolistic firm, which must decrease the price of its product if more is to be sold.

A profit-maximizing monopsonist will employ a variable production input such as labor up to the point at which the marginal input cost equals its marginal revenue product. The price of the input is determined by the corresponding point on its supply curve. Marginal revenue product is the addition to total revenue attributed to the addition of one unit of the variable input. Marginal revenue product is equal to the marginal revenue times the marginal physical product, which is the change in output from a change in input use. So as long as marginal revenue product exceeds marginal input cost, profit will rise with increasing input use. On the other hand, if marginal input cost exceeds marginal revenue product, profit will rise with decreasing input usage. Therefore, profit is maximized by employing the level of variable input where marginal input cost equals marginal revenue product.

This situation is depicted graphically in Figure 9-8. The profit-maximizing level of input usage by the monopsonist is given by Q_M, and the price per unit paid is P_M. Here the firm is equating marginal input cost and marginal revenue product at point A and paying the input price P_M given by the supply curve at point C. This situation differs from the profit-maximizing level of input usage under perfect competition given by Q_{PC} and price per unit Q_{PC}. Point B is the point at which demand and supply forces jointly determine price. Under perfect competition, the marginal revenue product curve is the same as the marginal value product because output price P equals average revenue and marginal revenue.

Note that the price paid for the resource is higher under conditions of perfect competition than under a monopsony. The level of resource use also is higher under perfect competition relative to monopsony. The difference between the prices paid for the input under perfect competition and monopsony is termed the **monopsonistic exploitation** of the input.

Finally, consider the situation in which the monopsonist is not only the sole buyer of a resource but also the sole seller of a product. This situation might

Figure 9-8

Monopsonists use less of a resource and consequently pay a lower price than they would under conditions of perfect competition.

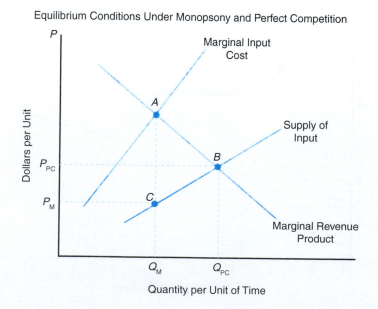

Equilibrium Conditions Under Monopsony and Perfect Competition

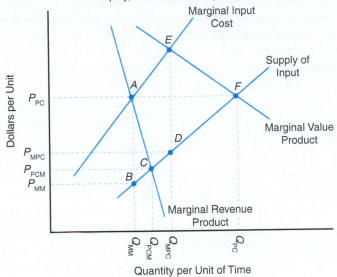

Equilibrium Conditions Under Alternative Combinations of Monopsony, Monopoly, and Perfect Competition

Figure 9-9

Each sole buyer or sole seller situation leads to lower levels of production than those given by perfect competition.

correspond to that of a meat packer who is the only purchaser of beef cattle and the only seller of wholesale dressed beef in the area. What is the firm's profit-maximizing level of input under this scenario? The answer to this question is illustrated, along with other market structure combinations, in Figure 9-9. The profit-maximizing level of input occurs at point A, where marginal input cost equals marginal revenue product. This point corresponds to input level Q_{MM} and price P_{MM}.

Under conditions of perfect competition in buying but under conditions of monopoly in selling in a particular market, the profit-maximizing level of input occurs at point C at which marginal revenue product equals the supply of the input. This situation corresponds to input level Q_{PCM} and price P_{PCM}.

Under conditions of perfect competition in selling and monopsony conditions in buying in a particular market, the profit-maximizing level of input is given where marginal input cost equals marginal value product (at point E), which corresponds to Q_{MPC} and P_{MPC} as the price paid for that input.

Under conditions of perfect competition for both selling and buying in a particular market, the profit-maximizing level of input usage is given where marginal value product equals the supply of the input (at point F). Thus, depending on whether imperfect or perfect competition prevails, profit-maximizing input levels and the prices paid for those inputs may vary considerably.

Oligopsony and Monopsonistic Competition

Similar considerations may prevail in buying activities in the case of oligopsony. When three meat-packing firms dominate a market, farmers often contend that prices are kept down because the respective firms follow a nonaggressive buying policy.

Oligopsony is defined as a market or industry comprising relatively few firms engaged in the purchase of resources. **Monopsonistic competition** is defined as an industry composed of many firms buying resources with the capacity of differentiating services. Differentiation of services to producers by buyers of resources may exist in terms of convenience of distribution and location of processing facilities, willingness to provide credit and/or technical assistance, and perhaps personal characteristics of the buyer. The quantity of a resource purchased at various prices by the oligopsonist or by the firm under monopsonistic competition to maximize profit is determined in the same manner as shown for the monopsonist.

Imperfect competition in buying Imperfect competition in buying activities can influence the market price for resources used in production. If there is only one firm buying a resource, the firm is said to be a monopsonist. The market structure is termed a monopsony. If there are just a few firms buying a resource, the firm is said to be an oligopsonist. The market structure is termed an oligopsony. If there are a relatively large number of firms buying a differentiated resource, the firm is said to be a monopsonistic competitor. The market structure is termed monopsonistic competition. The profit-maximizing level of input usage for any firm is determined where marginal revenue product equals marginal input cost.

The amount of profit earned under conditions of oligopsony or monopsonistic competition will depend upon the elasticity of supply for the resource. The more inelastic or steeper the supply curve for the resources, the larger the amount of profit earned. Under oligopsony, supply would most likely be less elastic than that under monopsonistic competition as a result of fewer firms and/or substitutable services. Note the symmetry between the increasing elasticity of supply and lower levels of profit in buying activities, and the increasing elasticity of demand and lower levels of profit in selling activities.

MARKET STRUCTURES IN LIVESTOCK INDUSTRY

To summarize imperfect competition from both a buying and a selling perspective, let's look at the market structures found in the U.S. livestock industry illustrated in Figure 9-10. Producers of feeder cattle, feeder pigs, and feeder lambs represent the original suppliers of the raw ingredient in the industry and closely resemble perfect competitors. These producers are so numerous that the actions of an individual producer have little or no effect on the market. They sell a relatively homogeneous product and have a fairly good knowledge of market opportunities and alternative prices. Because there are some differences in cattle, hog, and sheep, and because they do not have perfect knowledge of all possible market conditions, these producers operate within a near-perfect competition market structure. They are basically price takers and are powerless to affect the prices at which they sell their livestock. The consequence is that these livestock producers must accept the price outcomes from the interactions of groups further up the market channel.

A relatively new number of commercial feeders purchase feeder livestock from livestock producers. Consequently, commercial feeders operate as oligopsonists when buying feeder livestock. However, because there are more feeders with less market clout than packers, commercial feeders typically must accept the role of price takers when selling fed animals to packers.

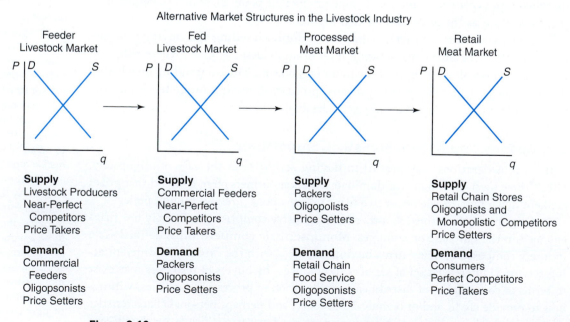

Figure 9-10

An examination of the market structures underlying demand and supply in the market channels of the livestock industries suggests the presence of several forms of imperfect competition.

Packers behave as oligopsonists in buying slaughter animals, but behave as oligopolists in attempting to sell their processed meat products. In fact, they sell their product facing the oligopsonistic behavior of large retail chain stores, large food service buyers, and many small wholesale and retail operations in specific geographical areas. This situation is known as bilateral oligopoly.

Large retail food stores and food service chains operate nationally as oligopolists in product markets, along with a fringe of monopolistic competitors in some local areas. They face a large number of consumers who purchase meat products as near-perfect competitors. The retail price is generally set through the price leadership of some dominant retailer or group of retailers in the oligopoly.

At the end of the market channel depicted in Figure 9-10 are consumers, the ultimate users of processed meat products prepared for retail sale. Consumers are price takers in a nearly perfect competition environment. They face the oligopolistic behavior of the large national retail chains and the more price-competitive behavior of local retailers.

GOVERNMENTAL REGULATORY MEASURES

Various measures may be employed to counteract possible adverse effects of imperfect competition in the marketplace. These measures include legislative acts, institution of maximum or ceiling prices received for output, use of lump-sum taxes, and institution of minimum prices paid for resources.

Legislative Acts

Historically, **legislative acts** have been passed by Congress in an attempt to minimize the social waste of imperfect competition. In 1890, Congress passed the Sherman Antitrust Act, prohibiting monopoly and other restrictive business practices. Section 1 of this historic act makes it illegal to act in restraint of trade by conspiring with other individuals or firms. The act forbids restraining trade through price-fixing arrangements or controlling and sharing industry output by collusive agreement. Section 2 of the act forbids the use of economic power to exclude competitors from any market.

In 1914, two further measures were enacted, namely, the Federal Trade Commission Act and the Clayton Act. The Federal Trade Commission was charged with the responsibility of investigating business organizations and practices and with carrying out the provisions of the Clayton Act. Although the Sherman Antitrust Act was general in identifying what actions were illegal, the Clayton Act was quite specific. The Clayton Act thus plugged loopholes and deficiencies in the Sherman Act.

A typical market structure situation in agriculture is one in which a large number of farmers face a single buyer (a monopsonist) for their products. Recall that the monopsonist has sufficient market power to be able to hold down the offered price. To offset this disparity in market strength, farmers historically have organized **agricultural bargaining associations**, or farmers' cooperatives. One example of a cooperative in the marketing and distribution of oranges is Sunkist Growers, Inc. Cooperatives are business organizations that are exempt from federal anti-trust legislation and whose chief goal is to improve producer income through either higher prices received for products or lower prices paid for inputs. The growth of farmers' cooperatives in the United States dates from the passage of the Capper–Volstead Act of 1922, which exempted cooperatives from certain restraints imposed by the **Sherman Antitrust Act of 1890** and the **Clayton Act of 1914**. The **Capper–Volstead Act of 1922** was the principal legislation exempting cooperatives from anti-trust laws.

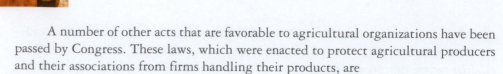

A number of other acts that are favorable to agricultural organizations have been passed by Congress. These laws, which were enacted to protect agricultural producers and their associations from firms handling their products, are

- **Packers and Stockyards Act of 1921**, which reinforced anti-trust laws regarding livestock marketing;
- **Cooperative Marketing Act of 1926**, which permitted farmers or their associations to acquire, exchange, and disseminate a variety of price and market information;
- **Robinson–Patman Act of 1936**, which primarily covered price discrimination practices; and
- **Agricultural Marketing Agreement Act of 1937**, which established agricultural marketing orders.

Marketing orders and agreements refer to arrangements among producers and processors of agricultural commodities. The chief goal of a marketing agreement or order is to improve producer income through the orderly marketing of a commodity by a group of producers and to avoid price fluctuations faced by an individual operating alone in the marketplace. Commodity prices may be controlled through negotiations with different groups involved in the marketing process or limitations on the supply placed on the market. Marketing orders have historically been used by the federal government to control the quantity coming into specific markets and, hence, support the market prices and incomes received by farmers.

If an agricultural bargaining association is to be effective in increasing farmer-members' incomes, it must gain control of, or influence, total supply. However, such control or influence is very difficult to achieve and almost impossible to maintain. Not all growers of a commodity are convinced that they should join the bargaining association. A further difficulty is due to the inability to control the supply of substitutes or imported quantities of that good. Finally, success in bargaining brings about a loss in incentive to maintain membership in the association. When product price is increased through the actions of the bargaining association, some of the group members may be tempted to leave the association. Under the scenario of increases in product prices, a strong incentive exists for each producer to expand production and consequently total market supply. Historically, bargaining associations have not been very successful. Nevertheless, they have been a force in improving the degree of competition in markets for agricultural commodities.

Countervailing actions as related to the agricultural sector have included the establishment of cooperatives and the development of marketing orders and agreements. However, other measures also may be employed to counteract the possible adverse effects of imperfect competition.

Ceiling Price

A regulatory agency such as the Federal Trade Commission may reduce profit and actually effect an increase in the output of a monopoly by instituting a maximum or ceiling price that can be charged. This countervailing action is illustrated graphically in Figure 9-11. Without assistance from the regulatory agency, the monopolist would produce Q_M, charge P_M per unit, and earn a profit that corresponds to area AP_MBC. If a governmental regulatory agency imposes a price ceiling of P_{MAX}—which of course is below P_M—the demand curve is given by $P_{MAX}ED$. Over the range in which the demand curve is horizontal, price equals marginal revenue because the monopoly can sell additional units of output without lowering

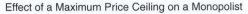

Effect of a Maximum Price Ceiling on a Monopolist

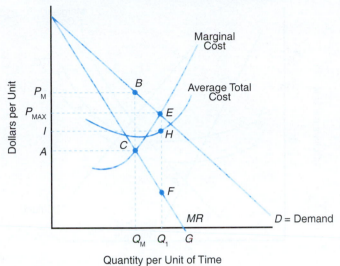

Quantity per Unit of Time

Figure 9-11
The profit earned by a monopolist will fall if a price ceiling is instituted.

price. Therefore, $P_{MAX}E$ is the marginal revenue curve up to Q_1 units of output. At output levels that exceed Q_1, the original demand curve is unchanged, so the FG segment of the original marginal revenue curve associated with the ED portion of the demand curve is still relevant. The entire marginal revenue curve is then P_{MAX}-EFG. The marginal revenue curve is discontinuous at Q_1. Therefore, the imposition of the price ceiling of P_{MAX} would cause the monopolist to produce Q_1, charge P_{MAX} per unit, and earn a profit equal to $P_{MAX}EHI$. Note that by instituting a price ceiling, the government would encourage the monopolist to produce more output $(Q_1 - Q_M)$, and charge a lower price $(P_{MAX} - P_M)$. The profit earned by the monopolist falls from AP_MBC to $P_{MAX}EHI$. Thus, this price regulation reduces monopoly profits and lowers price.

Lump-Sum Tax

Alternatively, a governmental regulatory agency may eliminate or reduce the profit of a monopoly by assessing a **lump-sum tax** on the firm's operation. This lump-sum tax may be a license fee or one-time charge. In essence, this countervailing action corresponds to a fixed tax, regardless of the level of output. This situation is exhibited in Figure 9-12. Without the lump-sum tax, the monopolist would produce Q_M, given by the intersection of marginal cost and marginal revenue at point F, charge P_M per unit, and earn a profit equal to area AP_MBC. With the imposition of the lump-sum tax, however, the firm's average total cost curve shifts upward from ATC_1 to ATC_2. Hence, the monopolist under this arrangement would still produce Q_M and charge P_M per unit. However, the profit now earned by the monopolist is reduced to area EP_MBT, which is less than the profit earned without the tax.

Minimum Price

In monopsony, the government could regulate the price of a resource purchased by imposing a minimum price that must be paid for the resource. In the case of the resource labor, this government regulation is the minimum wage law. This countervailing action is depicted graphically in Figure 9-13.

Legislative acts to counteract possible adverse effects of imperfect competition
Legislative acts passed by Congress such as the Clayton Act of 1914, the Packers and Stockyards Act of 1921, and the Capper–Volstead Act of 1922 are vehicles designed to minimize the social costs of imperfect competition. Countervailing measures including price regulation and taxation, as well as marketing orders and marketing agreements, also exist for offsetting potential adverse effects of imperfect competition.

Figure 9-12

Imposition of a lump-sum tax would lower profits but leave the level of output and product price unchanged.

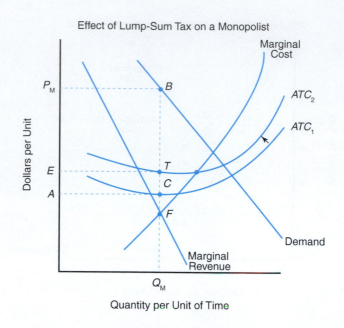

Effect of Lump-Sum Tax on a Monopolist

Figure 9-13

The impact of this government regulation is to raise the level of resource use and cost per unit.

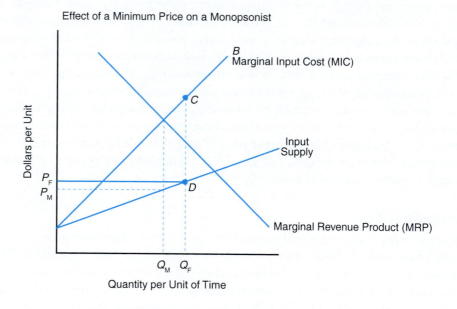

Effect of a Minimum Price on a Monopsonist

Without this minimum price regulation, the monopsonist would employ Q_M units of the resource and would pay P_M per unit. With the minimum price regulation, that is, imposing a price P_F that is higher than P_M, the firm's marginal input cost curve becomes $P_F DCB$ with a discontinuous portion equal to the vertical distance CD. If the firm employs a level of input less than Q_F, marginal input cost is less than marginal revenue product, and the firm will be able to expand profits by employing more units of the input. At an input level greater than Q_F, marginal revenue product is less than marginal input cost, so the firm would not employ inputs at this level. This reasoning suggests that the firm will employ an input level of Q_F to achieve profit maximization. Under this regulation, the monopsonist would employ Q_F units of the resource and pay P_F per unit. Interestingly, the minimum price regulation causes the monopsonist to employ more units ($Q_F - Q_M$) and pay a higher price per unit ($P_F - P_M$) for the resource.

SUMMARY

The purpose of this chapter is to discuss the determination of the price and quantity experienced under conditions of imperfect competition. The major points made in this chapter are summarized as follows:

1. Imperfectly competitive forms of market structure can be classified by
   - the number of firms and size distribution in the market,
   - the number of buyers in the market,
   - the degree of product differentiation,
   - the extent of barriers to entry, and
   - the economic environment within which the industry operates.

2. If there are a relatively large number of firms selling (buying) a differentiated product, monopolistic (monopsonistic) competition is said to exist.

3. If there are just a few firms selling (buying) a particular product, an oligopoly (oligopsony) is said to exist.

4. If there is only one firm selling (buying) a product, the firm is said to be a monopolist (monopsonist).

5. The profit-maximizing level of output for any firm is determined where marginal revenue equals marginal cost.

6. The profit-maximizing level of input usage for any firm is determined where marginal revenue product equals marginal input cost.

7. Social costs of imperfect competition exist in the marketplace. These social costs are known as dead-weight loss.

8. Several countervailing measures exist for offsetting the adverse effects of imperfect competition, including price regulation and taxation, and marketing orders and agreements.

KEY TERMS

Absolute unit-cost advantages
Agricultural bargaining associations
Barriers to entry
Capital access and cost
Cartels
Collusion
Countervailing actions
Dead-weight loss
Differentiated product
Economies of scale

Imperfect competition
Legislative acts (Sherman Antitrust Act of 1890; Clayton Act of 1914; Capper–Volstead Act of 1922; Packers and Stockyards Act of 1921; Cooperative Marketing Act of 1926; Robinson–Patman Act of 1936; Agricultural Marketing Agreement Act of 1937)
Lump-sum tax

Marginal input cost
Marketing orders
Monopolistic competition
Monopoly
Monopsonistic competition
Monopsonistic exploitation
Monopsony
Nonprice competition
Oligopoly
Oligopsony
Preferential government policies

TESTING YOUR ECONOMIC QUOTIENT

1. Plot the marginal revenue curve associated with the following demand curve faced by a monopolist or a monopolistic competitor.

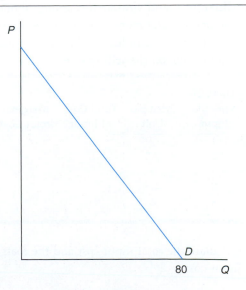

2. Use the following graph to answer the following questions:

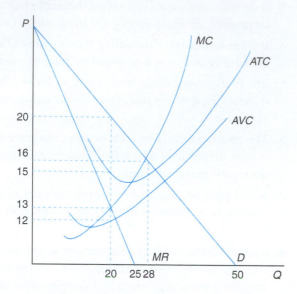

a. What price is charged by the monopolist in order to maximize profits?

b. Calculate the total revenue accruing to the monopolist at the profit-maximizing output.

c. Calculate the total cost to the monopolist at the profit-maximizing output.

d. Calculate the profit for the monopolist.

e. Calculate the total variable and fixed costs of the monopolist at the profit-maximizing output.

f. Now assume the MC curve represents market supply for a perfectly competitive market. What would the equilibrium price and quantity be for perfect competition? Are consumers better off or worse off with perfect competition or monopoly?

3. List differences and similarities among monopolies, oligopolies, and monopolistic competition. Be prepared to give examples of each form of imperfect competition on the selling side.

4.

Units of Variable Input	Price per Unit	Total Cost of Input	Marginal Input Cost
1	2		
2	2.5		
3	3		
5	4.5		
8	6		

a. Calculate the total input cost and the marginal input cost.

b. If the marginal value or marginal revenue products were 4, what would be the profit-maximizing level of input?

5. a. Find the equilibrium price and quantity for a monopsonist in the following graph.

b. Find the equilibrium price and quantity under perfect competition in the following graph.

c. What is the magnitude of monopsonistic exploitation?

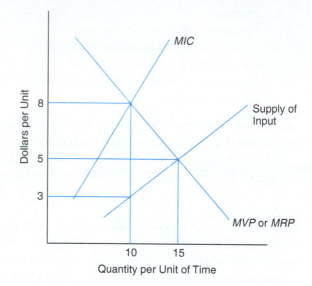

6. Explain the significance of the following acts:
   a. Clayton Act
   b. Capper–Volstead Act
   c. Packers and Stockyards Act

7. List and explain the various measures that may be employed to counteract possible adverse effects of imperfect competition in the marketplace.

8. On the following graph, show the effect of a lump-sum tax on a monopolist.

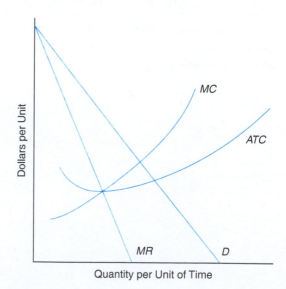

9. Using the following graph, answer questions (a) through (d).

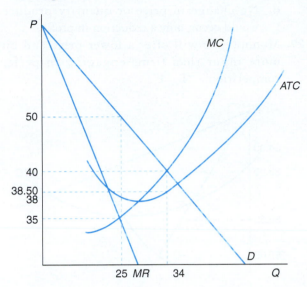

a. What are the profit-maximizing price and quantity levels for the monopolist?
b. Calculate profit.
c. Suppose the government imposes a price ceiling of $40. Now what is the optimal price and quantity combination?
d. Calculate the new level of profit.

10. This graph pertains to a firm labeled as a monopolistic competitor.

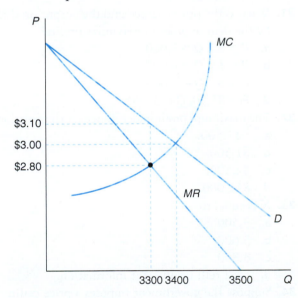

If the demand curve were to intersect the quantity axis, the quantity demanded would be equal to _____ units. This monopolistic competitor could produce _____ units and charge a price of _____. Under perfect competition, what output would be produced? What would be the corresponding market price?

Which of the following is true about monopolistic competition? (Circle all that apply.)
a. There is product differentiation.
b. There are few sellers.
c. The food retailing industry is an example of this market structure.
d. No firm can influence market prices.

11. Select the correct answers for conditions of oligopoly.
a. Number of sellers
    i. many
    ii. few
    iii. one
d. Product differentiation
    i. yes
    ii. no

Which of the following illustrates an oligopoly? (You may select more than one.)
a. Retail food industry
b. Airline industry
c. A utility (electric) company
d. Farm machinery manufacturers

12. Marginal input cost (MIC) refers to which of the following? (Circle the correct answer.)
a. The change in total cost when producing one more unit of output
b. The change in total revenue when using one more unit of input
c. The change in total cost when using one more unit of input
d. The change in total revenue when producing one more unit of output

13. The economic analysis of imperfect competition was originated by _____ and _____.

14. Measures to lessen the possible adverse effects of imperfect competition are called _____.

15. The arrangement among producers and processors of agricultural commodities in which the chief goal is to improve income is called a(n) _____.

16. In the United States, the government agency charged with the responsibility of investigating business organizations and practices is the _____.

17. In the beef processing industry, three firms—Conagra, Excel, and IBP—comprise roughly 85% of the buying side of the market. This situation characterizes a(n) _____.

18. The social costs of imperfect competition are known as _____.

19. A situation designed to increase or maintain profits through price fixing and/or to restrict entry of new firms in an industry is called _____.

20. A cooperative pool formed by oligopolists to set prices at artificially high levels is known as a(n) _____.

21. Attempts to increase the demand for a product or service via product differentiation are known as _____.

22. A market structure in which there are many buyers with the capacity of differentiating services (i.e., location of processing facilities; willingness to provide credit) is called a(n) _____.

23. Federal marketing orders were created in 1937 by way of which legislative act? _____.

For questions 24 through 28, circle the correct answer.

24. Which of the following is/are a common barrier(s) to entry?
  a. Economies of scale
  b. Absolute unit-cost advantages
  c. Capital access and costs
  d. All of the above

25. The first piece of legislation prohibiting monopoly and other restrictive business practices was the
  a. Clayton Act of 1914.
  b. Packers and Stockyards Act of 1921.
  c. Capper–Volstead Act of 1922.
  d. None of the above.

26. The Capper–Volstead Act of 1922
  a. reinforced anti-trust laws regarding livestock marketing.
  b. was the principal legislation exempting cooperatives from anti-trust laws.
  c. plugged loopholes in the Sherman Antitrust Act of 1890.
  d. all of the above.

27. The piece of legislation that not only plugged loopholes in the Sherman Antitrust Act of 1890 but also created the Federal Trade Commission was
  a. the Packers and Stockyards Act of 1921.
  b. the Clayton Act of 1914.
  c. the Capper–Volstead Act of 1922.
  d. none of the above.

28. If the government were to impose a lump-sum tax on a monopolist, what is likely to happen to the quantity produced of a commodity and the price charged relative to the situation where no lump-sum tax is imposed?
  a. The price would fall, but the quantity produced would rise.
  b. The price would fall and the quantity produced would fall.

  c. The price would remain the same and the quantity produced would fall.
  d. No change in price or quantity produced would occur, only a reduction in profit.

29. Monopolists will offer a lower price and buy more input than firms engaging in perfect competition.   T   F

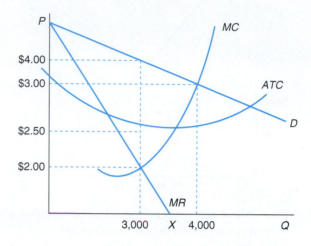

Use the following graph to answer questions 30 through 34.

30. Which of the following does this graph illustrate?
  a. Monopoly in the short run
  b. Monopolistic competition in the short run
  c. Perfect competition
  d. Both (a) and (b)

31. What is the price charged and the output produced by the firm in order to maximize profits?
  a. $P = \$2, Q = 3,000$
  b. $P = \$3, Q = 4,000$
  c. $P = \$4, Q = 3,000$
  d. $P = \$2.50, Q = 3,000$

32. The maximum profit under imperfect competition is
  a. −$1,500.
  b. $1,500.
  c. $4,500.
  d. $12,000.

33. X is equal to
  a. 3,500.
  b. 3,600.
  c. 3,800.
  d. can't tell; insufficient information.

34. Suppose the government imposes a price ceiling of $3. Which of the following is (are) *false*?
  a. In this situation, the price cannot exceed $3.
  b. Under this situation, $P = \$3, Q = 4,000$.
  c. The profit accruing to this firm is $12,000.
  d. None of the above.

Use the graph below to answer questions 35 through 37.

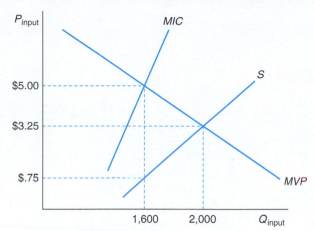

35. What is the equilibrium quantity of input used and what is the price paid per unit under the condition of perfect competition?
    a. $Q_{input} = 1,600, P_{input} = 5$
    b. $Q_{input} = 1,600, P_{input} = 2.75$
    c. $Q_{input} = 2,000, P_{input} = 3.25$
    d. None of the above

36. What is the equilibrium quantity of input used and what is the price paid per unit under the condition of monopsony?
    a. $Q_{input} = 1,600, P_{input} = 5$
    b. $Q_{input} = 1,600, P_{input} = 2.75$
    c. $Q_{input} = 2,000, P_{input} = 3.25$
    d. None of the above

37. What is the magnitude of monopsonistic exploitation on a per-unit basis?
    a. $2.25
    b. $1.75
    c. $0.50
    d. None of the above

REFERENCES

Chamberlin EH: *The theory of monopolistic competition*, 1933.

Pindyck RS, P Rubinfled: *Microeconomics*, Boston, 2009, AAE, Pearson Publishing.

Purcell WD: *Agricultural marketing: systems, coordination, cash, and future prices*, Neston, Va, 1979, Neston Publishing.

Robinson J: *The economics of imperfect competition*, London, 1933, Macmillan.

🍎10

Natural Resources, the Environment, and Agriculture

Perhaps more than any other industry in our economy, agriculture is intimately tied to natural resources and the environment. Despite all the modern technology that one finds on today's farms, soil, water, and sunlight are still three of the most critical inputs into the agricultural enterprise. Also, just as agriculture depends on the environment, it also affects the environment. Emissions from tractors and equipment pollute the air, and chemicals that farmers apply run off the fields and seep into the ground. There is increasing recognition that the environment and our natural resources are scarce assets and require careful management. In this chapter, we will study the economics of these important parts of our economy.

In Chapter 8, we found that in an ideal economy, the free market will maximize social welfare. However, we also saw that it doesn't always work out so well. If monopoly power exists in the economy, then the free market is not quite free and is not socially optimal. In this chapter, we find this inefficiency of market forces to be the case again. In this case, the problems arise because markets do not exist for all of the benefits provided by the agricultural sector and some of the costs of keeping the sector running are not paid by the farmers. Furthermore, we will also find that in some instances, the natural assets on which agriculture depends are not well managed due to poor incentives or weak understanding of those assets.

Some of the multiple interrelationships between agriculture and the environment are presented in Figure 10-1. The relationships portrayed in Figure 10-1 fall into two categories—environmental issues and natural resource issues. Since natural resources are part of the environment and the environment is a natural resource, this division is somewhat artificial, but it can be helpful. *Environmental economics* refers to the study of flows, such as pollution, that affect others. *Natural resource economics* is the study of natural assets that are valued for their productive capacity.

AGRICULTURE AND THE ENVIRONMENT

Water Pollution

By the middle of the last century, many U.S. waterways were severely polluted. The Cuyahoga River, which runs through Cleveland, Ohio, actually erupted in flames a number of times between 1936 and 1969. Environmental problems such as these inspired the 1972 passage of the **Clean Water Act**, which sought not only to clean up every major waterway in the United States but also to actually eliminate the release of "toxics in toxic amounts" by 1985. Although this overly ambitious goal was not achieved, there is no doubt that the act has dramatically improved the quality of the nation's water since its passage. Rivers and lakes that were once dangerous hazards are now attractive sites for fishing, boating, and even swimming.

Despite the improvements, water pollution in the United States remains an important environmental concern. In Figure 10-2, we see that there remains much

The Clean Water Act is the primary piece of federal law that seeks to restore and maintain the quality of U.S. water resources.

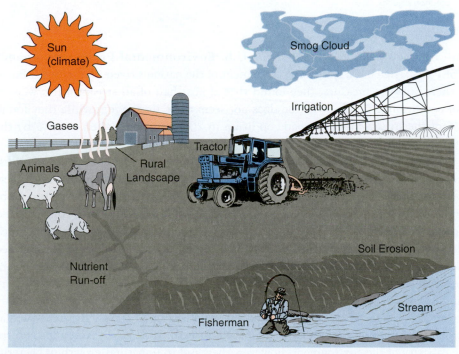

Figure 10-1

Agriculture interacts with natural resources and the environment in many ways. Water, climate, and the soil are all critical to the sector. Farming leads to both water and air pollution, but also can suffer as a result of such pollution. Farmland also is increasingly valued for the environmental services it provides, from rural amenities to habitat for some animals.

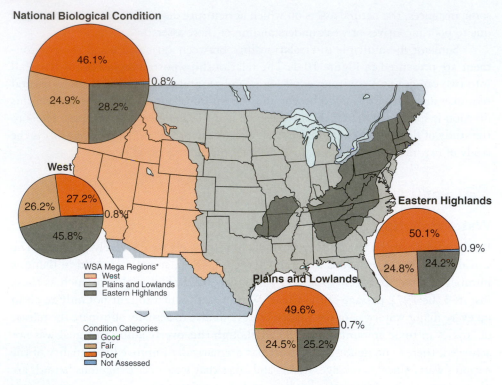

Figure 10-2

Despite years of water quality regulations, the most recent assessment of the quality of the water in U.S. streams found that only 28% of the length of the nation's rivers and streams had a biological condition that was qualified as good, while the condition of 46% was poor.

Source: From *National Rivers and Streams Assessment 2008–2009: A Collaborative Survey,* U.S. Environmental Protection Agency, 2016.

The U.S. Environmental Protection Agency (USEPA) is the federal agency in the United States that is responsible for enforcing or overseeing the federal environmental laws.

work to be done. In 2008–2009, the **U.S. Environmental Protection Agency (USEPA)** carried out a systematic assessment of the nation's rivers and streams. Based on their broadest measure, they found that almost half of all stream miles were in poor condition. And the situation does not seem to be improving; while they found 37% of the stream miles in good condition in 2004, this had fallen to 28% by the 2008–2009 assessment.

Although the Clean Water Act has eliminated the worst pollutants from industrial and municipal sources, it has done little to reduce nutrients that flow into rivers and lakes. Increasingly, it is not toxins that are at the root of U.S. water quality problems. As seen in Figure 10-3, there is a wide range of pollutants, and many lakes and rivers suffer from multiple types of pollution. Some pollutants, such as metals and pesticides, can have toxic consequences, leading to environmental impacts on fish or increased risks of birth defects or cancer. But others have primarily economic impacts. For example, sediment can clog hydroelectric facilities and passageways for boats, and salinity degrades the quality of water for irrigation and increases the costs of water treatment.

Excess nutrients are the most prominent pollutants in lakes. Nutrients such as nitrogen and phosphorus are essential for plant growth. When they exceed the absorptive capacity of the environment, however, they lead to unnatural growth of bacteria. These bacteria use up the available oxygen, leaving little for fish and other species in the ecosystem. These excess bacteria can also be toxic and, in extreme cases, can result in waterways becoming clogged with slimy green algal blooms.

Agriculture is the leading contributor to water pollution in the United States. In 2000, it was identified as a source of problems in 41% of the impaired waterways

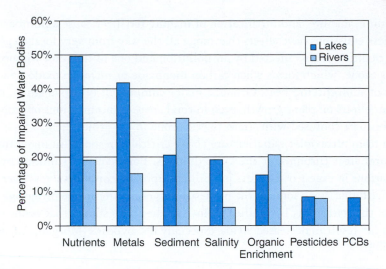

Figure 10-3

Water pollutants range from nutrients, which can lead to harmful algal blooms, to polychlorinated biphenyls (PCBs), a known cause of cancer. This figure presents the percentage of the impaired water bodies that are affected by various pollutants.

Source: From *National Water Quality Inventory Report to Congress*, U.S. Environmental Protection Agency, 2000.

in the country. It contributes to many of the pollutants listed in Figure 10-3, especially the excess nutrients. Fertilizer that is not absorbed by crops or incorporated into the soil runs off to streams, rivers, and lakes. Sediment, pesticides, and salinity all can be traced in large part to agricultural sources.

Air Pollution

Agriculture is also closely linked to air pollution, although the relationship is not nearly as significant as it is with water pollution. In this case, however, agriculture not only causes air pollution but also suffers from it.

The USEPA is required by the 1990 Clean Air Act Amendments to regulate air pollutants that "are known to cause or may reasonably be anticipated to cause adverse effects to human health or adverse environmental effects." The primary pollutants that are regulated include carbon monoxide, lead, mercury, nitrogen dioxide, particulate matter, ozone, and sulfur oxides. As we will discuss later, in 2007, the U.S. Supreme Court ruled that the USEPA must also regulate greenhouse gases that have been shown to cause global climate change. Some air pollutants are transported across great distances, while others fall near their source, causing localized air quality problems. The health impacts of air pollution are not trivial. For example, even though air quality had already improved greatly since the 1970s, researchers at Harvard School of Public Health found that improvements in air quality between 2000 and 2007 led to an increase in the average life expectancy in the United States by about one-third of a year (Correia et al., 2013).

Tractors and other agricultural equipment emit pollutants from the combustion of fuel. Although these emissions are tiny compared to those from the industry and transportation sectors, they can be locally important. For other types of emissions, however, agriculture is a leading source nationally. A 2003 report published by the National Academy of Science found that agriculture accounted for 89% of all emissions of ammonia (NH_3), 50% of nitrous oxide (N_2O), 9% of nitric oxide (NO), and 22% of methane (CH_4) (National Research Council, 2003). Emissions from animals,

Ozone is a gaseous molecule that contains three oxygen atoms (O_3). High in the atmosphere it plays a positive role by shielding the earth from harmful ultraviolet rays from the sun, but close to the ground it is a harmful component of smog.

crop burning, manure, and nitrification of manure fertilizer cause a number of problems. These emissions have effects that range all the way from very local to global.

Agriculture is also affected by air pollution and one of the most important pollutants is ozone, which forms when carbon monoxide or nitrogen oxides react in the presence of sunlight and heat. Ozone, a major component of urban smog, can have significant effects on plant growth, even in rural areas. Concentrations of ground-level ozone (not to be confused with stratospheric ozone in the atmosphere, which protects the earth from ultraviolet radiation) are above critical thresholds in many important agricultural areas throughout the United States. For example, summertime ozone concentrations in excess of 60 parts per billion (ppb) are common across over half the state of California, with some of the highest levels occurring in the prime agricultural regions of the state. This is significant since studies have shown that ozone exposure on plants can result in decreased yields, especially for soybeans (Morgan, Ainsworth, & Long, 2003).

Global Climate Change

Greenhouse gases are those gases that reflect radiation back to Earth. The most abundant of these is carbon dioxide (CO_2), which increases when fossil fuels such as oil or coal are burned.

Global climate change is a problem that has received a great deal of attention in recent years, yet remains confusing and contentious. The basic theory underlying global climate change, often referred to as the *greenhouse effect*, is presented in Figure 10-4. Greenhouse gases reflect radiation back to Earth. As the concentration of these gases increases, so does the greenhouse effect, warming the planet.

Although some aspects of the global climate change problem are subject to debate, there are important facts that are beyond dispute. First, atmospheric concentrations of greenhouse gases are increasing as a result of human activities. Carbon dioxide (CO_2), which is created whenever fossil fuels like coal, oil, and natural gas are burned, is the most abundant of the greenhouse gases. The concentration of CO_2 in the atmosphere increased from about 250 parts per million (ppm) in 1900 to 320 in 1960, and over 405 in 2016. To put these numbers in perspective, it is estimated that in the 650,000 years prior to 1900, the concentration of CO_2 in the atmosphere had

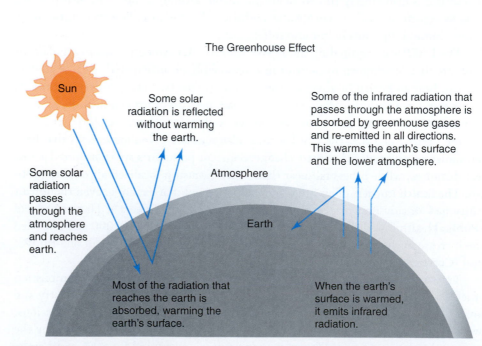

Figure 10-4

The greenhouse effect is caused by an increase in atmospheric gases that trap solar radiation, gradually warming the earth's surface.

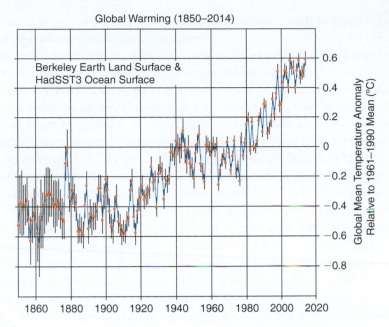

Figure 10-5

The average surface temperature for the planet, shown as the difference from average during the 1960s, clearly shows a warming trend for the planet over the last 100 years.

Source: From The Average Temperature of 2014 Results from Berkeley Earth by Robert Rohde and Richard Muller. Published by Berkeley Earth, © 2014.

never exceeded 300 ppm. Second, the planet has certainly warmed over the last 100 years; on average, from 2005 to 2015 the planet's global mean temperature was 1.7 degrees Fahrenheit warmer than it was from 1905 to 1915 (see Figure 10-5). Finally, since it is known that greenhouse gases trap heat in the earth's atmosphere, most atmospheric scientists believe that human production of greenhouse gases has contributed to the warming of the planet and, therefore, that continued burning of fossil fuels will lead to further warming.

There are, however, some important areas of uncertainty about global climate change. First, the magnitude, timing, and global distribution of climate change remain subject to significant debate. Modeling the climate is difficult. Second, we know little that is definite about the actual impacts of climate change. For example, agriculture in some regions of the United States is predicted to benefit from increases in temperature and a lengthening of the growing season, while in other regions drought and heat stress are expected. Most models also predict that the sea level will rise as the water expands in warmer temperatures and global ice caps melt, but the extent and speed of the rise are uncertain. Will sea levels rise by 1 foot or 4 feet? Will pest infestations be a serious problem in a warmer planet? How will natural ecosystems respond and what will be the consequences of those responses? If other nations are severely affected, what will be the consequences for the United States in terms of trade and immigration? These and other questions remain, making the development of national and international policies to address global climate change very difficult.

Agriculture is closely tied to the problem of global climate change. As the economic sector that is most intimately related to the weather, it is subject to the vagaries of the climate. Climate change may lengthen the growing season in the north, but might reduce yields in the south. Changes in precipitation patterns also will affect the sector. On the plus side, however, a fertilization effect is also likely because plants grow more rapidly in air with higher concentrations of CO_2.

Biofuels are fuels that are derived from biological sources, from corn kernels to trees. They include, but are not limited to, ethanol from corn, sugarcane, or other sources, and biodiesel, which is refined from vegetable oil.

Like the rest of the economy, agriculture emits CO_2, the principal greenhouse gas. In addition to CO_2, agriculture also emits methane and nitrous oxide, each unit of which traps 25 and 298 times more heat than CO_2, respectively, over 100 years. On the other hand, agriculture is also viewed as potentially being part of the solution. First, through changes in soil management, it is possible to increase the levels of carbon in the soil, eliminating CO_2 from the atmosphere. More significantly today, agriculture is now seen as a source of fuel, offsetting the burning of fossil fuels. Most prominently, the U.S. Energy Information Administration reported that ethanol represented 10% of U.S. gasoline consumption by volume in 2012. In principle, if *biofuels* are used instead of fossil fuels, then energy could be created without adding any CO_2 to the atmosphere since for every carbon atom that is emitted, another carbon atom is turned into plant matter as it grows. In the United States, however, government-subsidized ethanol from corn is the main biofuel currently in use and this form of biofuel does little to reduce CO_2 emissions since nearly as much energy is needed to produce the ethanol as is contained in the fuel itself. In net, therefore, nearly as much carbon might be emitted through the production of ethanol as if we used the fossil fuels directly. There are more promising alternatives such as ethanol from sugarcane and *cellulosic ethanol*, which uses not only the corn kernels but entire plants. However, sugarcane can be grown only in warm, wet climates, and the technology for production of cellulosic ethanol is still under development. Hence, though agriculture's potential to reduce the net CO_2 emissions of the country is great, that potential is still not being achieved.

Other Environmental Impacts

As seen in Figure 10-1, agriculture also generates a wide range of additional impacts on the environment. Among these we highlight odor and the impacts on the habitat for species.

Anyone who has ever driven past a feedlot will attest to the fact that the agricultural sector does not always smell like a rose. This stench is not just a nuisance. The odors wafting off large concentrated animal feeding operations (CAFOs) can be a health hazard and can depress nearby property values. Hence, these odors impose costs on people in the vicinity of the farm.

Another environmental impact of agriculture is on species of plants and animals. As of 2016, over 1,600 species of animals and plants were listed as endangered or threatened by the U.S. Fish and Wildlife Service. From the humpback whale to the Cumberland elktoe (a clam), these plants and animals are protected from harm by the **Endangered Species Act (ESA)**. Agriculture is one of many forces that can threaten species by encroaching on or contaminating their habitat. When a species vanishes from the planet, it is often impossible to know exactly what has been lost. Could it have led to the cure for cancer? Was it critical to its ecosystem? These concerns, and the awareness that the loss of a species is permanent and irreversible, have led the U.S. government to take strong steps to protect species from extinction.

Finally, agriculture generates valuable environmental services that are not captured by markets. For example, the open spaces and traditional landscape that farms provide are of increasing importance. City dwellers drive out to the comforting rural landscapes sprinkled with barns, farm animals, and crops. Other people move to these areas and pay premium prices for lots surrounded by farms. The farming sector, therefore, is part of the environment itself and something from which society benefits. But just as farmers do not pay the costs associated with their pollution, so they are not compensated for creating the open spaces that others enjoy. Because these costs and benefits are not covered by the market, the market fails to lead to a completely efficient outcome. This issue is covered in the next section.

ECONOMICS OF THE ENVIRONMENT

As we saw in Chapter 8, the market has a remarkable ability to allocate goods and services. If the economy is fully efficient, then the private actions of consumers and firms will maximize the economy's surplus, which is a *Pareto efficient* outcome. We now will examine whether this same remarkable result will occur for the environmental impacts that we have discussed earlier.

The first question we must ask is whether the environment has value. The answer is clearly *yes*. Consider the case of water pollution from agriculture. When water quality falls, there is a wide range of impacts downstream, from the decline in aesthetic value for sightseers to reduced recreational and commercial fish harvests. Users of the waterway, therefore, would be *willing to pay* (WTP) to decrease or *abate* the water pollution. In short, there is demand for environmental improvements much as there is for market commodities. Furthermore, there is clearly also a supply curve for environmental improvements. Reducing pollution is usually not free: farmers must either implement new *Best Management Practices* or change their operations in other ways. As pollution abatement increases, each additional reduction is usually more costly. Hence, as seen in Figure 10-6, there is also a *marginal cost* (MC) of abatement.

In Figure 10-6, we consider the pollution abatement problem. If farmers have already abated a_1 units, then marginally increasing abatement would cost c_1 dollars per unit. On the other hand, according to the graph, the public would be willing to pay p_1 dollars for a marginal increase in abatement. As long as the marginal WTP is greater than the MC, society as a whole can benefit by an increase in the level of abatement. We see, therefore, that $a*$ is the socially efficient level of pollution abatement. If the actual level of cleanup is less than $a*$, then increasing pollution abatement will increase net benefits; but if the actual level of abatement is greater than $a*$, then the benefits to the public of the last units of abatement will be less than the costs of achieving them.

We should be careful to avoid oversimplification. Estimating the amount that society would be willing to pay to reduce pollution is very difficult. When we are interested in the demand for potato chips, we can use market data. There is no such market for pollution abatement. So, estimating the WTP curve in Figure 10-6 can be quite difficult. Economists have developed a variety of methods to estimate such *nonmarket valuation*. A detailed discussion of these methods is beyond the scope of this chapter, but many studies have definitely shown that the environment provides

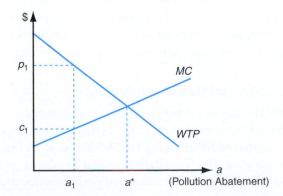

Figure 10-6

In this figure, we show the marginal cost to abate pollution and the willingness to pay for such abatement. The efficient level of abatement would be where the marginal cost (MC) is equal to the willingness to pay (WTP).

significant real benefits to the public, and people are definitely willing to pay for these benefits.

Efficient Property Rights

Clearly, Figure 10-6 looks remarkably like a standard supply and demand diagram. So, will a market actually arise for environmental improvements leading to a socially optimal outcome? Unfortunately, the answer is *no*. The reason is that the characteristics of efficient **property rights** usually are not satisfied for environmental goods. These characteristics are

- *Enforceability:* enforceability is satisfied if an individual's rights are secure—if you buy something, then it will not be taken from you without your permission.
- *Transferability:* a good or service satisfies transferability if it can be transferred from one individual to another.
- *Exclusivity:* a good or service satisfies exclusivity if all benefits and costs associated with it are received by only one individual at a time.

Let's consider why each of these characteristics is important for market efficiency. First, suppose that enforceability is not satisfied for a particular good (i.e., there is nothing that stops someone from taking the good from its owner). In such a world, would anyone pay the costs to produce the good? No, because they could not be assured that they would be paid. Would anyone buy the good? No, because the property could be taken from them. In short, even if people were *willing* to pay for the good, no market would arise for its production, and efficiency would not be achieved. This is why laws protecting property are so critical to economic prosperity.

Second, suppose that transferability is not satisfied (i.e., a good or service cannot be transferred from one individual to the next). For example, there may be laws limiting the sale of certain goods or services. In this case, no market can arise for the production of these goods or services because their sale is not allowed and their efficient transfer from one individual to the next cannot occur.

Finally, and most importantly in the context of the environment, *exclusivity* must be satisfied for a market to yield a fully efficient outcome. The classic example here is the case in which some of the costs of producing a good are borne not by the producer of the good but by the public at large. For example, farmers produce agricultural commodities and pay for their inputs of capital, labor, and supplies. But they do not pay the cost that might be imposed on fishermen downstream who suffer reduced catch because of declining water quality. This environmental cost is called an **externality**, because it is *external* to the market—neither the farmer nor the buyer of the farmer's crops pays these costs. Externalities can be positive or negative, but regardless of their sign, when externalities are present, the market typically will not lead to a socially efficient outcome. We now consider this problem in more detail.

Efficiency and Externalities in Agriculture

Figure 10-7 presents the case of a negative externality. The graph represents the aggregate supply and demand for a market good, Q. As discussed in Chapter 8, the market supply curve, S_M, is obtained by summing the MC curves of all individual firms. The market demand for the good, D_M, is similarly obtained by adding up the demand curves for all the buyers in the market. Market forces will push the price to p_M, at which point the willingness to pay for the last unit is equal to the marginal cost to produce the unit.

If the market establishes the price p_M, then producers will supply Q_M units at a cost equal to the area under the MC_M curve, area A. Producer surplus is equal to area B (i.e., total revenue, $B + A$, minus total cost, A). Consumer surplus is the triangle

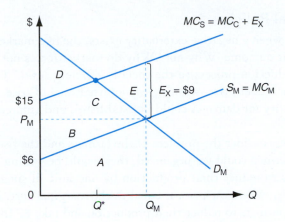

Figure 10-7

An externality is represented in this figure. We assume that for each unit of Q that is produced, external damages of E are inflicted on society. The social marginal cost, MC_S, includes both the marginal cost seen in the market, MC_M, and the external cost, E_X.

$C + D$ (i.e., the total willingness to pay for the good, $A + B + C + D$, minus the amount that they pay to the suppliers, $A + B$). Producer surplus plus consumer surplus is, therefore, the familiar triangle $B + C + D$.

In Figure 10-7, however, we assume that producing Q also causes pollution. We assume that this pollution imposes costs on others because it leads to the degradation of downstream fisheries. Neither the producers of the good nor the consumers of the good take these costs into account—they are external to the market. For simplicity, we assume that each unit of production causes external costs worth $9 that are borne by the downstream fishermen. The social marginal cost for each unit of production, therefore, is equal to $MC_M + E_X = MC_S$. If Q_M units are produced, there is an additional cost equal to $Q_M \times E_X$, the area $B + C + E$ in Figure 10-7.

From the market equilibrium, therefore, we see that social net benefits equal producer surplus, plus consumer surplus, minus external costs (i.e., $B + C + D -$ [$B + C + E$]). Noting that areas B and C enter on the benefits side and the external costs side, we find that the net social benefits created by the production and sale of Q_M units are equal to $D - E$. It can easily be seen that net social benefits could be increased by simply supplying less of Q. At Q^*, the net benefits are equal to area D, and the negative portion, E, is absent, so net social benefits are greater than those at Q_M. Hence, the quantity provided by the market is inefficiently high relative to the socially optimal level.

The inefficiency of the market in this case can also be seen at the margin. Consider the marginal benefits and marginal costs of the very last unit produced at the market level of production, Q_M. The consumers would be willing to pay p_M for this unit. The cost to producers is also p_M. So, the net benefits to producers and consumers of the Q_Mth unit are essentially zero. But there is also an additional cost of E_X to the fishermen downstream. Hence, after adding up benefits and costs, the benefits of this unit are less than the costs. The marginal net benefits (WTP $- MC_M - E_X$) are negative for every unit beyond Q^*.

There are two important things to note here. First, we see that in the presence of externalities, the free market will not achieve the socially efficient level of production. This is why an externality is frequently referred to as an example of *market failure*. Second, we also see that although the production of Q causes pollution and external costs, this does not mean that this activity should be halted entirely. On the contrary, it would be socially inefficient to set production at zero since in this case production of Q could yield positive net benefits, even after subtracting the external costs, E_X.

Environmental Policies

As we have seen, when a negative externality exists, the free market will not lead to a socially efficient outcome. Why not? Why do market forces not develop in order to bring the production process to the socially efficient level? This question was considered by the Nobel Prize–winning economist Ronald Coase. Coase concluded that if responsibility for damages could be established and enforced, then a market could arise.

For example, consider the problem of the farmers and the fishermen discussed earlier. If the fishermen could get organized, they would be willing to pay $9 to convince the farmers to reduce their production by one unit. If such a payment were offered, then at Q_M, where the price is equal to the marginal cost, it would be more profitable for the farmers to reduce their production and take $9 from the fishermen than to produce that last unit that just covers the marginal cost. As the supply of Q falls, the market clearing price would increase, leading to marginal profits for the farmers, but until Q^* is reached, farmers would be better off taking the money from the fishermen than they would be selling their crops. Hence, this privately negotiated agreement could lead to the socially efficient level of production.

Coase's market-based approach would also work if the farmers were required to pay the fishermen for the right to pollute. Until production reached Q^*, the farmers would be willing to pay the fishermen at least $9 for the right to produce a unit of Q. In this case, the external cost (i.e., the cost imposed on the fishermen) is *internalized*. Now that the farmers bear all costs of production, including the environmental costs, the market leads to a socially efficient outcome.

Although Coase's approach to resolving the problem of externalities is enticing, it has not been widely adopted. The most significant problem with this approach is the difficulty of coordinating the many firms and individuals that are involved. Most critically, a *free-rider* problem often arises. For example, suppose that a fishermen's association forms to pay the farmers to reduce pollution. Although only members of the association pay, all anglers, whether members or not, benefit from the decreased pollution. There can be a strong incentive to not join the association and let others take care of the problem. This phenomenon is known as *free riding*.

As a result of the free-rider problem, the government, with its ability to tax and regulate, is frequently involved when externality problems arise. The question that we now consider is this: when the government becomes involved, how should its policy be designed?

In all of the approaches we consider, the government chooses the total level of abatement. If efficiency was desired, then the government would choose the level of abatement where the marginal WTP to abate pollution is equal to the marginal cost. Because of the difficulty in establishing the true WTP for environmental improvements, however, environmental standards are rarely set at a level that is deemed socially efficient. Instead, most environmental targets are set based on scientific standards. Taking this level as given, economists still have much to offer by helping to reach an environmental target in the most efficient way possible.

Command-and-Control Policies   Many environmental policies consist of regulations on technology or restrictions on practices. Economists have given this approach the ignominious title of *command-and-control* (CAC) regulations. In the prototypical CAC approach, all businesses are treated more or less equally. All firms are required to install the same equipment or to reduce their pollution load to the same level.

The main economic problem with the CAC approach is that it fails to take into account the diversity that naturally exists in the economy. For example, if one farmer is required to reduce pollution, all are required to do the same. But Farmer John may have old equipment, so for him the implementation of the practice could be quite costly. Farmer Sue, on the other hand, is about to buy a new tractor, and buying one that pollutes less than the required level might be relatively inexpensive. If the two farmers are located near each other, then the same environmental improvements could be achieved at lower cost if Farmer Sue reduced her pollution more than the standard and Farmer John reduced less.

This principle is demonstrated in Figure 10-8. Figures 10-8A and B represent the marginal cost curves of two firms, for which the marginal cost of abatement increases as pollution abatement goes from 0 to 10. Figure 10-8C presents the marginal cost curves of both firms together. A point on the horizontal axis in Figure 10-8C identifies an allocation of pollution abatement responsibilities for both firms. For example, starting at the point where $a_1 = a_2 = 5$, movement to the right would increase the abatement of firm 1 and decrease that of firm 2. Movement to the left would decrease the abatement of firm 1 and increase firm 2's share. No matter what point on the horizontal axis is chosen, total abatement equals 10.

The problem with a CAC approach can be seen in Figure 10-8. The total cost of achieving 10 units of abatement with each firm abating 5 units is shown by the matched area under the MC curves in Figure 10-8C. When both firms are required to abate 5 units, the last unit of abatement by firm 1 costs substantially more than the last unit of abatement by firm 2. Hence, the total cost could be reduced by moving to the left, where firm 1 abates less and firm 2 abates more. In principle, the government could make such an efficient allocation but would have to know the MC curves of each polluter. Hence, it is very difficult for a CAC-type policy to reach an environmental goal in a cost-effective manner.

Taxes and Subsidies Economists have long endorsed *incentive-based* approaches to pollution control as an alternative to the CAC approach. The first such approach is either a tax on pollution or a subsidy for abatement. In this case, the tax or subsidy serves to correct a market failure, improving the economy's economic efficiency. This is different from distortionary taxes that decrease welfare by creating a dead-weight loss. Consider first the case of a subsidy of t dollars per unit of abatement as presented in Figure 10-9. In the case presented, firm 1 would abate 3 units, because for the first 3 units the cost of abatement is less than the subsidy; the firm would not abate more because beyond 3 the marginal abatement cost is more than the subsidy. Similarly, firm 2 would abate 7 units, but not more. If the subsidy is chosen just right, therefore, 10 units of abatement will be achieved.

Incentive-based policies are those that use economic incentives to achieve their objectives. Taxes, subsidies, or transferable permits can be used to achieve environmental goals, and all of these have the potential to be less costly than command-and-control policies in which the government dictates what should be done.

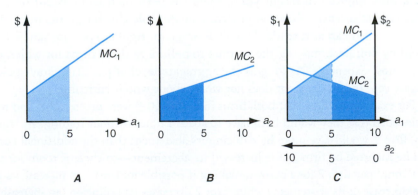

Figure 10-8

In these figures, we present the marginal cost of abatement for two firms: firm 1 in A, firm 2 in B, and both firms in C. In C, firm 1's abatement increases from left to right, while firm 2's abatement increases from right to left. The vertically matched areas indicate the cost to the firms of 5 units of abatement each.

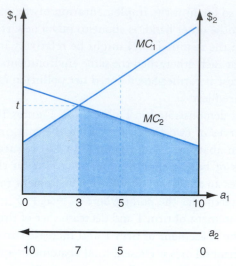

Figure 10-9

A tax or subsidy of $t per unit will induce the firms from Figure 10-8 to abate pollution. Firm 1 will abate 3 units, where its MC is equal to t. Firm 2 will abate 7 units. If the MC of abatement is equal across firms, then the allocation of abatement responsibilities is cost-effective.

A tax on pollution can work just like a subsidy for abatement. In this case, the tax of t dollars per unit of pollution would mean that for each unit of abatement the firm saves t. As far as the firm is concerned, therefore, the choices will be the same as in the subsidy case—the firm will continue to abate as long as the tax savings are greater than or equal to the cost of abating. Again, if the rate is chosen correctly, 10 units of abatement can be achieved.

The important advantage of taxes, subsidies, and all forms of incentive-based approaches is that whatever level of abatement is finally achieved, it will be done at the lowest possible cost. In Figure 10-9, the total cost is the shaded area, which is less than the costs under the CAC allocation in Figure 10-8. A disadvantage of a tax or subsidy program is that unless the *MC* curves are known, the government will not know with certainty what level of abatement will be achieved. If t is too low, then too little abatement will be achieved. If t is too high, too much abatement will be achieved. On the other hand, if the government does not have a clear target level of abatement but does know the social cost of each unit of pollution, then the tax or subsidy approach can be socially efficient even if the final level of abatement cannot be predicted.

Transferable Rights   In recent years, there has been increasing attention paid to another type of incentive-based approach, **transferable discharge permits (TDP)**, commonly called **cap and trade**. In a TDP program, rights to pollute can be bought and sold by polluters, moving the permits to pollute to those firms for which abatement is most expensive. As long as the aggregate level of pollution stays below an aggregate cap, the government does not worry about who is emitting.

For example, assume that both firms in Figure 10-9 were given rights and responsibilities such that they would both have to abate 5 units without trading. At this allocation, firm 1 would save more by reducing its abatement than the additional cost that would be incurred by firm 2 if it increased its abatement—so there is room for a deal. Firm 1 could pay firm 2 for a permit, making it possible for firm 1 to increase its pollution (a decrease in its abatement) while firm 2 decreases its pollution (an increase in its abatement). Such trading could benefit both firms. Permits could continue to move from firm 2 to firm 1 until they reach the point where the *MC* curves intersect.

As with taxes and subsidies, TDP programs are cost-effective, minimizing the cost of achieving a pollution abatement goal. The advantage of TDP programs is that the government can control the overall level of pollution, which is often what they are most interested in, and leave the allocation up to the market.

Note that although cap and trade approaches rely on a market, that market is created by a government intervention. Unless there is a government-established cap requiring some abatement, there will be no market and no abatement. Such programs, therefore, are not voluntary in the traditional sense of the word.

THE ECONOMICS OF THE RESOURCES OF AGRICULTURE

We now turn to the related set of issues surrounding agriculture and natural resources. The central distinction here between environmental issues and resource issues is the extent to which externalities are present. An issue falls in the category of environmental economics if there are important externalities and in the realm of natural resource economics if the costs and benefits of using a natural resource accrue primarily to the user. Here we look specifically at two important resources for agriculture: soil and water. When a farmer takes steps to prevent soil erosion, this protects the quality of the soil on his or her farm and benefits him or her directly. When a farmer uses water, this increases his or her yields.

As revealed in the story of the Dust Bowl in the 1930s, the agricultural sector does not always take care of its most precious resources. Why is this? Do markets fail to give the right signals to producers, or is it simply a case of ignorance? (The answer is a combination of these two.) Water is increasingly scarce, and competition for this resource is increasing. Do the existing rules governing the use of water lead to an efficient outcome? These are the questions we consider here.

SOIL QUALITY AND QUANTITY

According to the USDA Natural Resource Conservation Service, "Soil quality is the capacity of a specific kind of soil to function, within natural or managed ecosystem boundaries, to sustain plant and animal productivity, maintain or enhance water and

It was conditions like this dust storm that forced many farmers to abandon this U.S. Resettlement Administration land use project and other areas in the spring of 1935.   Credit: kwest/Shutterstock.

An asset is something that generates benefits to the user, typically over a period of time. Machinery is an asset. So is soil quality. In placing a value on future benefits that can be generated, we use discounting, taking into account that future returns are not equivalent to benefits you receive today.

air quality, and support human health and habitation." In short, soil quality is the most fundamental asset of the agricultural sector.

The major source of decline in soil quality is erosion, which can occur either as a result of rainfall, which leads to sheet and rill erosion, or by wind. Over time, erosion can lead to a decline in soil characteristics by washing away productive soil or by degrading features of the soil that are essential for crop productivity. Although farm management and the replacement of nutrients using fertilizer can restore the productivity of eroded soils, this comes at a cost.

As shown in Figure 10-10, the problem of soil erosion diminished dramatically from 1982 to 2012, though progress has been limited since 2002. Today it is widely held that in the United States, the on-farm consequences of soil erosion from declining productivity are of less public concern than are the off-farm consequences. Eroded soil clogs irrigation and hydroelectric equipment and carries nutrients into rivers. It is not surprising that farmers have done much to reduce soil erosion for, as we discuss now, soil is a valuable asset that deserves conserving.

THE ECONOMICS OF SOIL CONSERVATION

Characteristics of Soil

Any farmer will tell you that soil quality is precious. It can make the difference between profit and loss and be the central determining factor in what crops a farmer chooses to grow. Depth, density, acidity, nutrient content, and microbial activity are some of the characteristics that make up soil quality. The quality of soil is a complex function of its physical, chemical, and biological characteristics. In the end, high-quality soil has the potential to produce bountiful crops for a long time.

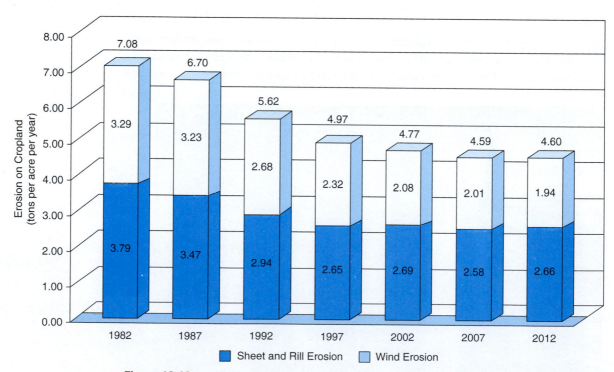

Figure 10-10

As seen in this figure, the soil erosion rate in the United States has declined significantly since the 1980s, but has remained fairly constant since 2000.

Source: From National Resources Inventory Annual NRI, Natural Resource Conservation Service, United States Department of Agriculture, 2015.

So what is the value of this precious resource and how much should be spent on preserving it? In the standard economic framework, a farmer values soil because its productive potential can generate profits over the long term. Imagine the case of Farmer Jones (FJ), who year after year has harvested an average of 100 bushels per acre from his farm. He spends time and effort to take care of his farm, and his profits are moderately low, just $10 per acre. For simplicity, we assume that FJ plans to continue to farm his land far into the future.

What is the future stream of income worth to FJ? Let's consider the next 5 years. In each year, he expects profits of $10 per acre, but it would be inaccurate to say that the value of the farm today is $50 per acre. The $10 that he will earn in 5 years is not the same as the $10 per acre that he will earn next year. The further out in the future that income is to be obtained, the less important that money is to FJ today. To add up values that occur at different times, economists use an approach called **discounting**.

Discounting and Present Value Analysis

To compare dollar values across time, economists use discounting to convert all dollar values to present values. The **present value** of money to be received in the future is equal to the amount of money that the individual would have to be given today in return for the future money. For example, consider the case of FJ's $10 profit in 5 years. How much would we have to give FJ today so that he would be willing to give us the $10 in 5 years? In other words, returning to the familiar concept of opportunity cost discussed in Chapter 1, what is the opportunity cost today of that future money?[1]

Let's start with an example. Suppose FJ bought a $5 certificate of deposit at his local bank today paying 5% interest annually. After 5 years, he would have $5 plus the compounded interest, a total of $6.38. So, the opportunity cost of $6.38 in 5 years is $5 today.

We can now see that the opportunity cost of $10 in 5 years is the amount of money, say $X, today that FJ could put in the bank and have exactly $10 in 5 years. If he took $X today and put it in the bank for 5 years at 5% interest, then he would have $X \times (1.05)^5$ in 5 years. So all we need to do is find out what X is such that $X \times (1.05)^5 = \$10$. Dividing both sides of this equation by $(1.05)^5$, we get $X = \$10/(1.05)^5 = \$7.83$. If we gave FJ $7.83 today and he put it in the bank at 5% interest, he would have exactly $10 after 5 years. Hence, $7.83 is the *present value* of $10 in 5 years, given the interest rate is equal to 5%.

The Present Value of Soil Resources

We can now return to the value of Farmer Jones's soil. Remember that his farm had traditionally earned profits of $10 per acre. The present value of the stream of productive services provided by his farm would be the amount of money that we could pay

> **The present value** refers to what money in the future is worth today. If you discount future returns at the rate of r per year, the present value of X to be received in n years is $\$X/(1 + r)^n$.

---

[1] To answer this question, we have to understand the principle of compounded interest. Suppose FJ puts $5 in the bank today and leaves it there for 1 year at 5% interest. He will earn 25¢ in interest, leaving him with $5.25 next year or

$$\$5.00 \times (1 + 5) = \$5.00 \times \left(1 + \frac{5}{100}\right) = \$5.00 \times 1.05 = \$5.25$$

If he left the money in the bank for a second year, the money would increase further to $5.25 \times 1.05 = \$5.00 \times (1.05)^2 = \$5.51$. After 5 years, FJ would have $5.00 \times (1.05)^2 = \$6.38$. The equation can easily be generalized: if you have $X today and put it in the bank at r% interest per year, then after n years you will have $X \times (1 + r/100)^n$.

Present value then works backward. It tells us how much money today is equivalent to a specified sum in the future. If we are to receive $Y in n years and the interest rate is r% per year, then the present value of Y(PV) can be found using the following equation: $PV = \$Y/(1 + r/100)^n$. Discounting and present value analyses are extremely useful for evaluating a decision with future consequences. This is true for issues ranging from business investments to personal finance.

him right now that would be equivalent to this stream of future income. The value of this stream of profits can be found by adding up the present value (PV) of profits next year, the following year, and so on. Assuming again an interest rate of 5% per year, we arrive at the sum

$$V = \frac{\$10}{(1.05)^1} + \frac{\$10}{(1.05)^2} + \frac{\$10}{(1.05)^3} + \cdots \tag{10.1}$$

If we add up the PV of profits over the first 5 years, it is equal to $43.29. Over 10 years, it reaches $77.22, and over 30 years $153.72. It turns out that no matter how far out in the future we look, the PV of the stream of profits will never pass $10/5% = $200. Hence, the PV of the stream of profits looking far out to the future is $200. In general, the **capitalized value** of a constant stream of $Y per year given an interest rate of $r\%$ is $\$Y/r\%$.

So the stream of profits from FJ's farm is originally worth $200 per acre. However, since those profits are a result of not only the soil but many other factors including FJ's ingenuity and the fertilizer and seed that he uses, we cannot meaningfully apply this full value of his farm to the soil. What we can do, however, is consider the marginal value of the soil.

Suppose that FJ allows his brother to manage part of his farm for 1 year. Because he works the land more aggressively and less carefully than FJ, significant erosion occurs during the year, leading to gullies and a loss of productive top soil. As a result, FJ's average profits decline by $1 per acre to $9. The capitalized value of the land, therefore, falls from $200 to $9/0.05 = $180. Hence, the decline in soil quality reduced the land's capitalized value by $20. If the only change from one year to the next was the soil quality, we can appropriately ascribe the decline in value to the soil. Or, looking at it another way, the value of FJ's annual soil conservation efforts was $20 per acre.

With this information, one can then evaluate the economic efficiency of soil quality management. For FJ, the benefit of avoiding the decline in productivity that followed from his brother's 1 year of management is $20. But how does this compare with the cost? If FJ's brother's profits were $35 per acre, a difference of $25, then the benefits of aggressive management exceeded the costs in terms of the decline in the value of the soil asset.

Practice: The Dust Bowl

From the viewpoint of soil as an economic asset, soil quality has clear economic value and, at least on the margin, can be quantified, although it may be a complicated concept and difficult to measure. From the story of the Dust Bowl, we know that at some times in history soil resources have been severely diminished.

From an economic standpoint, soil quality is a resource for which the characteristics of efficient property rights are satisfied. The on-farm benefits and costs of soil management fall on the owner of the land, the soil's characteristics can be sold to others in the sale of land, and the right to one's land is usually quite secure. Hence, markets should lead to the efficient management of soil resources. The beautifully terraced fields on mountainsides throughout Asia remind us how marvelously effective such incentives can be.

So what happened during the Dust Bowl? How did erosion get so excessive on such a wide scale? Of course, one part of the problem was just bad luck—drought conditions occurred. But other factors were at the root of the widespread erosion that took place during this period. First, there was simply the lack of knowledge; farmers at that time were uninformed of the basics of soil conservation, so they did not know how to protect their valuable resource. Second, land itself was not seen as scarce at that time. Farmers often operated as if they could use up their land and then easily move on to the next field. Together, these and other factors resulted in the loss of substantial top soil, and huge economic and social costs for the nation.

WATER AS AN ECONOMIC ASSET

"Water, water, every where, Nor any drop to drink."
SAMUEL TAYLOR COLERIDGE, "THE RIME OF THE ANCIENT MARINER", THE EDUCATION PUBLISHING COMPANY, 1906

Water covers two-thirds of the globe. It is so abundant that one might wonder how it could ever be considered scarce in an economically meaningful sense. Yet Coleridge's familiar rhyme is actually a telling portrayal of the world's predicament. Although we certainly can't say there is literally not a drop to drink, it is really not much more than that. If all the world's water were fit into a gallon jug, the amount of fresh water available for consumption or irrigation would make up only about one tablespoon, less than 0.5%. Hence, it is not surprising that fresh water is a scarce resource.

Fresh water is the primary interest to agriculture because salt water destroys crops. In some countries where fresh water is extremely scarce, desalination plants process ocean water for irrigation. But desalination, which is costly and energy intensive, is currently uneconomical in the United States and most of the world. Hence, we restrict ourselves here to considering fresh water in its two main forms: surface water in rivers and lakes and groundwater stored in underground aquifers.

Surface Water   Subject to the vagaries of the weather, lakes and rivers can change from dry beds to overflowing hazards in a matter of hours. Nonetheless, during most of the year, most rivers and lakes carry water that can be distributed among the many competing uses, or are left in their place to support ecosystems and provide recreational and aesthetic values to humankind.

A distinguishing characteristic of rivers and lakes is that, with few exceptions, they are renewed throughout the year through precipitation and runoff. Hence, one of the most important questions that economics can help us understand is how these resources are and should be allocated between competing uses. In the early history of the United States, water was so abundant that it seemed there would never be a limit on the amount an individual could extract; it just wasn't always in the right place at the right time. The fallacy of that myth has now been revealed in the trickles of water that flow through numerous rivers severely overexploited throughout the West.

The overextraction of water is not only environmentally tragic but often economically inefficient. What we mean by efficient use of water can be illustrated by considering a stylized case of a river where farmers are competing for a flow of 100 acre-feet. In Figure 10-11, we assume that for each farmer, increasing water improves crop yields. The marginal revenue of water and marginal cost of pumping the water to their farms are such that each farmer would like to use 80 acre-feet, where the marginal cost is equal to the marginal revenue.

Consider a situation in which there are two such farmers, but one is upstream of the other. The upstream farmer has unlimited access to the water, and he or she will use 80 acre-feet, leaving only 20 acre-feet for the downstream farmer. We can see that this allocation is economically inefficient. The value of the last marginal unit of water to the upstream farmer was essentially zero since MR = MC at 80 acre-feet. For the downstream farmer, left with only 20 acre-feet, a marginal increase in the water would yield profits of *a–c*. Total net benefits could be increased by taking water away from the upstream farmer and giving it to the downstream farmer. Total

Figure 10-11

This figure presents the marginal revenue (MR) and marginal cost for farmers who share access to a river. A farmer's profits would be maximized where MR = MC, at 80 acre-feet of water.

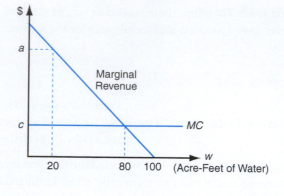

Prior appropriation property rights is a system in which individuals have rights to specific quantities of water and these rights can be sold to others. This system can lead to an efficient allocation in which water is used by those who value it most.

net benefits would be maximized where all farmers have the same marginal net benefits (MR = MC), which in this case is achieved by an equitable allocation of the water.

As the simple example shows, a system in which upstream users have preference over downstream users can be inefficient; some users have more than enough water and others go thirsty. As more and more people wanted access to scarce water resources, these inefficiencies led to the creation of an alternative system of water rights in water-scarce areas called **prior appropriation water rights**. Under a prior-appropriation system, water rights can be held by people throughout a region, and any person who does not hold such a right cannot extract water from the system. The key advantage of prior appropriation arises when transferability is allowed. This allows market forces to move water from low-value uses to high-value uses.

For example, suppose now that the upstream farmer in Figure 10-11 has an initial right to 80 acre-feet and the downstream farmer has a right to 20 acre-feet. If the rights were transferable, the downstream farmer would be willing to pay more for an increase in his or her water than the upstream farmer would demand; a deal could be made. Under ideal circumstances, the farmers could bargain back and forth until the equal distribution of 50 acre-feet each was reached.

But what about the fishermen, the rafters, and the birds? Even where water rights are conferred under a prior-appropriation system, some groups may not hold rights. In many regions, water rights are conferred only to those who put the water to "productive use." This typically would include agricultural, municipal, and industrial users, but would exclude groups who value the water in situ (i.e., in the lake or river itself). This arrangement also leads to an inefficient allocation because some groups that place a high value on the water are denied the opportunity to obtain water. An allocation of surface water is truly efficient only if the marginal value of the water to all uses is equal, including both extractive and nonextractive uses.

Groundwater   Unlike surface water, groundwater reserves are often recharged very slowly. There are many kinds of aquifers, from those that are recharged annually with water flowing through them like an underwater river, to fossil aquifers that store water that can be thousands of years old and recharge at rates that are so slow that they are negligible on a human time scale.

As with surface water, the efficient use of groundwater resources requires that the marginal net benefits to competing uses be equal. This is particularly challenging for groundwater since monitoring use is difficult. If water metering is not required, then the resource falls into the category of the "rule of capture," in which a user gets what a user takes. Like the upstream farmer taking as much river water as he or she desires, the user of groundwater who has the "biggest straw" and sucks the hardest is able to take the most. This leads to an inefficient allocation of water across users.

A second issue arises in groundwater reserves that recharge slowly. In this case, water withdrawals diminish not only the water available for other uses today but also the water available in the future. Hence, an efficient allocation must also strike a balance between current and future uses of the water resource. As we discussed for soil resources earlier, the present value criterion is typically used when considering future consequences. The allocation of the water is *inter-temporally* inefficient (i.e., across time) if the present value of a marginal increase in future supply is greater than the cost of a marginal decrease in consumption today.

If rights are not clearly defined, then users of an aquifer will realize that they only partially control the water that will be available in the future. So even if one user conserves carefully, overuse by his or her neighbors could still lead to rapidly declining water stocks. This results in an incentive for everyone to take as much as they can, as fast as they can. In the end, all the users are worse off, a pattern of inefficient resource use that Garret Hardin referred to in his 1968 article as "The Tragedy of the Commons."

Again, property rights are one way to move toward a more efficient use of groundwater resources. For example, establishing an aquifer-wide authority that sets a cap on aggregate annual consumption levels could balance between current and future uses of the watershed. Markets then could be used to reallocate the rights across competing uses.

Unfortunately, aquifers rarely lie neatly within political boundaries, usually extending across county, state, and even national borders. As a result, even if clear rights are established in one jurisdiction, competition across borders can lead to over-exploitation. Until scarcity is such that cooperation is seen as the only way to avoid a crisis, achieving efficient management of groundwater resources is often difficult.

GOVERNMENT POLICIES FOR AGRICULTURE, NATURAL RESOURCES, AND THE ENVIRONMENT

So far in this chapter, we have reviewed some of the most important issues in the area of environment and natural resources as they relate to agriculture. We have also seen why market failure sometimes arises, providing a justification for government involvement. We now provide a brief discussion of the most significant environmental policies in the United States that affect agriculture.

Soil Erosion Policies and the Conservation Reserve Program

In response to the crisis of the Dust Bowl, the U.S. government began taking steps to reduce soil erosion in the early 1930s. The agency that is now called the Natural Resource Conservation Service began addressing soil management problems through educational efforts, establishing demonstration plots and disseminating information about soil conservation. By the 1950s, the government started changing the incentives of farmers by paying them to take land out of production. This effort eventually became the Conservation Reserve Program (CRP) in 1985.

The CRP pays farmers to take their land out of production, compensating them with a rental payment and paying up to 50% of the cost of establishing a cover crop to diminish erosion. The program's focus has shifted over time as farmers have increasingly recognized the importance of protecting their soil assets and learned how to do so. In 2015, total payments amounted to over $1.8 billion, covering over 24 million acres.

The benefits of the CRP are much broader than simply maintaining soil for the farmers. Keeping soil on the land also keeps it out of the rivers and lakes. Moreover,

The Conservation Reserve Program has traditionally been the largest conservation program in the United States. Originally part of the national emphasis on stopping soil erosion, the program pays farmers to take fields out of production. Today, the CRP is only one of a number of conservation programs that seek a variety of environmental objectives.

CRP lands provide habitat for birds and other animals. As we have noted, a farmer's economic incentives should lead to efficient maintenance of his or her soil asset, but the external costs of erosion and the external benefits of conservation mean that market forces alone will probably not provide incentives for socially efficient land management.

Indirectly, the CRP reduces the supplies of agricultural commodities. This causes farmers to look for other fields to plow and pushes farm prices higher, encouraging more intensive use of non-CRP land and the conversion of nonqualifying lands into production. This slippage is estimated to be on the order of 20% and offsets about 10% of the erosion reductions brought about by the CRP (Wu, 2000).

Other Federal Incentive Programs for Agricultural Conservation

Roughly every 5 years there is a new incarnation of the Farm Bill, changing priorities and introducing new programs to foster conservation. As seen in Figure 10-12, conservation programs increased from a little over $3 billion in the 1990s to $4 to $5 billion in the last decade. Since 2006, working-lands programs have been the largest category of USDA-funded conservation program and in recent years these programs have represented over half of all conservation funding. Working-lands programs include the Environmental Quality Incentives Program (EQIP) and the Conservation Stewardship Program (CSP). These programs advance a range of national, regional, and local environmental priorities such as improving water quality and providing habitat for migrating birds.

Environmental Regulations

Unlike incentive-based programs, most regulations faced by agriculture are far from voluntary. Farmers must comply with many rules governing workplace conditions, handling of chemicals, machinery, treatment of animals, and many other aspects of a

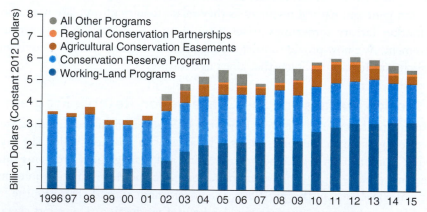

Major USDA Conservation Program Expenditures, 1996–2015

Note: Working-land programs include the Environmental Quality Incentives Program, the Conservation Stewardship program, Conservation Technical Assistance, and predecessor programs. Predecessors of the Agricultural Conservation Easement Program include the Wetland Reserve Program, Farmland Protection Programs, and part of the Grassland Reserve Program. Other programs include Voluntary Public Access, Healthy Forest Reserve Program, Agricultural Management Assistance, and watershed programs. Source: USDA, Economic Research Service using data from USDA, Office of Budget and policy Analysis budget summary data.

Figure 10-12

Under the Farm Bill, Congress provides support for a wide range of programs designed to support natural resource conservation.

Source: From Economic Research Service, United States Department of Agriculture.

farmer's day-to-day operations. Some of these regulations address the increasingly large-scale nature of many agricultural operations. Concentrated animal feeding operations (CAFOs) have been especially tightly regulated. Since 1976, the Clean Water Act has regulated CAFOs, treating them much the same as a chemical factory or a sewage treatment facility, and, over time, the number of operations categorized as CAFOs has increased.

Except for CAFOs, however, most farms have not been tightly regulated under the Clean Water Act. That is changing. Across the United States, the USEPA is implementing **Total Maximum Daily Load (TMDL)** programs, which seek to control pollution from all sources in a watershed and bring all pollutants below a cap. Some of these programs seek voluntary approaches to encourage pollution abatement by agriculture using incentives and transferable discharge permits. The USEPA is legally required to achieve the standards, however, and if voluntary approaches are not sufficient, there will be pressure to begin more forceful actions to achieve water quality goals.

As we have seen, pollution is an externality, so it is typically assumed that government regulation is required to overcome this market failure. But regulation may not always be necessary. Because of social pressures or concern for their neighbors, farmers frequently take measures to reduce the offsite impacts of their farms, including environmentally responsible farming. Efficient government intervention can provide additional incentives to address environmental externalities without unnecessarily tying hands or imposing unnecessary costs. But some regulations have high costs and result in poor environmental performance. This phenomenon is frequently referred to as *government failure*.

The Endangered Species Act

The Endangered Species Act (ESA) may be the most notorious environmental law in the United States. It makes it illegal to "harass, harm, pursue, hunt, shoot, wound, kill, catch, capture, or collect" any species identified as endangered by the U.S. Fish and Wildlife Service. Although most landowners are unaffected by this law, for farmers who own land with habitat for an endangered species, the act can fundamentally change the way the land is managed and dramatically alter the farm's economic potential. The ESA prohibits any activity that might threaten an endangered species, no matter how typical or seemingly innocuous.

The economic justification for the ESA is that all species provide benefits to all of society, a positive externality. The value of some species, like the humpback whale, comes from the simple knowledge that the species exists. Other species provide tangible benefits in the form of medicines or the support of ecosystems that provide services to humankind. When a species is pushed to extinction, therefore, this imposes costs on society that typically are not internalized by those who have caused the extinction.

The ESA has been widely criticized for its harsh approach that can turn simple fence-building or land-clearing into federal offences. Landowners and others are frequently irate about the enormous costs that the act can impose when it limits their ability to use their land as they like. In some cases, this results in a huge reduction in the land's value.

Originally passed in 1973, the ESA was amended in 1978, 1979, and 1982, giving some flexibility to a generally inflexible law. Nonetheless, the law continues to restrict activities on lands where endangered species are found, essentially punishing landowners that have endangered species. This has created the *perverse incentive* to illegally "shoot, shovel, and shut up" to eliminate species before they are found by a regulator. Although the ESA has had many important successes and definitely has reduced species extinctions in the United States, it will never be fully successful in protecting U.S. biological diversity until the perverse incentives problems can be resolved.

SUMMARY

The purpose of this chapter is to introduce the basic economics of the interaction between agriculture and the natural environment. The major points of the chapter are summarized as follows:

1. Agriculture affects and is affected by a wide range of environmental problems, from local air and water quality to global change.

2. The environment is like many other economic assets; people are willing to pay for a better environment, and there is a cost to providing a clean environment. In other words, there is a demand curve for environmental quality and a supply curve.

3. However, unlike most other goods and services discussed in this book, the market economy typically does *not* provide the efficient level of environmental services. The reason for this is that the three conditions that are necessary for the efficient provision of a good or service are not all satisfied for most environmental services. In particular, when we consider environmental pollution, it is clear that some of the costs associated with many production processes do not exclusively fall upon the individuals or firms that reap the benefits—everyone in the path of the pollution can suffer the environmental costs.

4. An externality is present when costs or benefits fall on people other than those who are directly involved in a market exchange. For example, when runoff from farms kills fish, harming recreational fishermen downstream, that is a negative externality. When farmers provide habitat for migratory birds that are enjoyed by birdwatchers, that is a positive externality. Usually when externalities are present, the unregulated market does not achieve a socially efficient outcome.

5. Soil represents a critical asset to the agricultural enterprise. It is valued because it provides the foundation for farm production for many years to come. The resource must be valued, therefore, in terms of both its present and future value. To do this, economists use the principle of discounting and present value.

6. Water is also a critical asset to agriculture, but in this case it is shared by many users. Whether the water is deep below the ground or lies at the surface in rivers or lakes, economic efficiency requires that the water be allocated between competing uses, both now and in the future. An efficient allocation is one in which the marginal net benefit to all users is the same.

7. Agriculture in the United States exists within a policy environment in which there is a range of regulations and incentives that encourage environmental and natural resource stewardship. Because of these many positive and negative incentives, farmers today must be aware that their operations have impacts on the environment that cannot be ignored.

KEY TERMS

Capitalized value
Cap and trade
Clean Water Act
Discounting
Endangered Species Act (ESA)
Externality

Global climate change
Present value
Prior appropriation water rights
Property rights
Total Maximum Daily Load
    (TMDL)

Transferable discharge permits
    (TDP)
U.S. Environmental Protection
    Agency (USEPA)

TESTING YOUR ECONOMIC QUOTIENT

1. The _____ is the primary law that governs water quality in the United States.

2. Agriculture contributes to air pollution through _____, but also is affected by air pollution since _____.

3. Concentrations of CO_2 in the atmosphere were _____% higher in 2016 than the highest estimated levels in the 650,000 years prior to 1900.

4. Agriculture can play a role in reducing CO_2 concentrations in the atmosphere through _____ and _____.

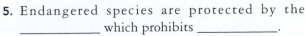

5. Endangered species are protected by the _____ which prohibits _____.

6. The socially efficient level of pollution abatement occurs when the _____ is equal to the _____.

7. An externality occurs when the benefits or costs of an economic activity are external and not felt by _____.

8. Externalities lead to market failure because the market fails to _____.

9. When a free-rider problem arises, it is often necessary for _____ to become involved.

10. Equal treatment of two polluting firms may not lead to a cost-effective outcome because _____.

11. A tax on pollution will cause a firm to reduce its pollution until the _____ is equal to the tax rate.

12. Soil erosion is a public concern because of the effects of erosion on _____.

13. If the interest rate is 5% per year, the present value of $1 million to be received in 20 years is $_____. If the money is to be received in 200 years, the present value is only $_____.

14. If the interest rate is 5%, then the capitalized value of an investment that pays $8 per year would be $_____.

15. The efficient allocation of water is achieved if the marginal value of water to all users is _____.

16. Groundwater resources suffer from the "Tragedy of the Commons" when _____.

17. Prior to 2006, the largest conservation program funded by the Farm Bill was _____, but since that time _____-lands programs have consumed a larger share of the total conservation budget.

18. The Endangered Species Act seeks to protect species from going extinct by prohibiting anything that might put a species in danger of extinction, but it can also create perverse incentives for landowners to s_____, s_____, and s_____.

19. Suppose that a farmer has two ways to produce his or her crop. He or she can use a low-polluting technology with the marginal cost curve MC_L or a high-polluting technology with the marginal cost curve MC_H. If the farmer uses the high-polluting technology, for each unit of Q produced, one unit of pollution is also produced. Pollution causes pollution damages that are valued at E per unit. The good produced can be sold in the market for P per unit.

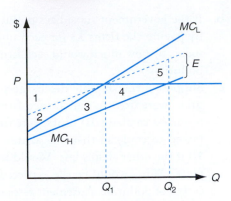

a. If there are no restrictions on the firm's choices, which technology will the farmer use and what quantity will he or she produce?

b. Given your response in part a, is it socially efficient for there to be no restrictions on production? Explain, referring to the areas identified in the figure.

c. If the government restricts production to Q_1, would this lead to a socially efficient outcome?

d. Now suppose the farmer is required to pay a tax of E per unit of production. What technology would the farmer choose? Would a socially efficient outcome be achieved?

20. Suppose that two firms have marginal costs of abatement as shown in the following figure.

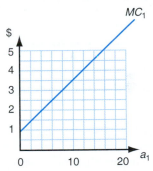

Firm 1

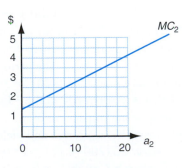

Firm 2

a. Would a government regulation requiring both firms to abate 10 units be cost-effective? Explain.

b. If the government subsidizes pollution control by paying the firms $3 for each unit of abatement, by how much would each firm reduce its pollution?

c. Suppose a transferable discharge permit program were used in which each firm is at first required to abate 10 units, but firms are allowed to trade so that one abates more than 10 if the other abates less. Would trading take place? What would be the expected final allocation of pollution abatement responsibilities?

21. Farmer Jones's farm is in a hilly section of the state. Erosion has always been a problem, but in recent years it has caused a decline in his yields. Over the next 5 years, he expects that his profits per acre will fall as seen in the following table. He is considering putting in terraces, which would stop the erosion and hold profits at $10 per acre. If he doesn't pay for terracing, he will put the money in the bank and earn 5% interest per year.

Year	Profits per Acre without Terracing	Profits per Acre with Terracing
1	10	10
2	9	10
3	8	10
4	7	10
5	6	10

a. What is the present value of profits per acre with and without terracing?

b. If building terraces costs $10 per acre and terraces are good for only 5 years, would building terraces be a good investment?

22. Two identical farmers are competing for water on a river. Their marginal revenue from using water and the marginal cost of pumping water are presented in the following figure.

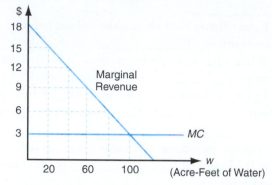

a. In wet years, there is 200 acre-feet of water in the river. If the upstream farmer takes as much as he wants and leaves the remainder for the downstream farmer, how much will each farmer use? Is this efficient?

b. In dry years, there is 100 acre-feet of water in the river. If the upstream farmer takes as much as he wants and leaves the remainder for the downstream farmer, how much will each farmer use? Is this efficient? How much would the downstream farmer be willing to pay to the upstream framer for a marginal increase in the water released in dry years?

REFERENCES

Coase RE: The problem of social cost, *Journal of Law and Economics* 3 (1960): 1–44.

Correia, Andrew W, C Arden III Pope, Douglas W Dockery, Yun Wang, Majid Ezzati, and Francesca Dominici: Effect of air pollution control on life expectancy in the United States: an analysis of 545 U.S. counties for the period from 2000 to 2007, *Epidemiology* 24, 2 (2013): 23–31.

Morgan, PB, EA Ainsworth, and SP Long: How does elevated ozone impact soybean? A meta-analysis of photosynthesis, growth and yield, *Plant, Cell & Environment* 26 (2003): 1317–1328.

National Research Council: *Air emissions from animal feeding operations*, Washington, D.C., 2003, National Academies Press.

U.S. Environmental Protection Agency: *National Rivers and Streams Assessment 2008–2009: A Collaborative Survey*, EPA/841/R-16/007, Washington, D.C., 2016, U.S. Environmental Protection Agency, Office of Water.

U.S. Environmental Protection Agency: *National water quality inventory report to Congress*, Washington, D.C., various issues, U.S. Government Printing Office.

United States Department of Agriculture, Natural Resource Conservation Service: *National resources inventory*, Washington, D.C., 2001, U.S. Government Printing Office.

United States Department of Agriculture, Natural Resource Conservation Service: What is soil quality?, http://soils.usda.gov//sqi/soil_quality/what_is/index.html, accessed 2004.

Wu J: Slippage effects of the conservation reserve program, *American Journal of Agricultural Economics* 82 (November 2000): 979–992.

Government Intervention in Agriculture

Chapter Outline

RATIONALE FOR GOVERNMENT INTERVENTION

Government provides many important functions, including the provision of law and order, a national defense, and infrastructure (highways, sanitation programs, scientific research programs, education programs, communication facilities, etc.) along with economic and social programs. The latter function includes providing the legal foundation and social environment necessary to have an efficient market system, maintaining competition in the marketplace, redistributing income and wealth to accomplish specific social objectives, monitoring the use of resources, and stabilizing the economy.

We made the distinction in Chapter 1 between capitalism, which in theory is free of government intervention, and socialism, which is characterized by massive government involvement. We concluded that most developed economies like the United States are mixed economic systems, or market economies in which specific sectors are regulated and/or protected by government. Chapter 10 focused on natural resources and the environment, including the role of government. Chapter 13 will focus on the functions of the Federal Reserve System and the conduct of monetary and fiscal policy. The last part of this book will focus on international trade and the role governments play in the global marketplace.

The rationale typically advanced for government intervention in the farm sector of an economy is the need to

- support and protect an infant industry,
- curb market powers of imperfect competitors when necessary to promote social good,
- provide for food security,
- provide for consumer health and safety, and
- provide for environmental quality.

The U.S. **food and fiber industry** clearly no longer qualifies as an infant industry. The farm sector is characterized by a large number of farming operations that produce crop and livestock products. It comes the closest of any sector in most economies to meeting the conditions for perfect competition we discussed in previous chapters. Other sectors in the U.S. food and fiber industry, on the other hand, are more closely characterized as imperfect competitors, a small number of firms selling a differentiated product and a small number of firms buying production from the farm sector. The government, for example, has taken action against what it perceived as imperfect competitive behavior by U.S. breakfast cereal manufacturers. The **Federal Trade Commission (FTC)**, an agency of the federal government, is charged with the responsibility of prohibiting companies from acting in concert to increase their market control using false and deceptive trade practices.

Many nations intervene in their food and fiber industry to ensure an adequate food supply from domestic sources. Although the U.S. secretary of agriculture watches stocks of specific commodities when conducting federal commodity programs, food security reasons are rarely cited as a rationale for federal government intervention in the U.S. food and fiber industry.

The food and fiber industry is one of the most highly regulated industries in the United States. Regulations that govern the use of inputs to produce crops and livestock, regulations that govern the processing and manufacturing of food and fiber products, and regulations that govern the markets in which these products are traded are just a few examples of how the government intervenes in the food and fiber industry. Government regulations cover everything from the regulation of chemical use in

Jamie L. Whitten Building main entrance, headquarters of the U.S. Department of Agriculture in Washington, D.C.   Credit: Mesut Dogan/Shutterstock.

crop production to the inspection of crops and livestock products before they reach the retail grocery store shelves.

The initial purpose of this chapter is to discuss the various approaches taken historically to support prices and incomes of producers of key commodities. This is followed by coverage of the current safety net provided by the federal government to farm producers. This chapter ends with coverage of consumer-related issues.

FARM ECONOMIC ISSUES

Politicians often debate the federal government's practice of making payments to farmers in an effort to support their farm incomes. This debate, which often pits rural politicians with their urban counterparts, has increased in intensity as the farm sector's relative contribution to national output has declined steadily over the post–World War II period. We will examine the nature of the issues involved and introduce the various ways in which the federal government intervenes to alter farm economic conditions.

Historical Perspective on the Farm Problem

What is the farm problem? Two symptoms typically associated with the farm problem are illustrated in Figure 11-1: (1) output fluctuations from one year to the next due to weather patterns, disease, and technological change, and (2) low net farm incomes in years of bumper crops on domestic and international markets. The inelastic nature of the own-price elasticity of demand for agricultural products, the lack of market power by farmers and ranchers, the interest sensitivity of the sector, and the fixity of farm assets represent the typical real roots of the farm problem. This has changed in several ways with the growth in demand for renewable fuels as discussed later in this chapter.

Inelastic Demand and a Bumper Crop   Any sector facing an inelastic demand for its products will suffer a decrease in its total revenue if supply expands faster than demand, causing market prices fall.

Crop and livestock production is subject to the vagaries of climatological and biological phenomena. Wet springs, dry summers, and early frosts, combined with

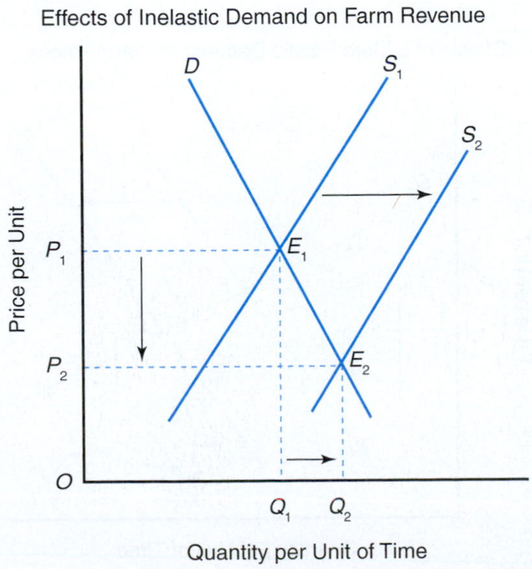

Effects of Inelastic Demand on Farm Revenue

Figure 11-1

This figure illustrates what happens to the total revenue received by farmers facing an inelastic demand for their products if the supply curve were to shift to the right. Total revenue would fall because area $OP_1E_1Q_1$ is greater than $OP_2E_2Q_2$.

things such as corn blight and cholera, will shift the sector's supply curve to the left. Bumper crops and technological change leading to increased productivity will shift supply curves to the right. A record crop, which is more often the rule than an exception, can lead to sharp declines in farm product prices and income levels.

Before the shift in the supply curve from S_1 to S_2, farmers were receiving a total revenue equal to the area formed by $OP_1E_1Q_1$. After the shift in the supply curve to the right, however, farmers were receiving a total revenue equal to the area formed by $OP_2E_2Q_2$. Obviously, the second area is smaller than the first, implying that the economic well-being of farmers would be diminished if supply increased. Observed another way, market prices would fall (P_1 to P_2) more than quantities marketed increased (Q_1 to Q_2).

Inelastic demand effects The inelastic nature of the demand curve for raw agricultural products, when combined with a bumper crop and rising imports, drives down the price of these commodities. Revenue falls sharply because of the steepness of the demand curve.

Would farm revenue fall by a smaller amount in years of bumper crops if the demand for the crop were more elastic? Figure 11-2 illustrates the effects of a more elastic demand on farm product prices. In Figure 11-2, we see that the market equilibrium price would decline from P_1 to P_3 if the demand curve facing farmers were represented by curve D_1 and the supply curve were to shift from S_1 to S_2. If the sector were instead confronted by a more elastic or flatter demand curve, represented by curve D_2, what would happen to the product price if the supply curve were to shift by the same amount (i.e., from S_1 to S_2)? The market equilibrium would be E_2 instead of E_1 and the equilibrium price would fall only to P_2 instead of all the way to P_3.

We can conclude from Figure 11-2 that the more inelastic or steeper the demand for farm products, the greater the decline (rise) in market prices will be for a given increase (decrease) in supply. A bumper crop, for example, could cause farm product prices to drop sharply under these conditions. Conversely, a drought that shifts the supply curve to the left could lead to a substantial increase in farm product prices.

Price takers Another characteristic besetting agriculture is the fact that farmers and ranchers are price takers in the market in which they buy their inputs and the markets in which they sell their production.

Lack of Market Power   Farmers and ranchers come close collectively to satisfying the conditions of perfect competition. There are a large number of producers producing a homogeneous product. No one farmer or rancher has sufficient market power to influence the market equilibrium price. Therefore, if Walt Wheaties suffers a sharp decline in wheat yields and does not have crop insurance, his revenue from wheat production will fall markedly, unless a large number of other wheat farmers experience the same decline in yields, and the sector supply curve shifts to the left.

Figure 11-2
This figure illustrates an important phenomenon of interest to farmers: the more inelastic the demand, the more prices will fall if supply increases.

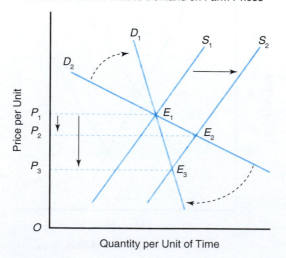

Effects of a More Elastic Demand on Farm Prices

Interest Sensitivity   The farm sector is one of the most highly capitalized sectors in the U.S. economy. There is more capital invested per worker than most other sectors in the economy. Farmers must borrow substantial amounts of money through short-, intermediate-, and long-term loans to purchase variable and fixed inputs. Thus, an increase in interest rates will increase farmers' interest expenses and, hence, their total production expenses. The high interest rates during a financial crisis in the farm sector in the early 1980s, for example, caused interest payments to reach over 16% of total production expenses, as compared to approximately 5% in 2016. Higher production expenses will lower net farm income, all other things constant.

In addition to their effects on farm production expenses, higher interest rates in the U.S. economy vis-à-vis the rest of the world can increase the value of the dollar in foreign exchange markets. The higher exchange rate between the dollar and other currencies will make it more expensive for other nations to import U.S. farm products. This will lower the demand for U.S. farm products, which will cause farm prices for products such as wheat and corn to drop, lowering farm revenue.

Asset Fixity   Another major problem facing farmers and ranchers is the notion of asset fixity. **Asset fixity** refers to the difficulty that farmers have in disposing of tractors, plows, and silos when downsizing or shutting down the business. These assets have little or no alternative uses and often sell for pennies on the dollar during hard times in the sector like the early and mid-1980s. A cotton combine, for example, has limited alternative uses. Farmers would have experienced a similar fate in the late 1990s to the early and mid-2000s were it not for substantial government subsidies.

Asset fixity   refers to the fact that many production assets employed in agriculture have little use in other sectors (e.g., a cotton picker). Furthermore, land, buildings, and machinery, all of which are fixed assets and have little liquidity value, dominate farm balance sheets.

Forms of Government Intervention

Government intervention in agriculture, required to improve farm economic and environmental conditions, has taken many forms. Chapter 10 covered the role of government in resources and the environment. We will discuss four additional forms in this chapter.

Adjusting Production to Market Demand   One approach to improving farm economic conditions is to reduce the number of resources employed to produce a product plagued by surplus conditions. This is achieved by the government's basically renting whole farms in the name of conservation or by paying farmers not to produce the product by requiring them to set aside part of the land normally used to plant this crop in order to qualify for farm program benefits. With less land planted to this crop, supply will decline and market prices will rise. If the demand curve is inelastic, farm revenue will increase.

Consider the opposite of the bumper crop case illustrated in Figure 11-1. Assume the original supply curve was S_2 and that policies that restrict resource use shift the supply curve back to S_1. Market equilibrium will now occur at E_1 instead of E_2. Because prices rise by a greater percentage (from P_2 to P_1) than quantity declines (from Q_2 to Q_1), total revenue represented by area $OP_1E_1Q_1$ will be greater than the total revenue represented by area $OP_2E_2Q_2$.

Price and Income Support Payments   Another general approach by government to improve farm economic conditions is to directly support farm prices and incomes, which can be achieved by establishing a price floor supported through government purchases of surplus commodities. If the level of production rises during the year, the government can step in and buy (and store) excess supply at the announced price floor. This would prevent farm revenue from falling below minimum desired levels. An alternative approach is to support farm incomes through direct transfer payments from the government to farmers. The **Federal Agriculture Improvement and Reform Act**, passed in 1996, for example, called for a number of efforts to help producers manage their exposure to risk. This included the establishment of the Risk Management Agency within the USDA and the implementation of pilot programs designed to provide revenue insurance. Thus, the nature of programs employed in the past to support pricing and/or income has taken several different forms. Since most payments under the 1996 FAIR Act were fixed, farm income could fluctuate more from one year to the next as market conditions changed. In response to the low commodity prices in the late 1990s, Congress passed a new farm bill in 2002 that returned a safety net concept to supporting farm prices and incomes. The 2008 Farm Bill continued this practice. The impact of the growth in demand for renewable fuels is discussed in the section titled Domestic Demand Expansion Programs. The 2014 bill introduced a major shift in farm policy by introducing crop insurance as the major safety net and eliminating direct payments. These programs will be discussed in more detail later in this chapter.

Foreign Trade Enhancements   A third approach to improving farm economic conditions involves the link between agriculture and foreign markets. There are essentially two general ways in which such enhancements can improve farm economic conditions in the United States. First, the government can institute **tariffs**, or a tax on imports, which make imported agricultural products more expensive to domestic consumers, or institute **quotas**, which limit the quantity of a particular good that can be imported. Both actions protect producers in the domestic agricultural sector. Quality restrictions on imports that grade imports or specify their sanitary conditions (pesticide and other chemical residues) can also effectively prohibit imports. These actions have the effect of limiting the supply coming into the market and raising the farm revenues of domestic producers. For example, a higher quota would shift the supply curve to the left and thus, like production controls, have the opposite effect of the bumper crop example illustrated in Figure 11-1. Prices will increase by a greater percentage (from P_2 to P_1) than will the decline in quantity (from Q_2 to Q_1), and farm revenue will increase from area $OP_2E_2Q_2$ to area $OP_1E_1Q_1$.

Second, the government can enhance the attractiveness of U.S.-produced agricultural products in foreign markets by subsidizing their purchase, thereby stimulating the export demand for U.S. agricultural products. Export credits help potential buyers finance the purchase of U.S. agricultural products. Subsidies are grants given by the government to private businesses to assist enterprises deemed advantageous to the public. Commodity assistance programs, such as P.L. 480, have historically promoted exports under direct food aid or subsidized concessionary sales.[1] These actions have the effect of shifting the demand curve to the right, and thus increase the farm revenue of domestic farmers.

The specifics of many of these and other foreign trade enhancements, along with the arguments for international agricultural trade, free trade, and multinational trade agreements, will be discussed in depth in Part 6 of this book.

---

[1] Public Law 480 is the name commonly given to the Agricultural Trade Development and Assistance Act of 1954, which seeks to expand foreign markets for U.S. agricultural products, combat hunger, and encourage economic development in developing countries. The act contains three titles: (1) Title I makes U.S. agricultural products available through long-term dollar credit sales at low interest rates for up to 40 years (often referred to as the Food for Peace Program), (2) Title II makes donations for emergency food relief abroad, and (3) Title III authorizes "food for development" grants. This program was unpopular with producers in recipient countries since they depressed local prices. The growth of other concessionary programs like the **Export Enhancement Program (EEP)** discussed later in Part 6 of the book explains the declining importance of PL 480's share of U.S. agricultural exports.

Other Forms of Intervention   The government has used a variety of other approaches to enhance economic conditions in agriculture. These approaches include credit subsidies to farm borrowers who cannot otherwise qualify for credit and loan guarantees for loans made to farmers and ranchers by commercial banks and other nongovernment lenders. The USDA provides direct loans to eligible farmers, who are primarily beginning farmers or socially disadvantaged farmers who cannot obtain credit from traditional lenders. These loans are available at subsidized rates from the **Farm Service Agency**. The USDA also guarantees loans made by traditional lenders used to finance operating expenses and buy farmland.

The USDA also helps farmers by providing subsidized crop insurance through private companies. Farmers, for example, can acquire highly subsidized coverage against catastrophic production losses (CAT insurance) as well as subsidized premiums for higher levels of coverage. Crop insurance has become a major feature of the safety net for participating producers and will be discussed in detail later in this chapter.

The federal government also offers programs such as the Animal and Plant Health Inspection Service to rid cropland of pests and diseases and make these crops more marketable. The Agricultural Marketing Service activities include a microbiological surveillance program on domestic fruits and vegetables through a food safety initiative. The "mad cow" scare in 2004, for example, led to more traceability of animals and further underscored consumer concerns over food safety issues.

Other forms The USDA's Farm Service Agency makes subsidized loans to qualified borrowers. Federal crop insurance policy premiums are another example of subsidized programs benefiting farmers.

HISTORICAL SUPPORT MECHANISMS

We made the point earlier in this chapter that the low own price of elasticity of demand for farm products, coupled with an increase in supply, can cause farm revenue to fall sharply. The historical problem of low returns to resources in agriculture has been dealt with in different ways over time by the federal government. The approach taken has depended in part on the nature of the conditions that existed at the time. The federal government has been involved in altering free market conditions in agriculture for almost 90 years. The roots of the price and income support mechanisms discussed in this section can be traced back to the federal legislation enacted in the 1930s and 1940s.

There are many excellent summaries of the history of federal government programs for agriculture.[2] Instead of reviewing the labyrinth of farm programs for the last 90 years, we will focus on five key features of more recent farm programs: (1) loan rate mechanism, (2) set-aside mechanism, (3) target price mechanism, (4) conservation reserve mechanism, and (5) subsidized crop insurance.

Loan Rate Mechanism

The U.S. Department of Agriculture has used the **commodity loan rate** mechanism since the early 1930s to support prices for commodities such as wheat, corn, and cotton. The loan rate essentially serves as a floor to farm prices for participating farmers.

Market Level Effects   To see how this mechanism works at the sector or market level, look at the market demand and supply curves for wheat in Figure 11-3. If competitive market forces were free to work, the market clearing price would be P_F and the quantity marketed would be Q_F. Assume that the federal government wished to

Loan rate The loan rate mechanism has long been a means of providing a price floor for selected program commodities like corn and wheat.

---

[2]For example, see Knutson, Penn, Flinchbaugh, and Outlaw (2006).

Figure 11-3

To achieve support prices of P_G, the federal government would purchase the surplus or excess supply marketed by farmers at the announced support price level. Farmers would be in a better position economically by the amount that area 4 exceeds area 2. Consumers would be in a worse position economically by the sum of areas 3 and 4. They would be paying a higher price (P_G) and consuming less (Q_D).

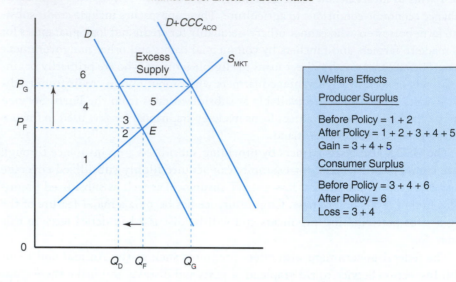

Market-Level Effects of Loan Rates

Welfare Effects

Producer Surplus

Before Policy = 1 + 2
After Policy = 1 + 2 + 3 + 4 + 5
Gain = 3 + 4 + 5

Consumer Surplus

Before Policy = 3 + 4 + 6
After Policy = 6
Loss = 3 + 4

CCC program The commodity acquisition/ storage programs run by the federal government actually led to more production and high operating costs for the federal government, and eventually were scrapped.

support prices at P_G, which lies above P_F. The quantity demanded by consumers at price P_G would be Q_D, and the quantity supplied would be Q_G.

The **Commodity Credit Corporation (CCC)**, a corporation operating within the U.S. Department of Agriculture, acts as the purchasing agent for the federal government in this instance. The CCC makes a nonrecourse loan to participating farmers at the loan rate, or at a fixed price. The loan, plus interest, can be repaid within 9 to 12 months from the proceeds the farmer receives from selling the commodity on the local market. If it is not profitable for these farmers to repay the loan, the CCC has no recourse but to accept the farmer's pledged collateral (the crop) as payment in full for the loan. The cost of this action to taxpayers in Figure 11-3 would be equal to P_G times $Q_G - Q_D$.

Participating farmers can either repay their loan if the market price is above the loan rate plus interest charges or forfeit the commodity pledged as collateral as payment in full for the loan if the market price is below the loan rate. It is the **nonrecourse loan** feature that provides the floor to the price received by the farmer.[3] Thus, CCC_{ACQ} in Figure 11-3 represents the additions to CCC stocks acquired by forfeits of commodities pledged by farmers under the nonrecourse loan program.

Firm-Level Effects   As illustrated in Figure 11-4, the individual farm would produce output q_G, where its supply curve (or marginal cost curve) equaled the perceived demand curve (or marginal cost curve) at P_G. The difference between Q_D and Q_G at the sector level in Figure 11-3 represents the excess supply or surplus at P_G that the federal government had to purchase and store for later distribution, if it desired to support prices at this level.

There are several problems associated with the loan rate mechanism when it is used alone. If the support price is higher than the free market price, it encourages higher production. The individual farmer will respond to the higher support price by producing q_G instead of q_F, as shown in Figure 11-4. This leads to greater government-held stocks and higher storage costs. Area 2 in Figure 11-4 represents the additional net income earned by the farmer under this farm program.

---

[3]The term *loan rate* dates back to the time when the purpose of the loan was to extend credit to farmers to help them store some of their production rather than sell it all at harvest time.

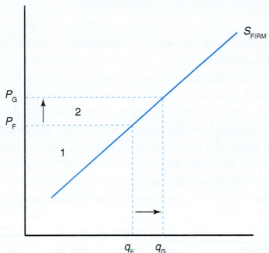

Firm-Level Effects of Loan Rates

Figure 11-4
At the higher announced support price, the individual farmer will produce a higher level of output. The farmer would be in a better position economically by the amount of area 2.

During the 1950s, the government acquired large quantities of surplus wheat in the form of forfeited nonrecourse loans when market prices were below the loan rate. The government spent more than $1 million a day for storage costs. These charges eventually reached several million dollars a day, which was big money at that time.

In addition, because price supports aid farmers in direct proportion to their level of production, the owners of large farming operations receive the bulk of the program's benefits. The value of expected future benefits from the loan rate mechanism is capitalized into the market value of farmland, resulting in an additional benefit to large landowners.

Set-Aside Mechanism

To combat the growing size and cost of government-held stocks, federal government policymakers adopted **set-aside** requirements in the **Food, Agriculture, Conservation, and Trade (FACT) Act of 1990** to support farm prices. The set-aside mechanism constrains the annual production levels of any single crop and thus avoids accumulating larger surplus stocks, which depress farm prices. Set-aside requirements call for farmers to remove a certain percentage of cropland from production as a condition for receiving farm program benefits. Set-aside requirements were used for most major food and feed grains as a means of reducing production of surplus crops such as corn and wheat. A major problem with the set-aside mechanism is that farmers will idle their poorest land first and crop their remaining acres more intensively, which can result in **slippage** or larger supply and lower prices than desired by policymakers.[4]

The **Acreage Reduction Program (ARP) percentages** spelling out these set-aside requirements were determined in part by the expected ratio of ending stocks to total use. There were individual ARP percentages set for corn, wheat, cotton, and other specific commodities that often varied dramatically from one year to the next depending upon demand and supply factors. The 1996 **FAIR Act** eliminated the

Set-asides The set-aside mechanism required participating farmers to idle a fraction of their farm land in exchange for government subsidies. While this lowered the quantity coming onto the market, it became extremely unpopular with farmers as well as input suppliers.

The ARP rates were the percentage of land that had to be idled to receive subsidies under the set-aside policy mechanism.

---

[4]Other production control mechanisms used either currently or since the end of World War II include acreage allotments, marketing quotas, land retirement programs, and paid land diversion programs. Farm programs calling for acreage allotments continue today for peanuts and tobacco. Marketing quotas have traditionally been implemented only if two-thirds of the farmers approve them in a referendum. The **Soil Bank Program** utilized in the 1950s led to whole farms being retired from production, eventually retiring approximately 30 million acres.

authority for this program, taking the USDA out of the commodity-level supply management game.

Market Level Effects   To illustrate how set-aside requirements work, study Figure 11-5. S_{MKT} represents the market supply curve for a commodity before acreage restrictions are implemented. The market would have cleared at equilibrium E_1, where price equaled P_F and quantity equaled Q_F. Assume that the government wants to support prices above P_F. At price P_G, the quantity demanded would be equal to Q_G, and the quantity supplied would be Q_S if farmers were free to produce all they desired. We know from the previous discussion of the commodity acquisition approach that this would have resulted in additional surpluses equal to $Q_S - Q_G$. The CCC, in the absence of set-aside requirements, would have had to accumulate these additional stocks if it wished to maintain the market price at P_G.

Firm-Level Effects   The provision for set-aside requirements in the 1990 FACT Act and earlier legislation restricted the amount of land the participating farmer can plant to a particular crop. This would result in a new firm-level supply curve S_{FIRM} for this crop that lies to the left of its original supply curve (its true marginal cost curve), as we see in Figure 11-6. The individual farmer operating under the set-aside mechanisms would produce q_G, where the backward-bent firm supply curve for this crop equals marginal revenue (P_G). This quantity is less than the quantity produced under free market conditions q_F and the quantity produced under the commodity acquisition approach (q_S).

The use of set-aside requirements during the 1980s, for example, called for wheat farmers at one point to retire approximately one-fourth of their land from wheat acreage. If a large percentage of these wheat farmers agreed to participate in the federal farm program, the market price would rise from P_F to P_G, as shown in

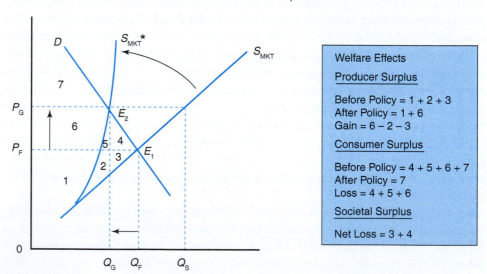

Figure 11-5

The set-aside mechanism at market level involves removing the inputs from production that were contributing to the accumulation of surplus stocks. Farmers would be in a better position economically by the amount area 6 exceeds the sum of areas 2 and 3. Consumers, on the other hand, are in a worse position economically by the sum of areas 4, 5, and 6. The net loss to society as a whole due to the reduced level of output would equal the sum of areas 3 and 4. Areas 2 and 5 represent the additional surplus received by firms supplying inputs to agriculture.

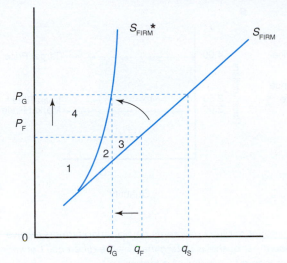

Firm-Level Effects of Set-Aside Requirements

Welfare Effects

Producer Surplus

Before Policy = 1 + 2 + 3
After Policy = 1 + 4
Gain = 4 − 2 − 3

Figure 11-6
With the set-aside mechanism at firm level, the individual farmer would produce less than he or she would under free market conditions and under the loan rate mechanism. Farmers will be in a better position economically as long as area 4 exceeds the sum of areas 2 and 3.

Figure 11-6. The goal is to set price support above loan rate levels to avoid having the CCC accumulate large stocks of surplus commodities by farmers forfeiting their output as payment in full for their CCC loan.

To convince farmers to participate in this form of production control program, the federal government offered income support benefits called **deficiency payments**, which are not available to nonparticipants.

Farmers naturally responded to acreage restrictions by idling their worst land first and cropping their remaining acres more intensively, which allowed them to produce output q_G, which may be more than the government expected them to produce.

Target Price Mechanism

A major feature of the 1990 FACT Act, ended in 1996 by the Fair Act but brought back in the 2002 Farm Bill and continued in the 2008 Farm Bill to determine the countercyclical payment, was the **target price mechanism**. The concept of target prices actually was first proposed by Secretary of Agriculture Charles F. Brannan of the Truman administration in 1949. Target prices help determine the level of **direct payments** to farmers. The payment per bushel is equal to the difference between the target price and the market price or loan rate, whichever is higher.

Consider the market conditions depicted in Figure 11-7, in which target prices have been augmented with both the commodity loan rate mechanism and the set-aside mechanism that limited output to Q_M. If the target price for wheat was \$4 per bushel, the market price was \$3 per bushel, and the loan rate was \$2.50, farmers participating in the program would receive a direct payment from the federal government of \$1 for every bushel of wheat produced, which is illustrated by the shaded area in Figure 11-7A. If the market price fell to \$2, the deficiency payment per bushel would be \$1.50, which is illustrated by the shaded area in Figure 11-7B.[5] Thus, the deficiency payment acted as an income support over and above the price support given by the loan rate mechanism.

Target prices Target prices, still in use today, established the ceiling or maximum price the government was willing to support for selected program commodities.

---

[5] Farmers participating in the program would also likely choose to relinquish title to their crop under the CCC nonrecourse loan program. This means that they would receive the \$2.50 loan rate from the CCC rather than only \$2 from the spot market. Thus, they receive the full \$4, or \$1.50 in direct deficiency payments and \$2.50 from the CCC nonrecourse loan.

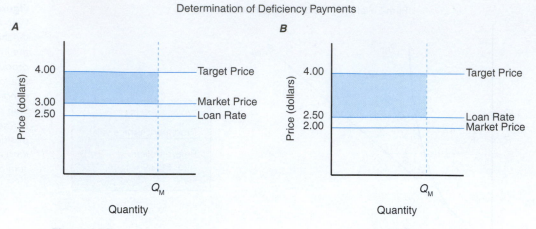

Figure 11-7

The target price helps determine the amount of direct payments participating farmers receive from the federal government. The amount per bushel is determined by the difference between the target price and the market price or loan rate, whichever is higher.

The target price mechanism used without the set-aside mechanism would stimulate production above free market levels much like the loan rate mechanism would. Farmers would produce to the point at which marginal costs equal the target price rather than the market price, unless otherwise constrained from doing so.

Countercyclical Payments Mechanism

The 2002 Farm Bill introduced the concept of countercyclical payments in an effort to restore the safety net to program commodity producers eliminated by the 1996 FAIR Act. The countercyclical payment, expressed on a dollar-per-unit (i.e., dollar-per-bushel) basis, is equal to the difference between the target price, discussed in the previous section, and the effective price. The effective price is the higher of the national market price over a 12-month period or the sum of the loan rate and a direct payment rate. This payment is based on the base acres for a commodity, which can be different from planted acres. In other words, the countercyclical payment is decoupled from production decisions. The 2008 Farm Bill made minor revisions, including eliminating benefits to farms with less than 10 acres. It also added a new countercyclical option—the **Average Crop Revenue Election**, or ACRE—which called for farmers to receive an ACRE payment based on the state-level difference between actual revenue and ACRE guarantee times a percentage of the farm's planted acres.

The nature of this mechanism means that the producer does not know for sure what the total payment will be until the end of the 12-month marketing year. For corn, this covers September of the year the crop is harvested through the next August. For wheat, this covers June of the year the crop is harvested through the next May. If the secretary of agriculture determines these payments are required, producers can elect to receive up to 35% of the projected countercyclical payment in October of the year the crop is harvested, an additional 35% in February of the following year, and the balance at the end of the 12-month marketing year.

Figure 11-8 illustrates the level of countercyclical payments to a program commodity producer if the market price (*PF*) is less than the announced target price. If the market price (*PF*) is equal to the target price (*TP*), there are no countercyclical payments. As illustrated in Figure 11-8, the fact that the target price is greater than the market price means that producers would receive countercyclical payments equal to *TP* – *PF* per bushel or ton in the case of cotton. Total countercyclical payments would be equal to area 3 + area 4 if the base acres and program yields translated into

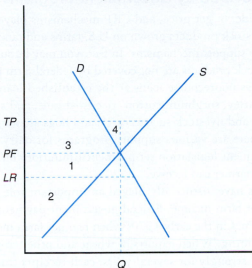

Countercyclical Payments

Figure 11-8
*Countercyclical payments
replaced deficiency
payments under the 2002
and 2008 Farm Bills.*

the quantity *Q*. Remember that payments are decoupled from the program commodities actually grown during the year. Producers can claim base acres in one commodity and grow another.

Conservation Reserve Mechanism

The **Food Security Act (FSA) of 1985** enacted the **Conservation Reserve Program (CRP)** to reduce acreage in production, reduce erosion, and improve water quality. Like the Soil Bank Program instituted in the 1950s, CRP is a voluntary long-term land retirement scheme that removes cropland from the production of program commodities such as wheat and corn. The retired land must be used for a soil-conserving cover crop such as grass or trees.

The 1985 Farm Bill authorized the retirement of up to 45 million acres of highly erosive land from production, with the caveat that no more than 25% of the land in any one county could enter the program. The 1990 Farm Bill expanded the CRP to include lands subject to water quality problems (i.e., a **Wetlands Reserve Program {WRP}**). Basically, farmers submit sealed bids that are then reviewed to determine which land to retire. If needed, the land can be put back into production.

Under the program, landowners who sign contracts agree to convert environmentally fragile land to approved conserving uses for 10 to 15 years. In exchange, the landowner receives a rental payment annually and cash or payment in kind to share up to 50% of the cost of establishing permanent vegetative cover. The FAIR Act in 1996 set the maximum CRP area at 36.4 million acres. The 2002 Farm Bill increased this to 39.2 million acres. New acreage enrolled in the CRP to replace acreage coming out at the end of their contract must meet stiffer criteria regarding soil erosion, water quality, or wildlife benefits. Pressure was exerted on the USDA to allow farmers to opt out of their program without penalty due to rising crop prices and the impact on food and fuel prices.

The Conservation Reserve Program calls for the retirement of additional acres of cropland, which means a further "bending back" of the market supply curve in Figure 11-5, lower levels of production for those crops previously grown on these lands, and higher prices in the short run. Farmers would be put in a better position economically, but consumers would be in a worse position economically. Society as a whole would be in a worse position because of the additional loss of output.

The features of the CRP program have been modified in subsequent farm bills, including reducing the number of acres enrolled in the program.

The Conservation Reserve Program was designed to idle "fragile" lands to prevent further erosion, pollution of water sources, and so on. This program is still in use today.

Commodities Covered by Government Programs

The loan rate, set-aside, target price, and CRP mechanisms played a significant role. Not all crop and livestock products grown on U.S. farms and ranches were covered by the price and income support mechanisms. In fact, you may be surprised to learn that most crop and livestock products are not covered by federal farm programs.

The CCC makes nonrecourse loans at the established loan rate to farmers for wheat, corn, oats, barley, sorghum, cotton, rice, soybeans, and sugar. Crops such as fruits and vegetables and livestock such as cattle and hogs are not covered by the loan rate mechanism. There are various support programs for dairy farmers. The 1990 Farm Bill and subsequent legislation set payment limitations on the amount of government payments a farmer can receive.

Farm programs have historically played an important role in determining the level of aggregate net farm income. Federal government payments to farmers helped to support farm incomes in the early 1970s when real net farm incomes from farming operations were declining. When exports of wheat and other crops rose markedly in 1972 and continued strongly for several years, cash receipts from farm marketings rose and the federal government provided less support. The same is true for more recent years as a result of the increased use of corn to manufacture ethanol.

Even with federal income supports, net farm income expressed in real terms fell in 1983 to a level not seen since the Great Depression. Higher interest rates, a recessionary economy, and declining export markets were all squeezing the profitability of farming operations. Figure 11-9 illustrates that direct payments from the federal government rose beginning in 1983, approaching $17 billion in nominal terms in the following years.

When annual federal budgets began to rise sharply during the late 1980s and early 1990s, annual expenditures for agricultural programs became increasingly criticized. As a consequence, the 1990 Farm Bill was tied specifically to measures taken to reduce the size of the federal budget deficit, which was expected to occur when the new Farm Bill was enacted. As a result of the growth in industrial use of corn for manufacturing renewable fuels, farm prices began to rise sharply in 2007 to 2012, reducing the need for the government to support farm incomes, and direct payments

Subsidy coverage Only a fraction of crop commodities are subsidized by federal government programs. These are primarily feed and food grains plus cotton.

Figure 11-9

Annual payments to participating farmers support the level of net farm income. The level of government payments rose dramatically as farm commodity prices fell and production costs rose in the 1980s and fell in the 2000s as prices rose. Government payments fell in recent years as the industrial demand for feed grains rose.

Source: From Economic Research Service, U.S. Department of Agriculture.

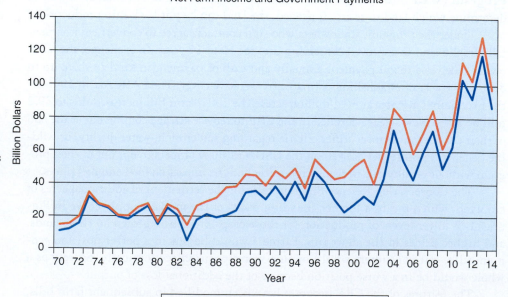

to farmers declined. Record income levels eventually led to direct payments being eliminated altogether in the 2014 farm bill.

Phasing Out of Supply Management

The 1996 Farm Bill, also known as the *Federal Agriculture Improvement and Reform Act* or simply FAIR Act, represented a transition to a market-driven agriculture. Under the 1990 Farm Bill, participants in farm programs had received deficiency payments reflecting the difference between market prices and target prices. Eligibility for payments was based on the acreage base history the farmer had established for a particular crop. Acreage Reduction Program percentages or ARPs were then used to tell farmers how much of the base acreage they had to set aside in a particular year to receive deficiency payments. The goal from a policy perspective was to manage supply, or the quantity coming onto the market.

The FAIR Act replaced target prices and deficiency payments with annual fixed transition or flexibility contract payments. The receipt of direct government payments was "decoupled" from planting decisions. Farmers were no longer restrained in their planting decisions by maintaining base acreage and annual ARP levels. They now had the flexibility to plant virtually whatever they wanted on their base acreage (referred to later as contract acres). Farmers who had planted cotton for generations, for example, were seen switching to corn in response to relative commodity price movements.

Crop Insurance

Federal crop insurance has been around for about 80 years. The **Federal Crop Insurance Corporation (FCIC)**, which administers the program, was created in 1938. Among the policies offered was **Multiple Peril Crop Insurance (MPCI)**, which protected against yield losses. This program however did not play a significant role until passage of the **Federal Crop Insurance Reform Act of 1994**. This legislation made participating in the crop insurance program mandatory if the producer wanted to be eligible for deficiency payments among other program features. Farmers could sign up for **Catastrophic Coverage (CAT)**, which covered losses exceeding 50% of an average yield and 60% of an established price. The premium on CAT coverage was completely subsidized. Premiums for higher coverage amounts were partially subsidized.

Congress repealed mandatory participation in 1966 but still required producers to purchase crop insurance if they wanted to be eligible for disaster assistance. The **Risk Management Agency (RMA)** was also created within the USDA to administer federal crop insurance programs and promote other risk management products and decision tools.

Participation in the federal crop insurance program was three times higher in 1998 than it was 10 years earlier, with more than 180 million acres under the program. By 2015, more than 298 million acres were insured representing 86% of planted cropland in the United States, covering 128 different crops with an insured value of $124 billion (Rain and Hail Insurance Society, 2016). Seventeen private sector insurance companies were selling and servicing the 1.2 million policies in force.

MPCI is the general name given to policies that provide coverage for a number of perils, including rain, hail, frost and drought, to name a few. The focus was largely on yield risk. Recent years have seen the development of a variety of products that provide both yield and price coverage. The **Revenue Protection (RP plan)**, for example, provides protection against the loss of revenue due to adverse fluctuations in both prices and yields. The producer can select coverage between 50% and 85%, with premiums increasing as the percentage coverage increases.

By 2015, there were 2,239,033 crop insurance contracts with 95% involving farmers buying up additional coverage levels. All plans involved revenue coverage

Program participation The decision to participate in a commodity program was decoupled from the production decision by the last two farm bills, thereby giving farmers more flexibility in their planting decisions.

Crop insurance The role played by crop insurance in supporting revenue has grown in recent years. Today is represents a major component of the safety net for many producers.

totaling over $82 billion. Producers paid $3.7 billion in premiums with the federal government paying $6.1 billion. Excessive moisture accounted for 48% of crop failures in 2015, with drought accounting for another 14%.[6]

CURRENT FARM BILL

The 2008 Farm Bill, due to expire in 2012, was extended. The U.S. Senate had passed legislation to replace the 2008 Farm Bill in 2012, but the U.S. House of Representatives did not bring a bill to the floor. The **Agricultural Act of 2014** was eventually passed some 2 years later. The 2014 Farm Bill modified many titles pertaining to health and nutrition, research and conservation programs to name a few. Of particular interest to us in this chapter is the safety net provided to crop and livestock producers. For example, the 2014 Farm Bill established a margin protection program for dairy producers and a permanent livestock disaster program.

The Agricultural Act of 2014 strengthened the federal crop insurance by making crop insurance more affordable for beginning producers and adding several new options. One is the **Supplemental Coverage Option (SCO)**, which enables producers to expand coverage against production losses and price declines. The other is the **Stacked Income Protection Plan (STAX)**, which accomplishes the same for cotton producers. Our focus in this section is on enhancements to support programs that supplement the federal crop insurance program benefits.

The 2014 farm bill introduced a major change to U.S. farm program policy. Direct payments to crop producers were eliminated. Producers no longer receive fixed direct payments per acre regardless of what happens to prices or what they plant or don't plant. Also eliminated were the countercyclical payment mechanism and the ACRE program. In their place, producers must make a one-time selection between two options they must operate under to 2018, when the 2014 Farm Bill expires. These two programs provide income support to producers under adverse yields and prices in addition to crop insurance coverage benefits. These two programs are the **Price Loss Coverage (PLC)** program and the **Agricultural Risk Coverage (ARC)** program.

Current farm bill The 2014 Farm Bill provides new options to producers to support farm revenue above insurance coverage.

Price Loss Coverage Program

The PLC program makes a payment to the producer when the market price for a covered crop falls below a fixed "reference" price. The reference price for selected program commodities are given below:

Commodity	Reference Price	Marketing Loan Rate
Wheat	$5.50 per bushel	$2.94 per bushel
Corn	$3.70 per bushel	$1.95 per bushed
Sorghum	$3.96 per bushel	$1.95 per bushel
Soybeans	$8.40 per bushel	$5.00 per bushel
Rice	$14.00 per cwt	$6.50 per cwt

This program covers losses in income from commodity price declines below these reference prices. The PLC payment rate is equal to the reference price minus the higher of the national average marketing year prices and the marketing loan rate. The

---

[6]There are also insurance programs for specific livestock commodities. They are designed to insure against declining market prices.

PLC payment is equal to the payment rate times the payment yield and 85% of the producer's base acres.[7]

If the PLC program option is selected, the producer can also choose the Supplemental Coverage Option (SCO), which allows producers to purchase additional crop insurance that could help offset the deductible on the producer's underlying insurance policy.

ARC Program

The ARC program makes a payment to the producer when either the producer's revenue for all crops (ARC-IC) or a county's revenue for a crop (ARC-CO) falls below 86% of a revenue benchmark. The maximum coverage band is 10 percentage points (76% to 86%) of the benchmark revenue. The ARC-IC program pays at the 65% level while the ARC-CO pays at 85% of base acres.

The producer must make a one-time choice between the two programs. Both programs are designed to supplement crop insurance by supporting revenue in a period of multiyear price declines and help producers cover the insurance policies' deductible. The PLC program makes a payment to a producer when the market price for a covered crop falls below a certain level while the ARC program makes a payment to a producer when either the revenue for all crops or the county's revenue for a crop falls below a benchmark level of revenue. Both programs are subject to payment limits. Payments to producers are limited to $125,000 per person and to those with an average adjusted gross income exceeding $900,000.

DOMESTIC DEMAND EXPANSION PROGRAMS

Raising the level of farm income can also be accomplished through domestic demand expansion, or by shifting the demand curve for farm products in the U.S. economy to the right, which is illustrated at the sector level in Figure 11-10.

When the market demand for a particular crop shifts from D_1 to D_2 and we move up the current market supply curve S, we see that the price of the commodity increases from P_F to P_G. Farmers are better off because the profits generated by their operations are rising (by an amount equal to area $P_F P_G E_2 E_1$).

Instead of the level of economic activity on farms and in rural communities, where farmers purchase farm inputs and market their output, being reduced directly or indirectly by production controls, domestic demand causes economic activity to rise. Farm output for this commodity would increase from Q_F to Q_G. The firm-level effects of an expanding demand from D_1 to D_2 are much the same as those illustrated in Figure 11-4 for the commodity acquisition approach. The domestic demand for farm products can be expanded in two ways:

1. The government can institute programs to promote the expansion of domestic demand through school feeding and other nutrition service programs, advertising and promotional programs, and so on.
2. The government can institute programs to subsidize the development of new uses for farm products (e.g., as an intermediate good in the production of other goods). A recent example is the use of state and federal subsidies for the production of ethanol, which is then blended with E10 gasoline and E85 fuel.

The first form of demand expansion program probably holds the least promise of the two in terms of its ability to alleviate the need for price supports in the farm

---

[7] A separate program was offered for cotton producers. The stacked income protection program or STAX is an area revenue plan which can be used alone or in conjunction with the cotton producer's underlying insurance plan. This program covers revenue losses of not less than 10% and not more than 30% of expected county revenue.

Figure 11-10
Under domestic demand expansion, farmers would be producing more than they would under free market conditions.

Market-Level Effects of Domestic Demand Expansion

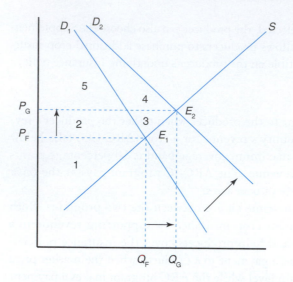

Welfare Effects

Producer Surplus

Before Shift = 1
After Shift = 1 + 2 + 3
Gain = 2 + 3

Consumer Surplus

Before Shift = 2 + 5
After Shift = 4 + 5
Gain = 4 − 2

Societal Surplus

Net Gain = 3 + 4

business sector. With a declining rate of population growth in the United States and the low-income elasticity for food, it is difficult to foresee how advertising and promotional programs can lead to a significant increase in the domestic demand for food and fiber products over the long run. Furthermore, food and nutrition service programs are already in place.

The effect of the second form of demand expansion programs has typically been minor given existing technologies, unless the demand–supply conditions for industrial uses make it more economical for nonfarm businesses to utilize farm products in their production processes. As the cost of imported crude oil began to rise dramatically beginning in 2005, the demand for corn, and to a lesser extent soybeans, used to produce renewable fuels grew as well. The price of corn rose from $2.00 a bushel in 2005 to $4.00 in 2007. The 2007 Energy Bill passed in late 2007 increased the mandate for ethanol well into the next decade and, coupled with expanded ethanol processing plant capacity, increased the demand for corn as a feedstock to ethanol production. Spot market prices at key locations around the nation rose to the mid $7.00 range, and the Chicago Board of Trade futures contracts for late 2008 delivery exceeded $8.00 at one point. While this was good news for corn farmers, it was not necessarily good for other components of demand, namely, livestock producers (rising feed costs) and consumers (rising food costs). The lesson here is that domestic demand expansion that benefits a particular commodity or group of commodities can have negative implications for other participants in the farm business sector and, ultimately, consumers.

IMPORTANCE OF EXPORT DEMAND

Export importance
Export demand is extremely important to many U.S. commodities. A prime example is wheat. The Asian financial crisis in the late 1990s, which hurt the ability of these countries to pay for U.S. imports, led to a sharp decline in the price of wheat.

Any phasing out of export subsidies to producers of export-sensitive crops like cotton places increased emphasis on strong export demand to keep prices at or above the old target price levels. Export subsidies represent one potential demand expansion device.[8]

---

[8]Public Law 480 (P.L. 480) is designed to export commodities to foreign countries under several different programs: long-term dollar and foreign currency sales (Title I sales), donations to other governments and various relief agencies to promote better diets and alleviate starvation (Title II donations), barter for strategic materials (Title III), or other foreign assistance programs. Almost 90% of P.L. 480 exports of wheat and flour are Title I sales. This and other programs designed to subsidize export sales have come under fire from the WTO and less-developed countries. More will be said about this in Part 5.

Other programs, such as those that extend export credit to importing nations, target subsidies to specific importing countries, and promote a "buy one, get one free" sale of farm commodities abroad, will be discussed in Part 5 of this book. The reduction of subsidies under the Uruguay Round and Doha rounds of the **General Agreement on Tariff and Trade (GATT)** affects this option, however, as discussed in Part 5. In later chapters, we will also emphasize two additional factors that have a direct impact on U.S. exports: (1) the purchasing power of national income in importing countries and (2) the relative cost of the U.S. dollar in foreign currency markets.

Monetary policies designed to combat inflation in this country that result in high interest rates lead to increases in foreign capital flowing into this country (which means that we are exporting recessionary pressures to other countries), and a higher price for U.S. dollars in foreign currency markets (which means that it would now take more units of a foreign currency to buy the dollars necessary to purchase a given quantity of U.S. products). Therefore, expansionary monetary policies will have a positive impact on export demand, and contractionary policies will retard export demand.

The demand curve for raw agricultural products facing U.S. farmers becomes more elastic (less inelastic) with a growing export demand for U.S. farm products, which shifts the total demand curve as illustrated in Figure 11-11.[9] E_0 reflects the market equilibrium at a price that almost halts foreign demand for a particular crop. E_1 represents the domestic market equilibrium at the price of P_{DD} if there were no foreign demand. E_2 represents the market equilibrium with foreign demand, which reflects both a higher product price (P_{TD}) and quantity needed (Q_{TD}).

To illustrate the importance of export demand, let us examine the real-world case of U.S. wheat farmers and how changing market conditions during the post–World War II period affected their total revenue.

Thomas Malthus, in the late 1700s, argued that the world would eventually suffer dramatic food shortages because population growth would exceed the growth in the food supply. Specifically, Malthus suggested that population grows at a geometric rate (e.g., doubles every 25 years), and food production, given a fixed amount of tillable land, grows at an arithmetic, or fairly constant, rate (Malthus, 1963).

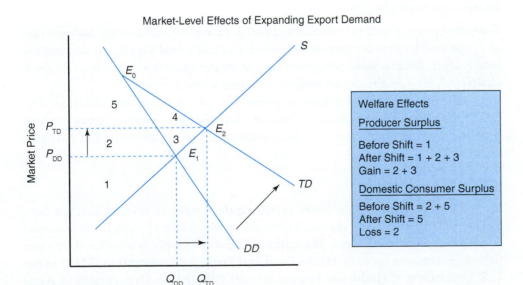

Market-Level Effects of Expanding Export Demand

Welfare Effects

Producer Surplus

Before Shift = 1
After Shift = 1 + 2 + 3
Gain = 2 + 3

Domestic Consumer Surplus

Before Shift = 2 + 5
After Shift = 5
Loss = 2

Figure 11-11
The domestic demand (DD) for farm products in general is highly inelastic. An expanding export demand at lower market prices will result in a more elastic total demand (TD) curve and increase the economic well-being of consumers and businesses.

---

[9] The quantity of U.S. farm products shipped abroad has more than tripled since 1950.

Government's involvement in preventing Malthus's prediction from coming true in the United States is reflected in the resources it commits to research and development activities, which has led to higher yields. The U.S. Department of Agriculture, along with researchers at land-grant universities around the country and researchers working for private agribusiness firms, has contributed to the yield-enhancing technologies embodied in a wide variety of agricultural inputs such as hybrid seeds and fertilizers that typically result in record crop levels for major food and feed grains in the United States.

CONSUMER ISSUES

The United States probably has the safest and most abundant supply of food and fiber products of any nation in the world. This is not by happenstance, but rather by design. The government is actively involved at various times during the production, processing, and delivery stages of the marketing channel to ensure a safe and nutritional food supply. There are numerous perspectives one can take to the discussion of consumer issues with respect to food and related products. The perspectives discussed in this chapter include the topics of food safety, an adequate and cheap food supply, nutrition and health, food subsidies, and consumer interests in animal rights.

Adequate and Cheap Food Supply

Consumers in the United States enjoy a relatively stable and cheap food supply. Consumers spend approximately 15% of their disposable income on food and rarely experience an interruption in the supply of food products.

Several studies of hunger in the United States over the post–World War II period indicate that, although progress has been made in reducing hunger in this country, hunger continues to be a serious problem. In the mid-1980s, one study estimated that 20 million persons, or almost one-tenth of the U.S. population, went hungry. These studies suggest that hunger is caused by a lack of income, poor knowledge of proper diets, and, in some geographical areas, poor access to nutritional food supplies.

Nutrition and Health

Consumer issues related to nutrition and health have many dimensions. Malnutrition and poor health can occur despite the existence of a safe food supply and an adequate food budget, because some consumers eat the wrong foods (too much fat, too much cholesterol, etc.) despite dietary recommendations to the contrary.

The government has played an important role in nutrition education. The USDA, for example, established the **Human Nutrition Information Service**, which distributes dietary guidelines that have contributed to improvements in nutrition and health.

Food Safety

Food safety   The safety of U.S. food supply is often taken for granted. Scares such as the mad cow disease outbreak in 2004 and 2005 led to calls for more traceability in the source of supply to the food manufacturing process.

Food safety covers a variety of issues, including the conditions under which raw farm products are produced and the conditions under which these products are processed and distributed to consumers. The safety of U.S. food supply is monitored by three federal government agencies: (1) the **Food and Drug Administration (FDA)** in the U.S. Department of Health and Human Services, (2) the **U.S. Department of Agriculture (USDA)**, and (3) the Environmental Protection Agency (EPA). The FDA has the broadest authority over monitoring the safety and wholesomeness of food and beverage products. The USDA inspects meat and poultry products. The EPA regulates the use of pesticides when they affect consumer health and safety.

Regulatory issues with respect to food safety often center on the testing that must take place before the product is cleared for public consumption; the tolerance limits associated with the presence of particular substances, such as rodent hair and insect parts in a food product; and product labeling. The FDA spends considerable resources in testing food and beverage products as required by federal legislation. Tolerance limits recognize that it is impossible to achieve a zero tolerance level, and reflect the concept of acceptable risk. Finally, food processors and manufacturers are typically very informative when it comes to labeling ingredients. The government has, in specific instances, stepped in and required health warnings on product labels. Perhaps the most widely known is the warning on cigarette packaging that says "Smoking is hazardous to your health." A similar health warning is attached to products containing saccharin.

Food Subsidies

The federal government conducts a food assistance program, which has beginnings tracing back to the 1930s. The current **Food Stamp Program (FSP)** was established by the Food Stamp Act of 1964 to aid needy households with food purchases. The FSP helps low-income households improve their diets by providing them with coupons to purchase food at authorized retail food stores. This program, along with educational programs, remains the federal government's major means of providing food assistance to those who qualify.

The **National School Lunch Program (NSLP)** is the oldest and largest child-feeding federal food assistance program in the United States. The USDA is the largest buyer of prepared foods in the country through the NSLP. This program provides financial and community assistance for meal service in public and nonprofit private high schools, intermediate schools, grade schools, and preschools, and also public and private licensed nonprofit residential childcare institutions. While all children may participate, a child may receive a free, reduced-price, or full-price meal based upon household income poverty guidelines.

The **Special Supplemental Food Program for Women, Infants, and Children (WIC)** was created to provide supplementary food assistance benefits to those individuals deemed by local health officials to be at nutritional risk due to their inadequate income and existing nutrition. Participants are given a voucher redeemable for specified foods at participating retail food stores. There are other programs that also provide food assistance to eligible individuals and households.

> **Food subsidies** There are a variety of federal programs designed to subsidize the provision of nutritious food products to children and others at risk.

SUMMARY

The purpose of this chapter is to illustrate the economic consequences of the inelastic nature of the demand for farm products and the continuing expansion of the farm business sector's capacity to produce, and to discuss the government programs that have been used to affect the market price farmers receive for their products. The major points made in this chapter are summarized as follows:

1. Changes in supply, coupled with the highly inelastic demand for raw food and fiber products, can translate into periods of booms and busts in farm income. The ever-expanding nature of annual farm output helps explain the low returns to resources observed historically in agriculture.

2. A short-run price elasticity of demand of -0.20 suggests a price flexibility (the reciprocal of the price elasticity) of -5.00. For farmers, this means that a 1% increase in the quantity they send to market will lower the market price they receive during the year by 5%, all other things constant.

3. One approach to dealing with the historical problem of low returns to resources is the implementation of commodity supply programs, based on federal policy mechanisms such as nonrecourse

loans, set-aside requirements, target prices and deficiency payments, and the Conservation Reserve Program.

4. Prior to passage of the FAIR Act in 1996, target prices were used to supplement the price supports under the CCC nonrecourse loan program. The target price concept called for direct deficiency payments to participating farmers, based on the difference between the target price and the market price when the target price is higher than the loan rate. This approach was eventually replaced by flexibility contract payments, which are fixed annually and decline in value until 2003, when they may disappear altogether.

5. The federal crop insurance program today represents a significant safety net for crop producers who can elect specific levels of coverage for losses stemming from adverse yields and falling commodity prices.

6. The 2014 farm bill, which extends farm programs to 2018, supplements the federal crop insurance program through two options available to producers: the Price Loss Coverage option and the Agricultural Risk Coverage option.

7. Raising the returns to resources in agriculture can also be accomplished by demand expansion programs. Domestic demand can be expanded by public (e.g., school lunch programs) and private (e.g., advertising) programs. The development of new uses for food and fiber products and expansion of export demand with export subsidies to importing countries also lead to higher domestic prices and farm income.

8. The level of export demand is important because it affects the elasticity of demand for certain farm products. Although farm programs may be needed in the short run to alleviate economic stress, the continuation of farm programs in which the elasticity of demand exceeds 1 may lower rather than raise farm revenue.

9. The rationale for government intervention in a nation's food and fiber industry typically includes the following: (1) to support/protect an infant industry, (2) to curb market power of imperfect competition when necessary to promote social good, (3) to provide for national food security, (4) to provide for consumer health and safety, and (5) to provide for environmental quality.

10. The U.S. food and fiber industry is one of the most highly regulated industries in the economy.

This includes regulations that govern the use of inputs such as agricultural chemicals, regulations that govern the processing and manufacturing of food and fiber products, and regulations that govern the markets in which these products are traded.

11. The inelastic own-price elasticity of demand for farm products, farmers, and ranchers, lack of market power, the interest sensitivity of the farm sector, and the fixity of farm assets and chronic excess capacity represent the roots of the farm problem.

12. A record crop, which is more the rule than the exception, can lead to sharp declines in farm product prices and income levels in absence of support programs.

13. The farm sector is one of the most highly capitalized sectors in the U.S. economy. There is more capital invested per worker than most other sectors in the economy.

14. Asset fixity can lead to a decline in farm asset values making it difficult for farmers and ranchers to scale back their operations to desired levels without selling excess assets at reduced prices.

15. Foreign trade enhancements that promote domestic farm economic conditions include tariffs and quotas on imports from other countries and export credits, subsidies, and commodity assistance programs to enhance exports of U.S. farm products.

16. Other forms of government intervention in agriculture include subsidized credit to farm borrowers, subsidized crop insurance, and support payments to farmers when crop yields and prices fall well below legislated levels.

17. The government also intervenes in the food and fiber industry on behalf of consumers. These efforts are to ensure food safety and to provide an adequate and cheap food supply. Other consumer issues include nutrition and health, the provision of food subsidies, and aid to rural communities.

18. The government also intervenes in the manner in which specific resources are used in the economy. This includes programs to control soil erosion, the adequacy (i.e., quantity and quality) of the nation's water supply, the minimum wages and safety of hired farm labor, and the conservation of energy and production of substitute fuels that use corn and other crops in their manufacturing.

KEY TERMS

Acreage Reduction Program (ARP)
  percentages
Agricultural Risk Coverage
  (ARC)
Agricultural Act of 2014
Asset fixity
Average Crop Revenue Election
  (ACRE)
Catastrophic Coverage (CAT)
Commodity Credit Corporation
  (CCC)
Commodity loan rate
Conservation Reserve Program
  (CRP)
Deficiency payments
Direct payment
Export Enhancement Program
  (EEP)
FAIR Act
Farm Service Agency
Farmer-Owned Reserve (FOR)
  Program

Federal Agriculture Improvement
  and Reform (FAIR) Act
Federal Crop Insurance
  Corporation (FCIC)
Federal Crop Insurance Reform Act
  of 1994
Federal Trade Commission (FTC)
Food, Agriculture, Conservation,
  and Trade (FACT) Act of 1990
Food and Drug Administration
  (FDA)
Food and fiber industry
Food Security Act (FSA) of 1985
Food Stamp Program (FSP)
General Agreement on Tariffs and
  Trade (GATT)
Human Nutrition Information
  Service
Multiple Peril Crop Insurance
  (MPCI)
National School Lunch Program
  (NSLP)

Nonrecourse loan
Payment rate
Price Loss Coverage (PLC)
Quotas
Revenue Protection Plan
Risk Management Agency
Set-aside
Slippage
Soil Bank Program
Special Supplemental Food
    Program for Women, Infants
    and Children (WIC)
Stacked Income Protection Plan
  (STAX)
Supplemental Coverage Option
  (SCO)
Target price mechanism
Tariffs
U.S. Department of Agriculture
  (USDA)
Wetlands Reserve Program
  (WRP)

TESTING YOUR ECONOMIC QUOTIENT

1. Based on the following graph, please place a T or F in the blank appearing by each statement:

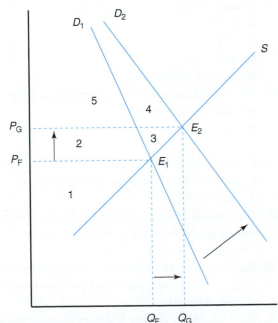

_____ There has been an increase in the quantity demanded.

_____ Consumers are better off economically if area 3 + area 4 is greater than area 2.

_____ Producers will be better off by the value of area 2 only.

_____ This figure illustrates the effects of domestic demand enhancement.

_____ Society as a whole will be better off by the value of area 3 + area 4.

2. Suppose the current market price for wheat is $3 per bushel and that 10 billion bushels are currently being marketed both domestically and abroad. Given the following market shares and own-price elasticities of demand for domestic and export markets, answer the following questions.

	Market Share (%)	Own-Price Elasticity
Domestic	40	−0.20
Foreign	60	−2.00

a. What would happen to U.S. wheat producers' total revenue if the quantity they supplied to the market *increased* by 3% (i.e., how much would their total revenue change from current levels)?

b. Would domestic consumers of wheat products be better off or worse off economically than they were before this increase in supply? Why? Illustrate your answer *graphically.*

3. The following graph depicts the market-level effects of using set-aside requirements to support farm prices and income.

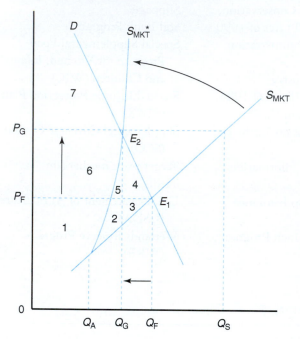

Enter a T or F in the following blanks to indicate whether the corresponding statements are true or false:

_____ Domestic consumers are made worse off economically in the current period as a result of the government's actions.

_____ The extent to which the government has to enter the market is unaffected by the elasticity of the demand curve.

_____ Q_F represents the quantity sold to foreign governments.

_____ Areas 3 and 4 represent the economic loss to society as a result of this policy action.

_____ Producers gain area 6 from consumers' willingness to pay price P_G for quantity Q_G.

4. Thomas Malthus observed some 200 years ago that
a. population grows faster at an arithmetic rate while food production grows at a geometric rate.
b. population grows at a geometric rate while food production grows at an arithmetic rate.

c. population grows at a decreasing rate while food production grows at an increasing rate.
d. none of the above.

5. A record crop can lead to
a. sharp declines in input use and cash receipts from sale of production.
b. sharp declines in farm product prices and cash receipts from sale of production.
c. sharp declines in cash receipts from sale of production and higher input prices.
d. none of the above.

6. The roots of the "farm problem" include _____, _____, _____, and _____.

7. The more inelastic the market demand curve for a commodity, the more prices will _____.

8. The 2013 Farm Bill will include the following "safety net":
a. Set-aside provisions for program commodities
b. Commodity storage programs for excess feed grain production
c. Direct payments
d. None of the above

9. The food and fiber industry includes which of the following:
a. Food and fiber consumers
b. Ethanol processing plants
c. Farmers and ranchers
d. All of the above

10. Commodities covered by federal government programs include
a. pork.
b. hay.
c. rice.
d. none of the above.

11. The role of government payments to supporting net farm income levels
a. was relatively important in the early to mid-1980s.
b. was relatively unimportant in the early 2010s.
c. varies directly with price–cost squeeze.
d. all of the above.

12. The more inelastic the market demand curve for a commodity is, the more prices will _____ if supply increases.

13. Excess capacity refers to the situation where _____ of a particular product is sufficient to meet demand.

14. The U.S. food and fiber industry is one of the least-regulated industries in the U.S. economy.   T   F

15. There is more capital per unit of output than all other sectors of the U.S. economy combined.   T   F

16. The rationale for the federal government's intervention in the U.S. food and fiber industry includes
    a. protecting an infant industry.
    b. keeping food prices low.
    c. providing for food safety.
    d. none of the above.

17. Examples of federal programs that expand demand include
    a. the National School Lunch Program.
    b. the Farmer-Owned Reserve Program.
    c. the Conservation Reserve Program.
    d. none of the above.

18. Examples of federal programs that pertain to crop insurance include
    a. the Risk Management Agency.
    b. indemnity payments.
    c. CAT.
    d. all of the above.

19. Features of the 2014 Farm Bill include
    a. the set-aside mechanism.
    b. the acreage reduction program.
    c. countercyclical payments mechanism.
    d. none of the above.

REFERENCES

Knutson RD, JB Penn, BL Flinchbaugh, and JL Outlaw: *Agricultural and food policy*, 6th ed., Englewood Cliffs, N.J., 2006, Prentice Hall.

Malthus T: *An essay on the principle of population*, Homewood, Ill, 1963, Richard D. Irwin.

Rain and Hail Insurance Society, *2016 Crop Insurance Update*, June 2016.

U.S. Department of Agriculture, Farm Service Agency, 2014 Farm Bill Fact Sheet, September 2014.

12

Product Markets and National Output

Chapter Outline

The discussion of market outcomes in the preceding chapters was conducted within a **partial equilibrium** framework; that is, we implicitly assumed when studying the events taking place in the market for unprocessed farm products that everything else in the economy would remain constant. We did not recognize the responses registered in all other markets to a new price for farm products, and vice versa.

If we were to analyze the impact of acreage controls placed on the production of wheat in a partial equilibrium framework, we would ignore the effects that this policy action would have on obvious things such as the price of bread and on less obvious things such as the federal budget deficit, national unemployment, and market interest rates. The effects this policy would have on the prices of other crops and livestock would also be ignored. Implicit in an analysis of the equilibrium in a single market is the assumption that the indirect or feedback effects of this action from all other markets are so small that they can be ignored.

The concept of **general equilibrium** regards all markets as being interdependent. Everything depends on everything else—to varying degrees, of course. The indirect impact on the price of wheat resulting from the effect that reduced wheat production has on the prices of other goods and services in the economy would not be ignored. Because the food and fiber industry represents almost one-fifth of U.S. output, a substantial shock to food and fiber markets can have a significant impact on the rest of the economy, and vice versa. In short, general equilibrium captures not only the equilibrium between producers and consumers in U.S. product markets but also the equilibrium between these products and the U.S. money market.

The purpose of this chapter is to illustrate how businesses and households are linked together through resource and product markets, to establish the conditions that must be satisfied for an equilibrium between consumers and producers for a given rate of interest, and to discuss the composition and measurement of national output. The assumption of a fixed interest rate will be relaxed in Chapter 13 when we examine money market equilibrium.

CIRCULAR FLOW OF PAYMENTS

Let us begin our discussion of multimarket relationships in the economy by assuming the existence of a **barter economy**, in which households and businesses exchange goods and services as a means for paying for their purchases. We will then relax this assumption by allowing for the presence of money as a medium of exchange.

Barter Economy

In a barter economy, there is no money to serve as a medium of exchange. Households own all the primary resources (i.e., land, labor, capital, and management), which they supply annually to businesses. Businesses need these resources to produce goods and services. As shown in Figure 12-1, households receive payments in kind; that is, their wages are in the form of the products they helped produce.

Households would have to **barter** among themselves to obtain the mix of goods and services they desire. Can you imagine the difficulties that consumers and producers would encounter in an economy as complex as ours if bargaining between all parties was necessary to satisfy the demands for goods and services? How would you like to work for a vegetable producer during the summer and be paid in heads of cabbage? This would mean that you would have to barter with the university when paying your tuition, and the university would have to decide how many heads of cabbage they would charge per course.

General versus partial equilibrium Partial equilibrium focuses on a single market, assuming everything else remains constant. General equilibrium, on the other hand, focuses on all markets in the economy.

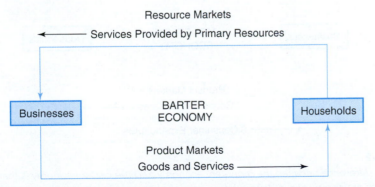

Figure 12-1

The business and household sectors in a barter economy interact with each other in the economy's resource and product markets. Businesses purchase the services provided by land, labor, capital, and management supplied by households in the resource market. Households buy the goods and services they need in product markets. Because there is no money in this economy, households barter among themselves and with businesses when exchanging goods and services. Bartering occurs even in a monetary economy. A plumber, for example, may do some plumbing work for a dentist in exchange for dental work of equal value. The value of these services would be considered income by the Internal Revenue Service.

Monetary Economy

Monetary economy
Parties trade goods and services in a barter economy when paying for other goods and services. Money typically changes hands in a monetary economy when paying for goods and services.

Now let's modify Figure 12-1 to reflect the fact that money is available for use as a medium of exchange. Continuing to assume that there are only households and businesses in our simplified economy, we see in Figure 12-2 that the household sector receives monetary remuneration in the form of rents, wages, salaries, and profits in exchange for providing the business sector with land, labor, capital, and management services. This figure also shows that businesses receive monetary remuneration in the form of consumer expenditures when they supply goods and services to households.[1] The sum of wages, rents, interest, and profits accruing to laborers and capital resource owners in the resource markets represents the economy's **national income**. The monetary value of the products flowing to households through the product markets represents **national product**.

You might have noticed that Figure 12-2 makes no mention of the economic role that government plays in the economy. This simplified flow diagram also ignores the possibility of savings and investment. There is also no recognition of the role financial markets play in the economy. To remedy these deficiencies, let us expand the simple flow diagram of a monetary economy in Figure 12-3 to include government and financial markets.

The expanded flow diagram in Figure 12-3 contains a financial market through which the net savings of households are passed on by depository institutions, the bond market, and the stock market to businesses that must either borrow, issue bonds, or sell stock to finance their expenditures. The term *net saving*, used to describe the flow of money from households to financial markets, and the term *net borrowing*, used to describe the flow of money from financial markets to businesses, signify the fact that money is flowing in both directions (i.e., households also borrow from financial

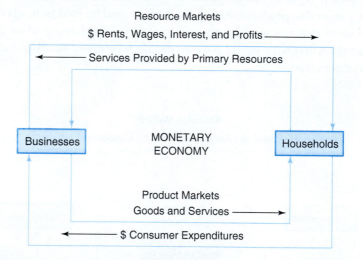

Figure 12-2
The major difference between the monetary economy and the barter economy is that transactions are completed in dollars. When households provide labor and other services to businesses, they are compensated in the form of rents, wages, interest, and profits. When businesses sell goods and services to households, they are compensated by the value of private expenditures. The sum of rents, wages, interest, and profits received by households in the economy's resource markets represents its national income. The value of total expenditures by households represents gross domestic product.

---

[1] The flow diagrams in Figures 12-1 and 12-2 assume that businesses do not save (i.e., retained earnings are zero). Instead, all profits are paid out in the form of dividends to households that are assumed to own all primary resources in the economy.

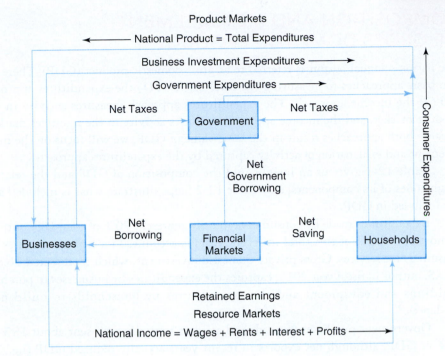

Figure 12-3

This figure adds a financial sector and a government sector. For example, households borrow and save. If their saving exceeds their borrowing, they are considered net savers. Businesses also borrow and save in financial markets. The government sector also borrows in the nation's financial markets to finance its budget deficits.

institutions, and businesses also save). Households are classified as net savers because they save more than they borrow, and businesses are classified as net borrowers because they borrow more than they save.

The production of goods and services in the monetary economy depicted in Figure 12-3 also generates income in the form of wages, rents, and interest. This income is captured at the bottom of this figure by the line drawn connecting the business and household sectors.[2] Businesses, of course, can decide to retain part of their profits (i.e., retained earnings) to help finance future expenditures.

The flow diagram in Figure 12-3 also captures the flow of goods and services and the flow of income and expenditures between government and businesses and households. For example, there is a flow of money from households to government in the form of tax payments. Households, in turn, receive money from the government in the form of social security payments, unemployment compensation, and other forms of payments (hence the use of the term *net taxes*).

Businesses also pay taxes and receive government payments. The government borrows in the nation's financial markets to finance its budget deficits (i.e., when government expenditures exceed tax revenues). It does this through the sale of government securities by the U.S. Treasury. When the government has a budget surplus, which last occurred during 1998–2001, it can buy back some of the securities it has issued. Finally, government expenditures for goods and services also represent a component of the **final demand** for the products produced by businesses. These expenditures for finished goods and services are included along with consumer expenditures and business expenditures when measuring the nation's **gross domestic product (GDP)**.

Measuring market activity
We can measure aggregate economic activity in product markets (national product) or resource markets (national income). Both values are identical in the national income and product accounts maintained by the federal government.

---

[2] For simplicity, only the money flows between the sectors are reflected in this figure, even though the physical flows of resources, services, and products continue to exist in the economy.

COMPOSITION AND MEASUREMENT OF GROSS DOMESTIC PRODUCT

The nation's annual output is referred to as gross domestic product (GDP). There are two basic approaches to measuring the level of GDP: (1) the expenditures approach and (2) the income approach. The expenditures approach measures activity in the product market, and the income approach measures activity in the resources market. Because both approaches result in the same value of GDP, we will focus on the measurement and explanation of activity captured by the expenditures approach.

Table 12-1 gives us an insight into the composition of GDP and the relative magnitudes of its components, and Table 12-2 helps illustrate what is included and not included in GDP.

Expenditures made by consumers typically represent 70% of total U.S. output. Almost two-thirds of this total in recent years is accounted for by expenditures by consumers for services. Gross private domestic investment, which represented 16.8% of U.S. output in fiscal year 2015, captures the expenditures by businesses to purchase buildings and equipment and the expenditures by households to build new residences.[3]

Government purchases of goods and services typically represent about 18% to 20% of GDP, although net exports in recent years actually reduced GDP (i.e., we have been importing more from other nations than we have been exporting). With the exception of net exports, these percentage shares of total GDP accounted for by

Table 12-1
U.S. GROSS DOMESTIC PRODUCT FISCAL YEAR 2015

	Amount (Billion Dollars)	Percentage of GDP
Consumption		
Durable goods	1,280	
Nondurable goods	2,668	
Services	7,918	
Total consumption	11,866	68.36%
Gross private domestic investment		
Nonresidential structures	506	
Producers' durable equipment	1,016	
Intellectual property	685	
Residential construction	559	
Total fixed investment	2,765	
Change in inventories	90	
Total investment	2,856	16.45%
Government purchases of goods and services		
Federal government	1,219	
State and local governments	1,956	
Total purchases	3,175	18.29%
Net exports of goods and services		
Exports	2,334	
Imports	2,872	
Total net exports	−538	−3.10%
GROSS DOMESTIC PRODUCT	17,358	100.00%

Source: Economic Report of the President, The Council of Economic Advisors, 2016

---

[3] The term *gross* here means that both annual expenditures to replace worn-out or obsolete machinery and net new investment expenditures are captured in this total.

Table 12-2
GDP: WHAT'S IN AND WHAT'S NOT

Type of Expenditure	Included in GDP?	Examples
Consumer expenditures:		
Durable goods	Yes	Autos, TVs, VCRs
Nondurable goods	Yes	Food, clothes
Services	Yes	Haircuts, airplane tickets
Gross private domestic investment:		
Change in business inventories	Yes	Corn stored on farm
Producers' durable equipment	Yes	Computers, tractors
Structures:		
Nonresidential structures	Yes	Factories, office buildings, shopping malls
Residential structures	Yes	Houses, condominiums
Net exports:		
Exports	Yes	Tractors, computers
Imports	Yes	Coffee, bananas, wine
Government purchases of goods and services:		
Intermediate services	Yes	Fire fighters, police officers
Consumption	Yes	City parks, street cleaners
Other activity:		
Government interest and transfer payments	No	Social security, welfare, unemployment benefits
Private intermediate goods	No	Wheat, iron ore, plastic, crude petroleum
Private purchases of used assets	No	Purchases of used home, used autos
Purchases of farmland	No	Cropland, grazing land

consumption, investment, and government expenditures have remained remarkably stable over the post–World War II period.

The value of GDP and its components in Table 12-1 are expressed in current dollars, which is commonly referred to as **nominal GDP**. No adjustments have been made for the effects that inflation has on the purchasing power of annual GDP. **Real GDP** reflects the effects of inflation, which is accomplished by dividing nominal GDP by the implicit GDP price deflator, a price index representing a weighted average of the prices of all the goods and services that are captured in GDP. Figure 12-4 illustrates the historical fluctuations in the real GDP growth rate from 1970 to 2014, which reflects the impact of the 2007–2009 Great Recession.

Newspaper articles reporting that the economy grew by say 3.2% during the year are referring to the percentage change in real GDP from the previous year. During the recession occurring in 1981, for example, nominal GDP grew 4.1%, but real GDP actually *declined* 1.9%. This particular difference is significant because nominal GDP growth gave a picture of a growing economy, while real GDP growth gave a picture of an economy in a recession. Therefore, here and in all other instances, it would be grossly misleading to discuss economic growth in nominal GDP terms.

Real GDP represents the value of goods and services produced after adjustments have been made for inflation.

CONSUMPTION, SAVINGS, AND INVESTMENT

There are two things that households can do with a dollar of **disposable income** (income after taxes): they can "consume" the dollar, or they can save it. If the dollar is used to finance consumer expenditures, it is gone forever. However, if the dollar is saved, it will be available to finance future consumption.

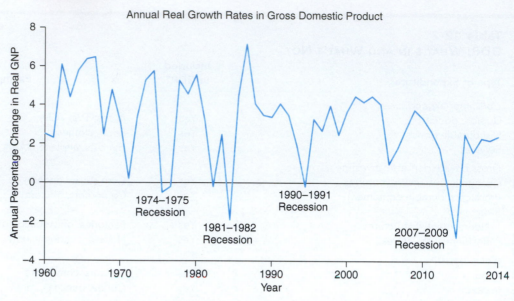

Figure 12-4

The percentage change in the real GDP of the United States shows sharp year-to-year variations during the past 100 years. The sharp decline beginning in 2006 marks the beginning of the "Great Recession," the longest and deepest decline in economic activity since the Great Depression in the 1930s.

Source: From Economic Report of the President, The Council of Economic Advisors.

Types of consumer expenditures Consumer expenditures are classified as durable goods, nondurable goods, and services. A new television is an example of a durable good, while food is an example of a nondurable good.

When households make expenditures, they are purchasing what are normally referred to as consumer goods and services. If the good has a life of less than 1 year, it is called a nondurable good. A service, by definition, is nondurable. If a good is consumed over a longer period of time, however, it is a durable good. Food is an example of a nondurable good, a haircut or an airplane trip is an example of a service, and a car is an example of a durable good.

Investment refers to expenditures by businesses on capital goods, such as a new machine or a new building; expenditures by households for new residences; and an increase in the inventories of businesses. We will identify the factors that influence the level of planned consumption expenditures by households and planned investment expenditures by businesses.

Determinants of Planned Consumption

In 1936, John Keynes identified the major determinant of planned **consumption** by households. Keynes suggested that people "increase their consumption as their income increases, but not by as much as the increase in their income." In other words, there is a relationship—but not a one-to-one correspondence—between the planned consumption expenditures by households and their current level of disposable income.

Figure 12-5 illustrates an annual consumption function and its break-even level of consumption for the hypothetical economy of Lower Slobovia.

Consumption Function When the household sector's disposable income goes up in Lower Slobovia, its level of planned consumption goes up also. This relationship represents the household's **consumption function.**[4] As Keynes suggests, planned consumption expenditures rarely change by the same amount as a change in consumer

---

[4] The term *function* refers to the fact that there is a causal relationship between income and consumption. If income increases, consumption will also increase. If consumption is said to be a function of income, this means that the level of consumption depends on the level of income.

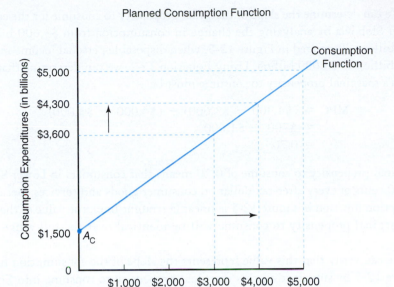

Figure 12-5

The relationship between actual disposable income and planned consumer expenditures by households suggests that households in Lower Slobovia will consume $1,500 billion ($1.5 trillion) per year if their incomes after taxes are actually zero. When disposable personal income increases, consumer expenditures will increase. The rate of this increase will depend upon the slope of this function.

disposable income. The hypothetical annual consumption function in Figure 12-5 suggests that planned consumption would be approximately $1,500 billion even if actual disposable income were zero. This level of consumption expenditures is referred to as autonomous consumption.

We can express the consumption function depicted in Figure 12-5 in general terms as

$$C = A_C + \text{MPC(DPI)} \qquad (12.1)$$

in which C represents the level of consumption expenditures for goods and services made by households, A_C represents the level of autonomous consumption expenditures, MPC represents the slope of the consumption function, and DPI represents the level of disposable personal income. Thus, when consumers' disposable personal income increases, consumer expenditures will increase. Assume the level of disposable personal income increased from $3,000 billion to $4,000 billion in Lower Slobovia. Figure 12-5 indicates that consumption expenditures would increase from $3,600 billion to $4,300 billion. Before we can use Equation 12.1 to replicate these graphical results, we must first calculate the slope of this function, or the MPC.

Slope of Consumption Function   The slope of the consumption function represents consumers' marginal propensity to consume (MPC); that is, it reflects the change in consumption expenditures associated with a change in disposable personal income, or

$$\text{MPC} = \Delta C \div \Delta \text{DPI} \qquad (12.2)$$

in which Δ represents the change in a variable, C represents the level of planned consumption expenditures, and DPI represents disposable or after-tax personal income.[5]

Marginal propensity to consume Marginal propensity to consume, the slope of the consumption function, tells us how much consumption will change if consumer disposable income changes.

---

[5] An alternative measure of income in the consumption function is given by the permanent income hypothesis. This theory implies that planned consumption expenditures do not depend on current income, but instead on a measure of expected permanent income over the next 3 to 5 years. In other words, current planned consumption will not change substantially if current actual disposable income changes, unless this change is expected to continue over time.

We can determine the slope or marginal propensity to consume for the economy of Lower Slobovia by analyzing the change in consumption from $3,600 billion to $4,300 billion, illustrated in Figure 12-5, when disposable personal income rose from $3,000 billion to $4,000 billion. Using Equation 12.2, we can determine that Lower Slobovia's marginal propensity to consume must be

$$\text{MPC} = (\$4,300 - \$3,600) \div (\$4,000 - \$3,000)$$
$$= \$700 \div \$1,000$$
$$= 0.70 \tag{12.3}$$

A marginal propensity to consume of 0.70 means that consumers in Lower Slobovia spend 70 cents of every after-tax dollar on consumer goods and services. Because the consumption function in Figure 12-5 is linear (a straight line), the value of the economy's marginal propensity to consume will be identical over the full range of this curve.

We can verify that this value represents the slope of the consumption function in Figure 12-5 by substituting this marginal propensity to consume into Equation 12.1 and solving for the level of consumption associated with the two levels of disposable personal income used in this example. Given the level of autonomous consumption in this economy of $1,500 billion and an MPC of 0.70, Equation 12.1 can be rewritten to read

$$C = \$1,500 + 0.70\,(\text{DPI}) \tag{12.4}$$

If the level of disposable income were $3,000 billion, then the level of consumption expenditures in the economy would be

$$C = \$1,500 + 0.70(\$3,000)$$
$$= \$1,500 + \$2,100$$
$$= \$3,600 \text{ billion} \tag{12.5}$$

And if the level of disposable income were instead equal to $4,000 billion, the level of consumption expenditures in this economy would be

$$C = \$1,500 + 0.70(\$4,000)$$
$$= \$1,500 + \$2,800$$
$$= \$4,300 \text{ billion} \tag{12.6}$$

Note these are the same levels of consumption expenditures associated with the two income levels illustrated in Figure 12-5.

Shifts in Consumption Function   While changes in the level of income correspond to movements along the consumption function illustrated in Figure 12-5, other determinants of consumption expenditures will shift the consumption function.

An increase in wealth of the U.S. household sector, for example, would shift its consumption function upward.[6] This implies that the household sector has a greater basis from which to spend for a given level of income than it does with less wealth. As Figure 12-6 illustrates, an increase in the wealth of the household sector in Lower Slobovia would increase consumption expenditures from $4,300 billion to $5,000 billion for a given level of disposable personal income ($4,000 billion) should the increase in wealth be used entirely for consumption. In terms of Equation 12.1,

Wealth effect   The consumption function will shift to the right (left) if consumer wealth increases (decreases).

---

[6] This measure of wealth should include the current market value of the household's physical and financial assets less debt outstanding, and should be adjusted for inflation.

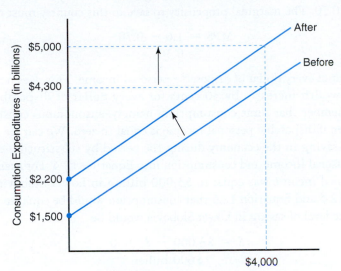

Effect of an Increase in Wealth

Figure 12-6

An increase in wealth will shift the consumption function upward, and a decrease in wealth will shift the consumption function downward.

autonomous consumption necessarily must have increased from $1,500 billion to $2,200 billion. A decrease in wealth would have the opposite effect.

A specific component of the household sector's wealth position—its holding of liquid assets—may be of special importance to the household sector when formulating its spending plans. A liquid asset is an asset that can be converted into money quickly with little or no loss in value. Cash is a perfectly liquid asset. Examples of other liquid assets include savings accounts at commercial banks.

Expectations also may affect how much households with a given level of current disposable income are willing to spend versus save, particularly in the short run. For example, if households expect a higher income in the near future, they may increase their current consumption expenditures. Conversely, the same households may decide to postpone consumption (and increase their savings) if they are pessimistic about the near future.

Determinants of Planned Saving

The current level of planned **saving** in the economy is equal to the difference between disposable personal income and planned consumption expenditures, or

$$S = \text{DPI} - C \qquad (12.7)$$

in which C represents the level of consumption expenditures and DPI represents the level of disposable personal income in the economy. This equation simply states that savings by households are equal to their personal income after taxes (or disposable income) and after household spending for goods and services.

The change in saving associated with a change in disposable personal income represents the marginal propensity to save, or

$$\text{MPS} = \Delta S \div \text{DPI} \qquad (12.8)$$

Because the level of savings also represents personal income after taxes not spent on consumption, we know that

$$\text{MPS} = 1.0 - \text{MPC} \qquad (12.9)$$

Marginal propensity concept   The marginal propensity to save is the change in saving as disposable income changes. If the marginal propensity to consume is 0.90 (90%), then the marginal propensity to save is 0.10 (10%).

We determined in Equation 12.3 that the marginal propensity to consume in Lower Slobovia was 0.70. The marginal propensity to save in this country must therefore be

$$MPS = 1.0 - 0.70$$
$$= 0.30 \qquad\qquad (12.10)$$

or 30 cents out of every dollar of disposable personal income. This does not mean that the nation's wealth increases by 30 cents for every dollar of disposable personal income. Remember that some consumption, namely, autonomous consumption, is carried on even if disposable personal income is equal to zero. We can see if there was saving or dis-saving in the economy during the period by substituting the levels of disposable personal income and consumption into Equation 12.4. For example, if disposable personal income was equal to $3,000 billion in Lower Slobovia, we know from Figure 12-5 and Equation 12.5 that consumption would be equal to $3,600 billion. Thus, the level of saving in Lower Slobovia would be

$$S = \$3,000 - \$3,600$$
$$= -\$600 \text{ billion} \qquad\qquad (12.11)$$

or dis-savings equal to $600 billion. In other words, consumption exceeded disposable income by $600 billion.

How high would the economy's disposable personal income have to be before dis-saving would be eliminated? A logical extension of this analysis is to determine the break-even level of disposable personal income, or the point at which consumption equals disposable personal income. We can determine this level of disposable personal income by substituting alternative levels of income into Equation 12.3 to determine the level of consumption and then use Equation 12.9 to determine the level of saving (see Table 12-3). Thus, a level of disposable personal income of $5,000 billion will be required if the economy's wealth position is not to be less at the end of the year than it was at the start of the year. If disposable personal income is less than $5,000 billion, dis-saving will occur. If disposable income is greater than $5,000 billion, saving will occur.

We can also determine the break-even level of disposable personal income graphically by extending a ray out of the origin in Figure 12-5 at a 45-degree angle. This ray would indicate all points in this figure at which consumption equals disposable personal income.

As Figure 12-5 indicates, the break-even level of consumption in Lower Slobovia's economy is $5,000 billion where consumption expenditures are equal to disposable personal income, the point at which a 45-degree ray would intersect the economy's consumption function.

As we now turn our attention to the economy's investment function, consider the following:

1. What would happen to the level of consumption expenditures in Lower Slobovia's economy if its federal government increased current income taxes by $10 billion?

Table 12-3
LEVEL OF DISPOSABLE PERSONAL INCOME IN BILLION DOLLARS

Disposable Income	Consumption	Saving
3,000	3,600	−600
4,000	4,300	−300
5,000	5,000	0
6,000	5,700	+300
7,000	6,400	+600

2. Can you answer question 1 graphically?
3. Does the consumption function in Figure 12-5 shift downward?
4. Can you answer question 3 using the consumption function equation?

Determinants of Planned Investment

Business investment is defined as business expenditures on new buildings and equipment plus net additions to their production inventories.[7] The level of business investment in the current period is determined in part by the market rate of interest. The level of household investment in new residences in the current period is determined in part by the market rate of interest. Not surprisingly, Keynes stated that the "inducement to invest depends partly on the investment demand schedule and partly on the rate of interest."

Investment Schedule The investment demand schedule that Keynes was referring to reflects the investment expenditure plans of businesses and households corresponding with specific rates of interest. Investment expenditures consist of fixed investments in business assets (factories, office buildings, shopping centers, trucks, tractors, silos, etc.) and changes in business inventories. Investment expenditures also include newly constructed houses and condominiums sold to individuals. A homeowner is treated as a business that owns the house as an asset and rents the newly constructed house to itself. The planned investment function for the economy reflects an inverse relationship between the rate of interest and the level of planned investment, or

$$I = A_I = \text{MIS}\,(R) \tag{12.12}$$

in which I represents the level of investment expenditures, A_I represents the autonomous level of investment, or investment that would take place if the market interest rate were equal to zero, MIS represents the marginal interest sensitivity of investment or slope of the investment function, and R represents the market rate of interest.

The interest sensitivity of investment in Equation 12.12 reflects the impact that a change in interest rates will have upon investment expenditures, and is given by

$$\text{MIS} = \Delta I \div \Delta R \tag{12.13}$$

Interest rate effect The slope of the investment function reflects the change in investment expenditures if the rate of interest changes.

or the change in investment expenditures associated with a change in the market rate of interest. We would expect the interest sensitivity of investment for businesses to be different from the interest sensitivity of investment for households. In general, the greater the value of MIS in Equation 12.12, the more sensitive the investment in Lower Slobovia's economy will be to changes in the market rate of interest.

Figure 12-7 presents the annual planned investment function for Lower Slobovia. This figure shows that if the market rate of interest in the current period is 7%, the level of investment expenditures by households and businesses will be $300 billion. But if the market rate of interest rose to 9%, investment expenditures in Lower Slobovia's economy would fall to $250 billion. We can calculate the marginal interest sensitivity of investment for Lower Slobovia based upon this observation using Equation 12.13 as

$$\begin{aligned}
\text{MIS} &= (\$250 - \$300) \div (9 - 7) \\
&= -\$50 \div 2 \\
&= -\$25.0 \text{ billion} \tag{12.14}
\end{aligned}$$

---

[7] Purchases of land or used capital goods at the aggregate level are not considered as investments because they do not represent capital goods formed in the current period.

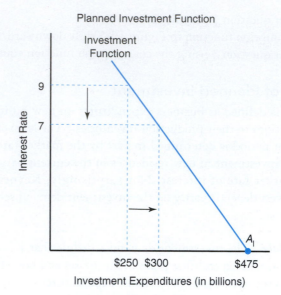

Planned Investment Function

Thus, a 1% increase in interest rates will depress investment expenditures in Lower Slobovia by $25 billion.

If we assume that the level of autonomous investment expenditures in the economy was $475 billion, we can solve Equation 12.12 for the two interest rates depicted in Figure 12-7 and verify the change in investment expenditures suggested by Equation 12.13. Substituting the values of A_I and MIS into Equation 12.12, we see that the level of investment expenditures when the market rate of interest is 7% would be

$$
\begin{aligned}
I &= \$475 - 25(7) \\
&= \$475 - \$175 \\
&= \$300 \text{ billion}
\end{aligned}
\tag{12.15}
$$

If the market rate of interest in Lower Slobovia increased from 7% to 9%, the investment expenditures economy-wide in Lower Slobovia would be

$$
\begin{aligned}
I &= \$475 - 25(9) \\
&= \$475 - \$225 \\
&= \$250 \text{ billion}
\end{aligned}
\tag{12.16}
$$

Note that the interest rate and investment expenditure combinations reported in Equations 12.15 and 12.16 represent the values depicted in Figure 12-7.[8]

Shifts in Investment Function   Other determinants of planned investment cause the investment expenditure function in Figure 12-7 to shift outward to the right or inward to the left. Like shifts in the consumption function, an outward shift in the investment function like that depicted in Figure 12-8 means that more investment will take place at a given market rate of interest (e.g., $350 billion instead of $300 billion at a 7% market rate of interest).

Businesses must project the profitability of an investment project over the life of the investment, and households must project the feasibility of a new home

---

[8] An increase in investment expenditures, by inducing additional output and income, *pulls up* the level of consumption expenditures in the economy. The result is an increase in the nation's output that is a multiple of the initial increase in investment expenditures. The size of this multiplier will depend on consumers' marginal propensity to save (MPS). The lower the MPS, the greater the multiplier effect of investment expenditures on output will be.

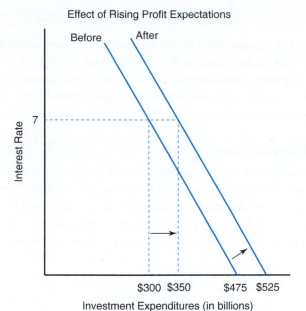

Effect of Rising Profit Expectations

Before   After

Interest Rate

7

$300 $350 $475 $525

Investment Expenditures (in billions)

Figure 12-8
*Favorable developments
with respect to expected
future incomes of
businesses and households,
prices of new investment
goods, technological
change, or taxes will shift
the investment expenditure
function to the right, and
vice versa.*

loan before initiating construction. Thus, their expectations about a wide range of factors that affect the future income of producers and consumers will affect decisions as to whether they should make an investment in the current period. If businesses are optimistic (pessimistic) about their future profit, the planned investment function will shift to the right (left) for each interest rate, as illustrated in Figure 12-8. If households are optimistic (pessimistic) about their future income, the planned investment function for new residences will shift to the right (left) for each interest rate.

The purchase price of new capital goods also represents a determinant of planned investment expenditures by businesses and households. If the cost of a new plant and equipment were to suddenly decrease (increase), we would expect the planned investment function in Figure 12-7 to shift to the right (left), as shown in Figure 12-8. If the cost of new homes were to suddenly decrease (increase) in Lower Slobovia, we would expect the planned investment function for new residences to shift to the right (left), as illustrated in Figure 12-8.

The technological change embodied in new capital goods also has an effect on planned investment expenditures. Improvements in the productive services, provided by equipment and innovations that affect the functions performed by capital, should shift the planned investment expenditure function to the right, as depicted in Figure 12-8. In other words, planned investment at each interest rate should now be greater than before as firms seek to become more efficient in their operations.

The level of taxes also will affect planned investment expenditures because businesses and households evaluate their expenditure decisions on the basis of their expected after-tax return. If there is a decrease in effective tax rates (i.e., deductions and credits are liberalized), we would expect businesses and households to expand their expenditures for new capital goods. The opposite would be true if tax rates were increased.

What would happen to the level of investment expenditures in Lower Slobovia if the market rate of interest fell by 2%? Can you answer this question graphically? Will the investment function shift to the right? Can you use Equation 12.12 to get a numerical answer to this question?

Future profit effect
Business expectations of rising (declining) future profits will shift the investment function to the right (left).

EQUILIBRIUM NATIONAL INCOME AND OUTPUT

To determine the equilibrium level of national income and output in the economy, we must employ the relationships discussed above for household and business expenditures. The total value of planned consumption, investment, government spending, and net exports in an economy is often referred to as aggregate expenditures. When aggregate expenditures equal aggregate output, the nation's product markets are in equilibrium.

Aggregate Expenditures

Table 12-1 shows that the nation's gross domestic product includes the demand for final goods and services in the economy by consumers, investors, governments, and foreign countries. Assume for ease of exposition that the economy of Lower Slobovia is a closed economy (i.e., it has no imports or exports). Also assume that the market rate of interest is constant. The levels of expenditures in Lower Slobovia can be summarized as

$$C = \$1,500 + 0.70(DPI)$$
$$I = \$475 - 25(R)$$
$$G = \$880 \tag{12.17}$$

in which the consumption expenditure function was developed earlier in Equation 12.4, the investment expenditure function was developed in Equation 12.15, and the government expenditure function here suggests that government expenditures on goods and services in Lower Slobovia are fixed at \$880 billion. If we assume that the market rate of interest for the moment is fixed at 7%, then the nation's aggregate expenditures would be given by

$$
\begin{aligned}
AE &= C + I + G \\
&= \$1,500 + 0.70(DPI) + \$475 - 25(7) + \$880 \\
&= \$2,680 + 0.70(DPI)
\end{aligned}
\tag{12.18}
$$

in which AE represents aggregate expenditures. Table 12-4 summarizes the levels of aggregate expenditures for alternative levels of disposable income and a market interest rate of 7%. The product market will be in equilibrium for a given market interest rate when aggregate expenditures and output are equal to \$8,000 billion.

The Keynesian cross illustrates the equilibrium level of aggregate expenditures in the economy, or where they equal aggregate output.

The Keynesian Cross

We can analyze the level of aggregate expenditures in Lower Slobovia in the context of what has come to be known as the Keynesian cross, using the observations for consumption (C), investment (I), and government (G) expenditures presented in Table 12-4. The 45-degree ray coming out of the origin in Figure 12-9 represents all

Table 12-4
TOTAL PRODUCT MARKET ACTIVITY IN LOWER SLOBOVIA IN BILLION DOLLARS

Planned Consumption Expenditures (C) $	Planned Investment Expenditures (I) $	Planned Government Expenditures (G) $	Aggregate Expenditures (C + I + G) $	Output (Y) $
5,420	300	880	6,600	6,000
6,120	300	880	7,300	7,000
6,820	300	880	8,000	8,000
7,520	300	880	8,700	9,000
8,220	300	880	9,400	10,000

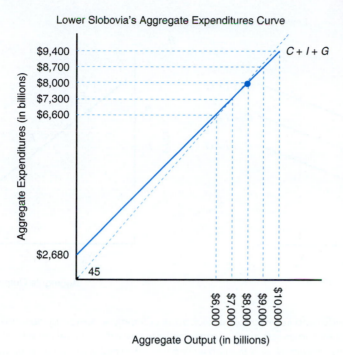

Lower Slobovia's Aggregate Expenditures Curve

Figure 12-9

The equilibrium level of aggregate output in the nation's product market in a closed economy occurs where $Y = C + I + G$. This occurs at a level of aggregate output and aggregate expenditures of $8,000 billion. This equilibrium assumes a given market interest rate and general price level.

points of equality between aggregate expenditures and aggregate output (gross domestic product also equals national income in the national income and product accounts). Figure 12-9 shows that equilibrium in the product market of the Lower Slobovia economy would occur at $8,000 billion. This is the point at which the aggregate expenditure curve $(C + I + G)$ for our hypothetical closed economy crosses the 45-degree ray. Table 12-4 also shows that aggregate spending in Lower Slobovia equals the nation's output at $8,000 billion.

Below this equilibrium, planned aggregate expenditures exceeded aggregate output, which should draw down unsold business inventories and increase pressures to expand output. After $8,000 billion, aggregate output exceeded planned aggregate expenditures, which should lead to increases in unsold business inventories and put downward pressure on future aggregate output levels.

Deriving Aggregate Demand Curve

The slope of Lower Slobovia's aggregate demand curve will depend upon the response of consumption expenditures and hence total expenditures to changes in real wealth (current wealth adjusted for changes in the general price level, which affects the purchasing power of wealth). The three aggregate expenditure curves in Figure 12-10A, are associated with three different general price levels. Let us assume equilibrium E_1 is associated with the level of aggregate expenditures illustrated in Figure 12-9.

The general price level P_1 is also associated with an aggregate output or real GDP equal to Y_1. **Equilibrium output E_3** would represent the level of aggregate output Y_3 that we would observe at a higher general price level P_3, and equilibrium E_2 would reflect a lower general price level P_2 and a higher aggregate output level Y_2. Figure 12-10B presents a plot of the price–quantity combinations associated with E_1, E_2, and E_3. A line drawn through these price–quantity coordinates gives the economy's aggregate demand curve (AD), a traditional downward-sloping demand curve- like those seen earlier in this book.

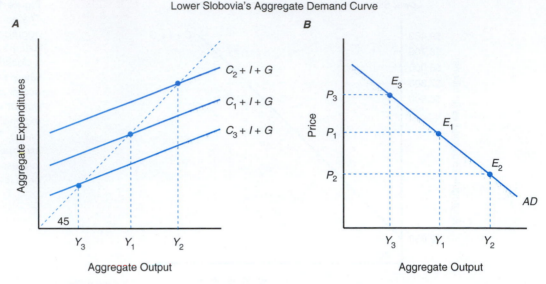

Figure 12-10
The aggregate demand curve for Lower Slobovia's economy is derived by observing the level of consumption spending and aggregate spending as we vary the general price level and hence real wealth of its consumers, and then observing what this means for the level of aggregate real output (real GDP).

Aggregate Supply and Full Employment

Aggregate supply represents the nation's aggregate output supplied to consumers, businesses, governments, and foreign countries. The aggregate supply curve, therefore, must necessarily represent the nation's aggregate output supplied to these groups for a full range of price levels. When aggregate demand increases, farm and nonfarm businesses in the economy will respond to the higher prices by increasing their output and/or reducing their current unsold business inventories. This can be seen as a movement up the economy's aggregate supply curve in the short run, because it takes time for current investment expenditures by businesses in buildings and other productive assets to expand productive capacity, which shifts the supply curve to the right.

If aggregate demand decreases and prices fall, businesses will respond by cutting production schedules, and business inventories will typically rise. This will result in a movement down the economy's aggregate supply curve.

The aggregate supply curve can take on a number of slopes depending on where the economy is on the aggregate supply curve. The curve has three distinct ranges: (1) the Keynesian depression range, (2) the normal range, and (3) the classical range. Each of these ranges is illustrated in Figure 12-11.

Supply curve ranges The aggregate supply curve for the economy has three ranges: the Keynesian or depression range, the normal range, and the classical range.

The depression range of the aggregate supply curve, sometimes referred to as the Keynesian range, after the British economist John Maynard Keynes, indicates that there is a perfectly elastic range of output over which increases in demand result in increases in supply unaccompanied by rising prices. The normal range of the aggregate supply curve takes the form of the firm- and market-level supply curves. The general price level will begin to rise as demand increases, more slowly at first and then more sharply when the curve becomes more inelastic. When the economy reaches its capacity to supply goods and services in the current period, the aggregate supply curve becomes perfectly inelastic. This range is referred to as the classical range because it takes on the properties of the aggregate supply curve found in the

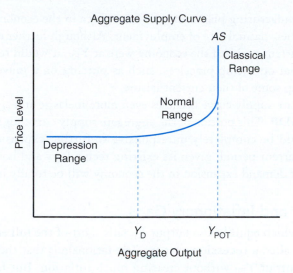

Figure 12-11
The economy's aggregate supply curve shows the depression, normal, and classical ranges.

writings of classical economists who were prominent before Keynes' theories surfaced in the 1930s.

The equilibrium between spending and output, illustrated in Figures 12-9 and 12-10B, represents one view of equilibrium in an economy's product market for a given market rate of interest. We can also illustrate the equilibrium in the product market by examining the intersection of the economy's aggregate demand curve (AD) and the aggregate supply curve (AS). Assume that the aggregate supply curve for Lower Slobovia takes the form illustrated in Figure 12-11 and repeated in Figure 12-12B.

The intersection of the AD and AS curves in Figure 12-12B identifies the equilibrium output level of the nation's product market, or Y_E. This corresponds with the point at which the aggregate spending curve ($C + I + G$) intersects a 45-degree ray from the origin, indicating that aggregate expenditures equal aggregate output. Figure 12-12 suggests that equilibrium output in the economy is less than full employment output (Y_{FE}) or the economy's maximum noninflationary level of goods

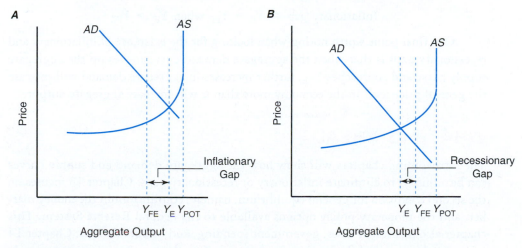

Figure 12-12
Let Y_{FE} represent the level of aggregate output at which the economy achieves **full employment GDP**, the point at which labor and capital resources are employed at a noninflationary or natural rate of employment. If the level of aggregate demand is less than what is necessary to be operating at Y_{FE}, the economy is suffering from a recessionary gap. If aggregate demand exceeds Y_{FE} and approaches Y_{POT}, the economy is suffering from an inflationary gap.

and services. Manufacturing plants and the labor force in the economy at Y_{FE} would be operating at their natural rate of employment. Although further output would be possible in the current period if the economy were at Y_{FE}, it would require businesses to adopt abnormal operating practices, such as putting on a graveyard shift, and laborers to give up some of their current leisure.

The aggregate supply curve becomes even more inelastic at Y_{POT}, or the economy's **potential GDP**. This portion of the aggregate supply curve suggests that aggregate output would be completely unresponsive to further expansion of aggregate demand in the current period, given its existing technology and normal production practices. Further demand expansion in the economy will be totally inflationary.

Recessionary and Inflationary Gaps

The amount by which equilibrium output Y_E falls short of the full employment output level Y_{FE} is called a **recessionary gap**. The rationale is that the economy could have produced output Y_{FE} without creating much inflation. But for one reason or another, the economy produced only output Y_E. The lower level of output also means that fewer people are employed (i.e., the unemployment rate for labor in the economy would be higher) than would be true at Y_{FE}. Figure 12-12 indicates that the recessionary gap in this instance would be equal to $Y_{FE} - Y_E$.

An **inflationary gap** will occur when planned aggregate expenditures are greater than the economy's full employment output. This economic situation can also be seen by reconsidering Figure 12-12A. An inflationary gap will occur when the aggregate demand curve AD intersects the aggregate supply curve AS to the right of Y_{FE}. The size of the inflationary gap would be the difference between the equilibrium level of output and Y_{FE}. This would correspond to the situation in which the aggregate expenditure curve in Figure 12-12A intersects the 45-degree ray from the origin to the right of Y_{FE}. Although not shown, Figure 12-12A and B would reflect a departure from the full employment level of output, Y_{FE}.

To summarize, departures from full employment output (Y_{FE}) are classified as either an inflationary gap or a recessionary gap. These gaps exist when the following conditions hold:

$$\text{Recessionary gap} = Y_{FE} - Y_E \text{ when } Y_E < Y_{FE}$$
$$\text{Inflationary gap} = Y_E - Y_{FE} \text{ when } Y_E > Y_{FE}$$

One final point worth noting when looking for the existence of inflationary and recessionary gaps is that when the aggregate demand curve moves up the aggregate supply curve and approaches Y_{FE}, further increases in aggregate demand will increase the general price level in the economy more than it will increase aggregate output.

WHAT LIES AHEAD?

The forthcoming chapters will show how the aggregate demand and supply curves can be adjusted to eliminate inflationary or recessionary gaps. Chapter 13 focuses on the second condition for general equilibrium, namely, equilibrium in the money market, and the monetary policy options available to the Federal Reserve System. This chapter also focuses on taxes, government spending, and budget deficits. Chapter 14 addresses the nature of fluctuations in the economy and what the government can do about them. Chapter 15 illustrates what all this means for the price of food and the economic performance of farmers and ranchers along with other segments of the nation's food and fiber industry.

SUMMARY

The purpose of this chapter is to introduce the concept of general equilibrium, the relationship between sector output and GDP, and the physical and financial links between the farm business sector and the rest of the economy. The major points made in this chapter are summarized as follows:

1. A partial equilibrium assumes other things remain unchanged. In many situations, however, it is necessary to take into account events taking place simultaneously in other markets. When we do this, we are focusing on a general equilibrium.

2. The circular flow of payments in a barter economy consists of a two-way flow of goods and services between businesses and households. A grower of cabbage selling his or her product to a plumber would have to negotiate an exchange of goods and services that is equally fair to both parties (e.g., 25 cabbages per hour of plumbing services). The circular flow of payments in a monetary economy has the benefit of permitting the cabbage grower to receive money in exchange for his or her product, which he or she can then allocate across his or her purchases of other goods and services.

3. When measured in current prices, economists refer to these aggregate expenditures as nominal gross domestic product (GDP). If deflated by the implicit GDP price deflator, however, economists instead use the term real gross domestic product (GDP), when referring to aggregate output of the economy.

4. Consumer expenditures amount to more than two-thirds of total aggregate expenditures in the economy. These expenditures can be disaggregated into the following categories:

   - Expenditures on food and related products
   - Expenditures on nonfood, nondurable goods
   - Expenditures on durable goods
   - Expenditures on services

   Durable goods are products that are consumed over a period of time, such as cars, television sets, and a good pair of shoes. Nondurable goods are products such as food, gasoline, and newspapers. Services include items such as haircuts and air travel.

5. The slope of the aggregate consumption function is called the marginal propensity to consume (MPC). This coefficient suggests what the change in consumption expenditures will be if disposable personal income changes. The marginal propensity to save (MPS) suggests what the change in saving will be if disposable personal income changes. Thus, $MPS = 1.0 - MPC$.

6. Other factors influencing the level of consumption expenditures, besides the level of income, are the level of taxes paid by consumers, the level of their real wealth, and their expectations about their future financial position.

7. The slope of the aggregate investment function is known as the interest sensitivity of investment, or the change in investment expenditures, given a change in the market rate of interest.

8. The determinants of planned investment expenditures in new business assets by businesses or new residences by households include factors such as the purchase price of new capital goods, the expectation of future income by businesses and households, technological change, and the level of taxes.

9. The *Keynesian cross* refers to the determination of the equilibrium level of output in the nation's product market by using a 45-degree ray out of the origin in a graph with aggregate expenditures on the vertical axis and aggregate output on the horizontal axis. The point at which this ray crosses the economy's aggregate expenditure curve indicates the equilibrium level of output.

10. The economy's aggregate supply curve reflects its willingness to supply goods and services at alternative levels of product prices. This curve has the following three ranges:

    a. A depression range, sometimes called the *Keynesian range*, after British economist John Maynard Keynes

    b. A normal range that looks like market supply curves

    c. A classical range, which is named after the classical school of thought that suggests that the aggregate supply curve is completely unresponsive to active demand-oriented macroeconomic policies

11. An inflationary gap occurs in an economy when planned aggregate expenditures exceed its full employment level of output. A recessionary gap occurs when the opposite is true; namely, aggregate expenditures are less than the economy's level of full employment output.

KEY TERMS

Barter
Barter economy
Consumption
Consumption function
Disposable income
Equilibrium output
Final demand

Full employment GDP
General equilibrium
Gross domestic product (GDP)
Inflationary gap
National income
National product
Nominal GDP

Partial equilibrium analysis
Potential GDP
Real GDP
Recessionary gap
Saving

TESTING YOUR ECONOMIC QUOTIENT

1. Referring to the following graph, place a T or F in the blank appearing next to each statement:

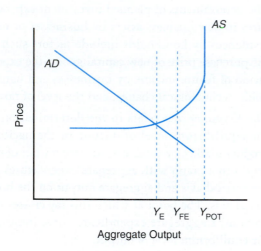

Aggregate Output

_____ Y_{FE} represents the economy's maximum output in the short run.

_____ This economy is experiencing a recessionary gap.

_____ Supply expansion policies are recommended to eliminate this gap in the short run.

_____ Demand expansion strategies could be used to eliminate this gap.

2. Assume that disposable income increases in the economy from $4,000 to $5,000 and causes consumer expenditures to increase from $4,300 to $5,200.

   a. What is the economy's marginal propensity to consume?

   b. Are households in this economy saving or dis-saving?

   c. What is the marginal propensity to save in the economy?

   d. If taxes were to increase by $500, what effect would this have on consumer expenditures?

3. Suppose the aggregate investment function for the economy is given by

$$I = \$400 - 40(R)$$

where I represents the level of investment expenditures and R represents the market rate of interest. Further assume the current rate of interest is 10% ($R = 10.0$).

   a. How much will investment expenditures in the economy change if interest rates fall by two percentage points?

   b. Would the level of autonomous investment change? Why?

   c. What factors might cause the investment curve given by the investment function mentioned earlier to shift to the left?

   d. What factors might cause the interest rate in this economy to decrease by two percentage points?

4. Given the following model for planned consumption expenditures, answer the following questions:

$$C_t = 140 + 0.80(Y_t - T_t)$$

where

C_t Total planned consumption expenditures in year t

Y_t Personal income in year t

T_t Total personal taxes due in year t

G_t Total government spending in year t

and where

$Y_t = 900$

$T_t = 50 + 0.20(Y_{t-1})$

   a. If total personal income last year was $800, what is the level of total planned consumption expenditures in the current year (year t)?

   b. What is the level of saving out of current income in year t? Are consumers saving or dis-saving?

   c. What is the value of the marginal propensity to consume in year t?

5. Carefully illustrate the difference between a recessionary and an inflationary gap in the nation's product market. Make sure you fully label all features of your graph.

6. Define or graphically illustrate each of the following terms. Make sure you correctly label all graphs used in your answers:
   Classical range of aggregate supply curve
   Marginal propensity to save
   Autonomous consumption
   Marginal efficiency of investment
   Equilibrium condition for the nation's product market

7. Given the following model for planned aggregate expenditures in a closed economy, answer the following questions:

   $C_t = 150 + 0.90(Y_t - T_t)$
   $I_t = 250 - 10(i_t)$
   $G_t = 250$
   $T_t = 50 + 0.10(Y_{t-1})$
   $D_t = 50$

   where

C_t	Total planned consumption expenditures in year t
Y_t	National income in year t
T_t	Total personal taxes due in year t
G_t	Total government spending in year t
D_t	Depreciation in year t
i_t	The current market rate of interest

   a. What is the equilibrium level of national output in the current year (year t) if the market rate of interest is 6% and national income in the previous year (year $t - 1$) was 3,000? (Hint: use 6.0 instead of 0.06 as the value for the market rate of interest.)
   b. What is the level of consumption in the current year (year t)?
   c. What is the level of savings in the economy in the current year?

8. The term *marginal propensity to consume*
   a. refers to the quantity of the last item consumed.
   b. is equal to consumption divided by disposable income.
   c. refers to the proportion of the marginal dollar of disposable income that is consumed.
   d. none of the above.

9. Given the following graph for the aggregate product market:

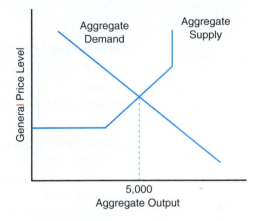

   a. If the full employment level is $6,000, is the economy experiencing a recessionary or an inflationary gap? Why?
   b. If the full employment level is $4,000, is the economy experiencing a recessionary or an inflationary gap? Why?
   c. If the full employment level is $5,000, is the economy experiencing a recessionary or an inflationary gap? Why?

10. Given the following information, what is the break-even level of disposable income?
    Marginal propensity to consume = 0.85
    Autonomous consumption = $1,500

11. Given the information from the previous question, what is the level of consumption and saving if the level of disposable income is $7,500?

12. How would your answer to the previous question change if the level of disposable income was $13,000?

REFERENCES

Economic Report of the President, Washington, D.C., various issues, U.S. Government Printing Office.

Keynes JM: *A general theory of employment, interest and money,* London, 1936, Macmillan Publishing.

13

Macroeconomic Policy Fundamentals

Chapter Outline

It is difficult to imagine a large, highly integrated economy like that of the United States without the existence of money. Money is perhaps the most important invention created by human beings. Bartering is both time consuming and costly. Can you imagine a manufacturer of combines trying to exchange its product for the goods and services needed to manufacture additional combines? Workers certainly would not want to be paid off in combines, nor would the utility company that is supplying electrical power to the firm.

Our willingness to accept money made of paper or coins made of specific metals allows households and businesses in various regions of the country to specialize in the goods or services they provide despite a local non-coincidence of wanted goods. Given the critical role played by money in modern economies, it is important that we understand the factors that influence its demand and supply.

The purpose of this chapter is to discuss the characteristics of money, including the functions it performs; the Federal Reserve System, which is commonly called the Fed, and its role in our monetary economy; and the determinants of equilibrium in the money market. We will also discuss the federal budget, the financing of annual federal budget deficits, and the growth of the national debt.

CHARACTERISTICS OF MONEY

If you stopped people on a street corner and asked them to define the term *money*, they would probably define it as the coins in their pocket and the paper bills in their wallet. The typical definition of money, however, includes more than just currency. At a minimum, it also includes checking account balances, which are referred to as demand deposits.[1]

Functions of Money

Money can be defined by the functions it performs. We have already discussed its role as a **medium of exchange**. Money allows us to specialize in endeavors in which we have a comparative advantage by facilitating payments to others for the goods and services they have provided and by facilitating payments from others for our labor.

Money also serves as a **unit of accounting**; that is, it provides a basis with which we can compare the relative value of goods and services in the economy. For example, we have discussed the level of gross domestic product and national income in terms of billions of dollars. Physical measures of aggregate statistics such as these would be confronted with the age-old problem of "adding apples and oranges." Money also serves as the unit of accounting ranging from assessing the profitability and net worth of a business to establishing a household budget.

Money is also an asset and, as such, has a **store of value**. You could store all your wealth in condominiums, for example. One problem with this is that condominiums are not liquid assets; that is, they cannot always be quickly converted to cash to pay for an unexpected bill without cost (realtor fees) or a loss in market value. Money is "cash" and can readily be used to meet an unexpected bill. It is partially for this reason that households and businesses store at least part of their wealth in money.

Money serves a variety of roles in a monetary economy, including a medium of exchange, a unit of accounting, and a store of value.

Money versus Near Monies

If we define money by the functions it performs, currency and demand deposits definitely would be considered as money. This—along with such things as traveler's checks—constitutes the narrowest definition of money, a definition labeled M1 by the Federal Reserve System. Until the 1980s, only commercial banks could offer demand deposits. Today there are a variety of checking-type accounts offered by many financial institutions. The 1980 Depository Institutions Deregulations and Monetary Control Act expanded the definition of M1 balances to include these additional checkable accounts and demand deposits at other institutions such as mutual savings banks. M1 balances are also referred to as "narrow money."

Other forms of assets may be considered **near monies** because of their relatively high liquidity, or how quickly they can be converted to cash without penalty. Small time deposits, which must be held for a specific time before they can be withdrawn without penalty, fall into this category. Savings accounts are particularly liquid. Although a 30-day notice of intent to withdraw funds from a savings account may be required, such notice in practice is rarely required. Shares in money market funds, which typically include holdings of government and corporate bonds, can also be

Near monies are those financial assets that can be converted to cash quickly with little or no penalties.

---

[1] Do not confuse money with a credit card. While use of a credit card may circumvent the need for paper money to complete a transaction, a credit card transaction is really a loan from the bank or other institution issuing the credit card. When you retire part or of your credit card balance, you are repaying the loan.

classified as near money. These funds typically allow individuals to write checks on the share they own in the fund.

The Federal Reserve System includes these near monies together with those assets previously captured in M1 in their M2 definition of money. Certain specialized overnight assets are also included in M2.[2] Certificates of deposit at all depository institutions issued in denominations of $100,000 or more are the major factors distinguishing the Federal Reserve System's M3 definition of money.[3]

Backing of Money

A fiduciary monetary system refers to monetary systems like the United States where the value of money is tied to the confidence the public has in its acceptability in completing transactions. Before 1957, silver certificates were backed by silver.

What makes a piece of paper with green ink and a few numbers on it acceptable to us as payment for our labor prices? Monetary economies such as that of the United States rest on what is termed a **fiduciary monetary system**; that is, the value of money rests on the public's belief that a piece of paper with green ink on it can be exchanged for goods and services. Take a dollar bill from your wallet or purse. Is there a promise to exchange the dollar for a specific quantity of gold? No.

There are three principal reasons why money in a fiduciary monetary system has a positive value. First, money is acceptable by others when purchasing goods and services. Second, demand deposits and currency have been designated as legal tender by the federal government. Most paper money will contain the phrase "this note is legal tender for all debts public and private" somewhere on the note; thus, it must be accepted for payment of debts. Third, money also is a predictable value in nominal terms. A dollar is worth a dollar, today and tomorrow. Although the purchasing power of a dollar will decline in periods of inflation, it will always be worth a dollar in nominal terms.

FEDERAL RESERVE SYSTEM

Congress passed the Federal Reserve Act, and it was signed into law by President Woodrow Wilson on December 23, 1914. The Federal Reserve System began operation in the fall of 1914 as the U.S. central bank. Its principal domestic goals were to encourage economic growth and combat both inflationary and recessionary tendencies in the domestic economy. The Federal Reserve accomplishes these goals by regulating the money supply. More recently, the Fed has targeted a specific interest rate known as the federal funds rate. The purpose of this section is to discuss the organization of the Federal Reserve System, the function it performs, its relationship to the banking system in this country, and the policy instruments that it has at its disposal.

Organization of the Federal Reserve System

The Federal Reserve System is organized around a Board of Governors located at the system's headquarters in Washington, D.C. The Board of Governors consists of seven full-time members appointed by the President and approved by the U.S. Senate. The Board of Governors sets monetary policy and exerts general supervision over the 12 district Federal Reserve Banks, which are located in selected major cities in the United States. The locations of these district banks and their 25 branches are shown in Figure 13-1.

The board determines reserve requirements, reviews and determines the discount rate, orders the buying or selling of government securities, determines margin

---

[2] Overnight assets include overnight repurchase agreements issued by commercial banks and overnight Eurodollar deposits held by U.S. nonbank residents at Caribbean branches of U.S. banks.

[3] Term repurchase agreements issued by commercial banks greater than 1 day and savings and loan associations are also included in M3. The Federal Reserve System also publishes a definition of liquid assets, which includes everything in M3 and financial instruments such as banker's acceptances, commercial paper, and liquid Treasury obligations. See Board of Governors of the Federal Reserve System, Federal Reserve Bulletin, Washington D.C., various issues, U.S. Government Printing Office.

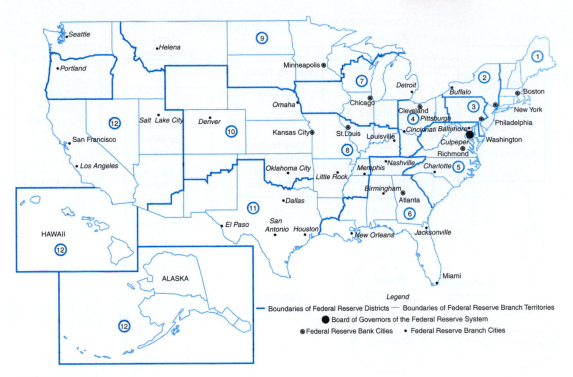

Figure 13-1
District bank boundaries in the Federal Reserve System.

requirements on stock market credit, and in the past determined the ceilings on the yields paid by banks to their depositors (these ceilings have now been phased out). Each governor on the board serves one 14-year term and cannot be associated in any other way with banking during his or her term. In addition, no two governors may come from the same Federal Reserve district. The board has a chairman and a vice-chairman. Both are appointed by the President of the United States to serve their 14-year terms. These officers may be reappointed over the course of their 14-year term on the board.

The **Federal Open Market Committee (FOMC)**, which is composed of the seven members of the Board of Governors and five representatives of the district banks serving on a rotating basis, meets periodically to determine the desired future growth of the money supply and other important issues. This committee issues directives to the manager of the System Open Market Account, which is located in New York and is responsible for seeing that the appropriate buying or selling actions are taken in the government securities market.

The **FOMC** plays an important role in the implementation of monetary policy within the Federal Reserve System.

Each of the 12 district Federal Reserve Banks is a federally chartered corporation. Its stockholders are the member banks in that district. These member banks pledge up to 6% of their own capital and surplus in reserve stock. The district banks are not profit-maximizing businesses, however. Instead, their objective is to service and discipline member commercial banks and other depository institutions in their districts. Each district bank is managed by a nine-member Board of Directors serving in staggered terms. The role of these directors is to appoint the president of the district bank (with the approval of the Board of Governors) and to serve in an advisory capacity to the bank on economic and financial conditions in the district.

There were 5,441 banks in 2015 in the United States, down from the 14,000 banks in 1934. Unlike 1934, however, there were roughly 100,000 bank branches in 2009, although this number dwindled to nearly 90,000 in 2015 as technology has streamlined banking relationships. **National banks**, which receive their charter from the Comptroller of the Currency, are required to be members. The remaining banks are **state banks**. They

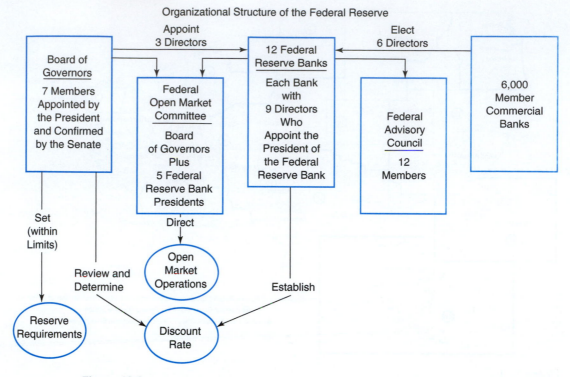

Figure 13-2

The Federal Reserve System consists of the Board of Governors, the Federal Open Market Committee, the 12 district Federal Reserve Banks, and the member banks located throughout the United States.

receive their charter from their respective state governments but have been elected to be members of the Federal Reserve System for one reason or another. Usually, this decision is based on weighing the cost of membership (e.g., the requirement of non-income-producing reserve requirements) and the returns (e.g., the ability to borrow from the Federal Reserve's discount window and the use of its check-clearing facilities and teletype wires to transfer funds).[4] Figure 13-2 illustrates the organizational structure of the Federal Reserve.

Functions of the Federal Reserve System

The Federal Reserve System's functions are to

- supply the economy with paper currency,
- supervise its member banks,
- provide check collection and clearing services,
- maintain the reserve balances of depository institutions,
- lend to depository institutions,
- act as the federal government's banker and fiscal agent, and
- regulate the money supply.

The first of the Fed's functions is to supply the economy with paper currency called Federal Reserve notes through the district banks. Every dollar bill has the words *Federal Reserve Note* printed on it. Each district Federal Reserve Bank must have enough paper currency to accommodate the demands for money in its district. Although this paper currency is printed in Washington by the Bureau of Printing and Engraving, it bears the code of the originating district bank.

---

[4] The Depository Institutions Deregulation and Monetary Control Act of 1980 virtually eliminated any differences between member and nonmember banks in the eyes of the Federal Reserve System. This act extended reserve requirements to *all* depository institutions (commercial banks, savings and loan associations, credit unions, etc.), and made its services available to all these institutions.

The Federal Reserve System also supervises its member banks, as do the Comptroller of the Currency and the Federal Deposit Insurance Corporation. Among the things that the examiners look at are the types of loans made; the backing, or collateral, for these loans; and who borrowed the funds.

A third function performed by the Federal Reserve System is to provide check collection and clearing services. All member banks and depository institutions can send deposited checks to their district Federal Reserve Bank, which competes with private clearinghouses for their business. Suppose that Ralph Rancher, who owns a farm in the Kansas City Fed's district, travels to San Francisco to attend a convention. While in San Francisco, Mr. Rancher writes a check to Wally's Wharf restaurant for $50. Assume that Wally deposits these funds in his checking account at his commercial bank in San Francisco. Wally's bank would then deposit the check in its reserve account at the San Francisco district Federal Reserve Bank. The San Francisco Fed would then send this check to the district Federal Reserve Bank in Kansas City, which would deduct $50 from the reserve account of Ralph's bank and then send this check to Ralph's bank. The final step sees Ralph's bank deducting $50 from his checking account and sending the canceled check to Ralph. He can look at the back of his canceled check to see this route his check has taken.

Another function performed by the Federal Reserve System is to maintain the reserve balances of depository institutions in each district bank as required by law. Depository institutions are required to keep a certain fraction of their deposits on reserve.

The Federal Reserve System also lends to depository institutions. The interest rate charged on these loans is called the **discount rate**. Banks with a heavy seasonal demand for loans such as rural commercial banks, which make a significant volume of loans to farm businesses, may qualify for special borrowing privileges.

The Federal Reserve System also acts as the federal government's banker and fiscal agent. The U.S. Treasury has a checking account with the Fed. In addition, the Fed helps the federal government collect tax revenues from businesses and aids in the purchase and sale of government bonds.

Finally, the Federal Reserve System attempts to regulate the money supply in an effort to promote economic growth and price stability. This function has received national attention in recent years of historically low interest rates.

The discount rate is the interest rate banks pay when they borrow reserves through the "discount window" at one of the 12 Federal Reserve District Banks.

Monetary Policy Instruments

The Federal Reserve System has three major monetary policy instruments it can employ to regulate the growth of U.S. money supply. These instruments are (1) changes in reserve requirements, (2) changes in the discount rate, and (3) changes in the direction or magnitude of its **open market operations**. Over the 2010–2014 period, the Fed also conducted additional operations to buy both government- and mortgage-backed securities under a program called *quantitative easing*, or simply QE.

Reserve Requirements The Federal Reserve System alters the supply of money by changing the amount of reserves in the banking system. One way to accomplish this is to change the required reserves at depository institutions, which are required to maintain a specific fraction of their customers' deposits as reserves. Total reserves can be divided into three categories.

1. **Legal reserves** for member banks consist of deposits held at the institution's district Federal Reserve Bank plus vault cash.[5]

---

[5] Nonmember banks and other depository institutions may treat as reserves their deposits with a correspondent depository institution holding required reserves, with the Federal Home Loan Bank (savings and loan associations), or with the National Credit Union Administration central liquidity facility (credit unions), as long as these reserves are passed on to a Federal Reserve bank.

2. **Required reserves** represent the minimum weekly average legal reserves that a depository institution must hold. This requirement is expressed as a ratio. The minimum and maximum reserve requirement ratios for depository institutions are changed infrequently.

3. **Excess reserves** represent the difference between total reserves and required reserves. The level of excess reserves determines the extent to which depository institutions can make loans.

Thus, manipulation of the reserve requirement ratio will either expand or contract the level of excess reserves at depository institutions and affect their ability to make new loans. We will see shortly how this translates into expansion or contraction of the money supply under normal economic conditions.

Discount Rate   As indicated earlier, the discount rate represents the interest rate the Fed charges for lending reserves to depository institutions. If a depository institution wants to increase its loans but does not have any excess reserves at the moment, it can borrow reserves from the Fed at its "discount window."[6] The Fed does not have to lend all the reserves these institutions need every time they want them, however.

As an alternative, banks can borrow funds on a short-term basis to meet reserve requirements. The **federal funds market** is an interbank market trafficking in reserves. Banks with excess reserves can lend to banks that are short on a 24-hour or 48-hour basis at what is known as the federal funds rate.

Open Market Operations   The third way the Federal Reserve System can alter the volume of reserves at depository institutions is by the sale or purchase of government securities. The directive by the Federal Open Market Committee (FOMC) to sell government securities results in a decrease in reserves at depository institutions. Deposits are withdrawn from these institutions to pay for the securities, which lowers the level of reserves in the banking system and hinders its ability to make loans.

When the FOMC issues the directive to buy government securities, for example, an increase in reserves in the banking system occurs. These open market operations are the most frequently used monetary policy instrument. Buying and selling of government securities influences fluctuations in the *federal funds rate*, or the rate of interest at which banks lend money to one another on a very short-term basis.

Other Instruments   The Fed also uses several other policy instruments to regulate the expansion of credit in the U.S. economy. The Fed, for example, determines the margin requirements (i.e., down payment required) for loans made by stockbrokers to customers desiring to purchase stock, and the maximum rates banks can pay depositors on specific types of deposits. Another action involves what is called quantitative easing.

The "Great Recession" during 2007–2009 discussed earlier in this chapter, which stemmed in part from the financial crisis beginning in 2006 and the bankruptcy of Lehman Brothers in 2008, was addressed by expansionary monetary policy actions undertaken by the Federal Reserve. Interest rates were already at historical lows. The real effective federal funds rate which banks use when lending overnight to one another was, for all intents and purposes, at near zero. The Federal Reserve had to use nonconventional methods to stimulate the economy, such as *quantitative easing* (QE). QE is used by central banks like the Federal Reserve to add liquidity in the economy when conventional tools targeting interest rates become ineffective. Such

---

[6] These reserves are referred to as *borrowed reserves*. Excess reserves minus these borrowed reserves are often referred to as *free reserves*.

actions are often considered a last-resort effort to stimulate a sluggish economy. This involved purchasing toxic mortgage-backed securities during 2010–2014 from commercial banks and other financial institutions. This can expand the quantity of money in the economy by increasing the amount of excess reserves in banks, if they are used to make loans. This also raises the prices of financial instruments like government bonds purchased by the Federal Reserve, which lowers their yield. The reduction in yields on longer term bonds and other financial instruments affects senior citizens who depend on interest income in their retirement programs.

The QE initiative undertaken by the Federal Reserve involved purchasing government bonds and mortgage-backed securities of a longer maturity with the goal in mind of lowering longer-term interest rates to bolster confidence and stimulate business investment and the housing market. The major risk in these operations is overshooting the needs of the economy, which can lead to inflation when the economy recovers. The challenge is how fast to mop up these excess reserves when this occurs.

CHANGING THE MONEY SUPPLY

A change in reserves in the banking system from the use of any one of the three monetary policy instruments will affect the ability of depository institutions to make loans. How does this affect the money supply? To address this question, we must first discuss the process of money creation. Then we can discuss how changes in the three major monetary policy instruments shift the supply curve for money.

Creation of Deposits

An important relationship between the level of reserves in the banking system and the money supply exists. In the discussion to follow, we will focus on the M1 definition of the money supply: coins and currency in circulation plus checkable deposits at depository institutions. Part of this money supply consists of coins and currency, which are physical units that can be carried around in a purse or wallet. The checkable deposits in M1, on the other hand, are merely entries in an account at depository institutions.

New deposits can be created, and the money supply expanded, by increasing the level of excess reserves in the banking system. If the level of excess reserves at depository institutions is zero, there can be no further expansion of the money supply. No new loans can be made at that point. To understand this, examine the balance sheet for a hypothetical bank, Bank Ag, in Table 13-1.

Table 13-1 shows how deposits can be expanded when excess reserves are greater than zero. Assume that the bank's assets consist entirely of reserves and loans, and its liabilities consist entirely of deposits. If the reserve requirement ratio is 0.20, this bank must hold at least $2 million of its $10 million of deposits in reserves. If Bank Ag already had outstanding loans of $8 million, as shown in Table 13-1, the bank

Table 13-1
BANK AG'S BALANCE SHEET BEFORE $1 MILLION DEPOSIT

Assets		Liabilities	
Reserves		Deposits	$10,000,000
required	$2,000,000		
Excess	0		
Total	$2,000,000		
Loans	8,000,000		
Total	$10,000,000	Total	$10,000,000

Table 13-2
BANK AG'S BALANCE SHEET AFTER $1 MILLION DEPOSIT

Assets		Liabilities	
Reserves		Deposits	$11,000,000
required	$2,200,000		
Excess	800,000		
Total	$3,000,000		
Loans	8,000,000		
Total	$11,000,000	Total	$11,000,000

would be fully "loaned up." The bank cannot increase its loan volume further because it has no excess reserves.

Assume that the depositors at this bank sell $1 million in government securities to the Federal Reserve and deposit the checks that they receive from the Federal Reserve System in Bank Ag (Table 13-2). As a result of these deposits, Bank Ag's total deposits would increase to $11 million, and its required reserves would have to increase to $2.2 million (i.e., its initial $2 million required reserves plus 20% of its new $1 million deposit). Assuming that it has not had time to make any new loans, the bank would have excess reserves of $800,000.

Banks under normal economic conditions desire to keep their excess reserves as low as possible for profit-maximizing reasons. Excess reserves represent idle balances that are earning no interest income.[7] Assume that Bank Ag lends its entire excess reserves of $800,000. The bank would then be fully loaned up again and would remain that way until new activities altered its balance sheet position.

Now assume that the proceeds of the loans made by Bank Ag become deposits in another bank. Assume the $800,000 lent by Bank Ag shows up as deposits in Bank B.[8] Given a required ratio of 20%, Bank B must now hold $160,000 in additional reserves (i.e., 0.20 × $800,000) and can now make loans of $640,000 (i.e., $800,000 − $160,000).

Table 13-3 shows how the initial deposit of $1 million in Bank Ag has been used by the banking system to expand the deposits that appear in the M1 definition of

Expansion of the money supply occurs as an initial purchase of government bonds by the Federal Reserve and gets "multiplied up" through the commercial banking system.

Table 13-3
CHANGE IN DEPOSITS IN THE BANKING SYSTEM AFTER $1 MILLION DEPOSIT

Bank	Change in Deposits	Change in Loans	Change in Reserves
Ag	$1,000,000	$800,000	$200,000
B	800,000	640,000	160,000
C	640,000	512,000	128,000
D	512,000	409,600	102,400
E	409,600	327,680	81,920
F	327,680	262,144	65,536
G	262,144	209,715	52,429
H	209,715	167,772	41,943
I	167,772	134,218	33,554
Subtotal	$4,328,911	$3,463,129	$865,782
Other banks	671,089	536,871	134,218
TOTAL	**$5,000,000**	**$4,000,000**	**$1,000,000**

---

[7] For reasons of safety and liquidity, banks normally hold some idle balances.
[8] This process could occur easily without all the loan proceeds showing up in different banks.

the money supply. If we look at the totals at the bottom of this table, we see that the deposits of the entire banking system have increased by a multiple of the initial $1 million change in reserves at these depository institutions brought about by the depositors of Bank Ag selling $1 million in government securities to the Federal Reserve.

If you divide the change in deposits ($5 million) by the change in reserves ($1 million) at the bottom of Table 13-3, you will see that the change in the money supply was five times greater than the change in reserves. From this example, we can make a generalization about the extent to which the money supply will increase when the banking system's reserves are increased.

If we assume that banks will minimize their holdings of excess reserves and that all the proceeds from loans are deposited in the banking system (none are hidden in tin cans in the backyard), we may assert that

$$\text{MM} = 1.0 \div \text{RR} \tag{13.1}$$

where MM represents the money multiplier and RR represents the fractional reserve requirement ratio. This equation suggests that if the **money multiplier** is equal to the reciprocal of the 20% required reserve ratio, the money multiplier would be 5 (i.e., $1.0 \div 0.20$).

The money multiplier tells us how much the change in reserves will change the money supply.

We may use the definition of the money multiplier in Equation 13.1 to calculate the level of U.S. money supply as

$$\begin{aligned} M_S &= (1.0 \div RR) \times \text{TR} \\ &= \text{MM} \times \text{TR} \end{aligned} \tag{13.2}$$

in which M_S represents the nominal money supply and TR represents the level of total reserves in the economy. The change in the money supply is given by

$$\Delta M_S = \text{MM} \times \Delta \text{TR} \tag{13.3}$$

Equation 13.3 suggests that if the reserve requirement ratio is 0.20 and there is an increase of $1 million in total reserves in the economy, U.S. money supply will change by

$$\begin{aligned} \Delta M_S &= \text{MM} \times \Delta \text{TR} \\ &= 5.0 \times \$1 \text{ million} \\ &= \$5 \text{ million} \end{aligned} \tag{13.4}$$

or a multiple of five.[9] This is identical to the total change in the money supply reported at the bottom of Table 13-3.

Under normal economic conditions, the money multiplier normally ranges between 2.0 and 3.0. This means a $1 million increase in reserves would increase the money supply by $2 million to $3 million. Several factors may cause the money multiplier to fall below the reciprocal of the required reserve ratio. These factors include **currency drains** (sometimes called leakages), which refers to the desire of individuals to hold currency rather than deposit funds in a bank or other depository institution. When this occurs, funds remain outside the banking system. Banks' desire to hold excess reserves seen since the 2007–2009 recession and global financial crisis also affects this multiplier. The bulk of the reserves resulting from the QE actions has been held in reserve rather than used to make loans.

If the depositors at Bank Ag received $1 million from the sale of securities to the FOMC and chose to bury the funds in the backyard, the level of deposits at Bank Ag would not change and the money supply would not expand by $4 million. Thus, the greater the incidence of currency drains, the smaller the multiplier will be. A bank's desire to hold idle excess reserves for reasons of safety and liquidity will also lower the

---

[9] The reverse of this conclusion is also true; that is, a $1 million decrease in reserves under normal conditions will reduce the money supply by $5 million, if the money multiplier is 5.0.

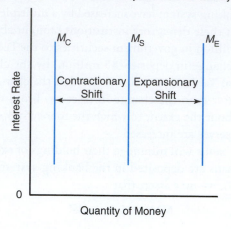

Figure 13-3

Expansionary monetary policy actions (i.e., the Federal Reserve buys government bonds, lowers the discount rate, or decreases the required reserve ratio) will lead to increases in the money supply, which is illustrated here by the outward shift in the money supply curve. Contractionary monetary policy actions (i.e., the Federal Reserve sells government bonds, raises the discount rate, or increases the required reserve ratio) will lead to decreases in the money supply, which will cause an inward shift in the money supply curve.

value of the money multiplier. The greater the level of desired excess reserves, the lower the money multiplier will be. The substantial excess reserves resulting largely from the Fed's QE operations compromise the money multiplier effect.

Monetary Policy and the Money Supply

Now that we understand how a change in reserves affects the money supply, let's return to the Fed's monetary policy instruments and examine their effect on the money supply curve. The money supply curve M_S pictured in Figure 13-3 reflects a policy that is invariant or unaffected by the level of interest rates; that is, a change in interest rates is assumed to have no effect on the money supply.

Expansionary Applications A change in open market operations, a change in reserve requirements, and a change in the discount rate can all affect the money supply. Table 13-4 lists the expansionary effects that monetary policy instruments have on U.S. money supply.

When the Fed purchases government securities in the open market, it increases the level of total reserves at depository institutions, which will cause the money supply curve in Figure 13-3 to shift to the right. The new money supply under normal economic conditions would be equal to the money multiplier times the new higher level of total reserves.

Lowering the discount rate will make it cheaper for depository institutions to borrow from the Fed and should increase the level of total reserves in the banking

Expansionary (contractionary) money policies will shift the money supply curve to the right (left).

Table 13-4
EXPANSIONARY EFFECTS ON U.S. MONEY SUPPLY

Expansionary Actions	Effects of Action
Fed purchases securities in the open market	Increases total reserves
Fed lowers the discount rate	Increases total reserves
Fed reduces the required reserve ratio	Increases the money multiplier

> **Table 13-5**
> **CONTRACTIONARY EFFECTS ON U.S. MONEY SUPPLY**
>
> | Contractionary Actions | Effects of Action |
> | --- | --- |
> | Fed sells securities in the open market | Reduces total reserves |
> | Fed increases the discount rate | Reduces total reserves |
> | Fed increases the required reserve ratio | Decreases the money multiplier |

system, which will also cause the money supply to shift outward to the right. The new money supply under normal economic conditions would again be found by multiplying the existing money multiplier by the new level of total reserves.

A reduction of reserve requirements forces depository institutions to hold a smaller fraction of their deposits in reserves, which increases the money multiplier under normal economic conditions (see Equation 13.1). A reduction of reserve requirements is also an expansionary application of monetary policy and would cause the money supply curve to shift from M_S outward toward M_E. The new money supply under normal economic conditions can be found by multiplying the existing level of total reserves times the new higher money multiplier.

Contractionary Applications   In contrast to the expansionary applications of the Fed's major monetary policy instruments, these same instruments can be used to contract the money supply. Table 13-5 lists the contractionary effects that monetary policy instruments have on the nation's money supply.

When the Fed sells securities in the open market to private investors, it reduces the total reserves in the banking system, which leads to a multiple decrease in the money supply. Therefore, the money supply curve in Figure 13-3 will shift back to the left, from M_S toward M_C.

An increase in the discount rate by the Federal Reserve makes it more costly for banks to borrow reserves through the Fed's discount window. An increase in the discount rate also reduces total reserves and leads to a multiple decrease in the money supply.

An increase in the reserve requirement ratio under normal economic conditions reduces the money multiplier (see Equation 13.1), which causes existing total reserves to expand by a smaller multiplier than before. If the Fed increases its reserve requirements at a time when all depository institutions are fully loaned up (i.e., each has zero excess reserves), these institutions will either have to call in loans or decrease their investments to be in compliance.

MONEY MARKET EQUILIBRIUM

Why do individuals prefer to hold part of their wealth in money rather than place all their wealth in income-earning assets? What happens to the interest rate if the demand for money increases or decreases? To answer these questions, we must understand the demand for money.

Demand for Money

Economists often identify three reasons why we hold money. The first represents our **transaction demand for money**. Households and businesses hold a certain amount of money because it is a widely accepted medium of exchange. Because their receipts of money income do not match the timing of their expenditures, households and businesses maintain some holdings of money to help finance these expenditures. The cost of not having transaction balances is the rate of interest households and businesses would have to pay when borrowing funds to finance these expenditures.

There are a number of reasons why we demand money, including the need for money to pay for goods and services.

The second reason we hold money represents our **precautionary demand for money**. Some households may wish to hold money when the prices of other assets are falling because the nominal value of money is fixed. One hundred dollars in cash will always be worth $100, but $100 invested in common stock could be worth less than that tomorrow. In a recessionary economy when asset prices are falling, the **speculative demand for money** should increase.

Thus, the demand for money reflects the liquidity preferences of households and businesses. The lower the rate of interest, the lower the opportunity costs of holding cash will be, and the larger the quantity of money demanded should be. The larger the level of disposable personal income, the greater the amount of cash balances held in the economy.

Assume that the general nature of the aggregate demand for money for Lower Slobovia's economy can be expressed in equation form as

$$M_D = c - d(i) + e(NI) \tag{13.5}$$

in which

$$d = \Delta M_D \div \Delta i \tag{13.6}$$

$$d = \Delta M_D \div \Delta NI \tag{13.7}$$

where c represents the autonomous quantity of money demanded, d reflects the interest sensitivity of the demand for money and the slope of the demand curve, i represents the market rate of interest, e reflects the income sensitivity of the demand for money, and NI represents the level of national income, which also equals the nation's gross domestic product in the national income and product accounts.

Figure 13-4A illustrates the inverse relationship between the demand for money and the rate of interest. When the returns on assets in general and interest rates in particular fall, households and businesses will attempt to substitute cash for such financial assets as stocks and bonds.

The Keynesian notion that the demand for money is inversely related to interest rates and positively related to national income is somewhat at odds with other theories of the demand for money. The Cambridge demand for money function, which was developed by classical economists at Cambridge, England, suggests that the

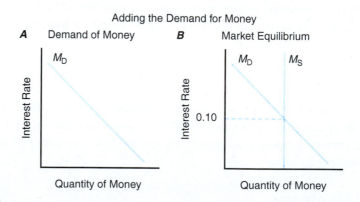

Figure 13-4

Before we can determine the market rate of interest in the economy, we must know the demand for money. (A) The demand for money function is downward sloping; that is, the higher (lower) the rate of interest, the lower (higher) the quantity of money demanded. This interest rate reflects the opportunity cost of holding money rather than an income-earning asset. The higher the expected income forgone by holding money, the lower the quantity of money demanded will be. (B) The market rate of interest is determined by the intersection of the money demand and supply curves. Thus, an increase in the money supply will lower interest rates in the short run, and a reduction in the money supply will raise interest rates.

demand for money is equal to some constant fraction of the nominal level of national income. This is the simplest of all monetarist positions. Modern monetarists such as Milton Friedman argue that the demand for money is a function of alternative rates of return, nominal national income, and the expected rate of inflation.

Equilibrium Conditions

An equilibrium will occur in the nation's money market when the supply curve given by Equation 13.2 intersects the demand curve given by Equation 13.5, or when

$$M_S = M_D \tag{13.8}$$

Figure 13-4B brings together the demand and supply curves for money. The intersection of the M_D curve and the M_S curve determines the equilibrium rate of interest, which would be 10%.

EFFECTS OF MONETARY POLICY ON THE ECONOMY

We now know that the money supply affects the level of interest rates. A shift to the right of the money supply curve in Figure 13-4B would lower the interest rate below 10%, and a shift to the left would raise the market interest rate.

Transmission of Policy

Figure 13-5 summarizes the mechanism through which changes in monetary policy are transmitted to the nation's aggregate demand. The change in policy first shows up in the money market, where it affects interest rates. This, in turn, affects the equilibrium in the product market through investment and, through the income multiplier, the equilibrium level of aggregate demand and national income. Because the demand for money is affected by changes in national income, we must account for the simultaneous interaction between the money and product markets when determining the general equilibrium level of income and interest rates.

The discussion of the aggregate demand curve for Lower Slobovia in figure 12-9 assumed a given market rate of interest. Changes in the market rate of interest resulting from the demand and supply forces operating in the nation's money market render the assumption of a fixed interest rate obsolete. **Expansionary monetary policy** actions that lower the rate of interest will spur investment expenditures, expand national income, and shift the economy's aggregate demand curve to the right. The opposite chain of events will occur if contractionary monetary policy actions are followed.

Expansionary monetary policy leads to lower interest rates, which stimulates investment and expands economic activity, and closes recessionary gaps.

Effects of Change in Monetary Policy on GDP

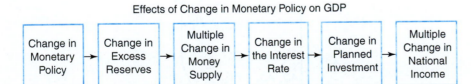

Figure 13-5

This figure illustrates the chain of events that takes place when a change in monetary policy is made. A change in monetary policy will directly affect the reserves that depository institutions such as banks are required to hold. This, in turn, will affect the banks' excess reserves and therefore their ability to make new loans, which ultimately affects the size of the money supply. An increase (decrease) in the money supply will put downward (upward) pressure on interest rates. A decline (rise) in interest rates will encourage (discourage) further investment expenditures by businesses in the economy.

Combating Recessionary Gaps

The monetary policies of the federal government can be used to reduce unemployment. To combat a sluggish economy and high unemployment, policymakers use expansionary monetary policies that lead to an increase in the money supply, lower interest rates, and an increase in credit being available to consumers and producers. Monetary policies that reduce unemployment include

- the Federal Reserve purchasing government securities in the open market, which increases the amount of reserves in the banking system,
- lowering the Federal Reserve's discount rate, which also increases the amount of reserves in the banking system, and
- lowering the Federal Reserve fractional reserve requirements, which under normal economic conditions increases the money multiplier.

An increase in total reserves in the banking system and the money multiplier will lead to an increase in the money supply over a period of time if the increased reserves are used to make loans. The increased credit availability and lower interest rates will expand consumption and investment expenditures, which will increase the need for more workers to help produce these additional goods and services.

To illustrate how expansionary demand-oriented macroeconomic policies can eliminate a recessionary gap, let's examine the product market equilibrium in Figure 13-6. As consumption, investment, and government expenditures increase in response to the expansionary monetary policies, the aggregate demand curve will shift outward to the right. If aggregate demand shifts from AD_1 to AD_2, aggregate output will increase from Y_1 to Y_2.

Additional workers will also be needed to produce this additional output. Because there was plenty of slack in the economy, these policies were not inflationary (i.e., prices remained at P_0 because the AS curve was perfectly elastic over this range of the AS curve). Finally, because of this increase in aggregate demand, the recessionary gap would be reduced from $Y_{FE} - Y_1$ to $Y_{FE} - Y_2$.

Expansionary monetary policy leaving the economy in the Keynesian range of the aggregate supply curve will not lead to inflation. A similar expansion in the classical range will lead to hyperinflation.

If aggregate demand is stimulated to the point at which the economy moves out to AD_3, aggregate output will increase to Y_3. The general price level would rise, reaching the level P_3. The rate of inflation resulting from this action would be the percentage change from P_0 to P_3, and the rate of growth in real GDP would be the percentage change from Y_2 to Y_3. The recessionary gap would be further reduced to $Y_{FE} - Y_3$, but not totally eliminated.

Figure 13-6

Monetary and fiscal policies can be used to expand or contract aggregate demand to promote economic growth and price stability.

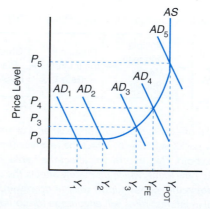

Elimination of Recessionary and Inflationary Gaps

Only if aggregate demand were expanded to AD_4 would aggregate output achieve full employment Y_{FE}, and the recessionary gap be completely eliminated. The general price level would reach P_4, which is somewhat inflationary, but does not swamp the growth in aggregate output from Y_3 to Y_{FE}. The increase in aggregate demand to AD_4 therefore expands the economy to full employment, or the point at which it achieves the lowest unemployment rate and highest capacity utilization rate without triggering a significant increase in inflation.

If aggregate demand was expanded to AD_5, the economy would reach its potential capacity to supply goods and services in the current period (Y_{POT}), and while additional output was forthcoming, any benefits would be swamped by the increase in the general price level from P_4 to P_5. This inflation would lower the real wealth of consumers and thus self-correct aggregate demand back toward AD_4, over time. The Federal Reserve would also likely pursue contractionary monetary policies to bring the AD curve back toward AD_4.

Combating Inflationary Gaps

Monetary policies, as suggested above, can be designed to reduce inflation. Contractionary monetary policies that lead to a decrease in the money supply, rise in interest rates, and decrease in credit available to consumers and producers combat inflation by lowering consumption and investment expenditures in the economy. Monetary policies that reduce inflation include

- the Federal Reserve selling government securities in the open market, which decreases the amount of reserves in the banking system,
- raising the Federal Reserve's discount rate, which also decreases the amount of reserves in the banking system, and
- raising the Federal Reserve fractional reserve requirements, which decreases the money multiplier under normal economic conditions.

This reduction in demand will put a downward pressure on prices in general. Unfortunately, these policies will also increase unemployment. This was precisely the tact taken by the Federal Reserve in 1980 when the rate of inflation reached double-digit levels. When inflation rates fell in the early 1980s, the annual unemployment rate began to rise, eventually reaching 10%. This inverse relationship and apparent sensitivity of employment to contractionary monetary policies raise the question of what would happen to unemployment if policymakers attempted to drive the inflation rate to zero, as some members of the Federal Reserve Board of Governors have argued in the past. Growth in productivity since the mid-1990s has led to the situations where *both* low inflation and low unemployment were achieved.

To illustrate, let's look again at Figure 13-6. Assume that aggregate demand in the economy is represented by the AD_5 curve. If contractionary monetary policies brought about a shift in this demand curve from AD_5 to AD_4, the aggregate demand curve would intersect the aggregate supply AS curve at full-employment output Y_{FE}. The general price level would fall substantially from P_5 to P_4 without causing a significant reduction in aggregate output. If policymakers unwittingly continued to dampen demand to the point where the demand curve shifted back to AD_3, the general price level would fall further to P_3. But aggregate output would fall below Y_{FE} to Y_3, thereby reducing the employment of the nation's resources to below full-employment levels and creating a recessionary gap.

Microeconomic Perspectives

Interest rates have an impact upon investments in new equipment and buildings by businesses and investments in new residences by households. Contractionary monetary policies that drive up interest rates will depress investment expenditures by

Combating inflationary gaps will shift the aggregate demand curve back down the aggregate supply curve, which is the opposite of what happens when the Federal Reserve follows an expansionary policy.

Table 13-6
EFFECTS OF A VARYING INTEREST RATE ON A 10-YEAR LOAN

Rate (%)	Annual Total Payment (%)	Annual Interest Payment ($)	Total Interest Payment ($)
8	22,354.69	7,354.69	73,546.90
14	28,757.67	13,757.67	137,576.68
20	35,782.44	20,782.44	207,824.40

businesses and households. Conversely, expansionary monetary policies that lower interest rates will stimulate investment expenditures in the economy. We can demonstrate how changes affect an individual business or household by examining the impact that specific interest rate levels have on the size of the loan payment and total interest expense for business and household loans.

Suppose a farmer is considering the purchase of a new combine and related equipment and seeks to finance $150,000 over a 10-year period in ten equal annual installments. Table 13-6 shows how varying interest rates affect the value of the annual loan payments and total interest paid over the entire loan period. The data in Table 13-6 suggest that the annual total loan payment (principal and interest payments) increases significantly as **nominal interest rates** rise significantly. These payments must be met annually from the revenue earned by the business and, hopefully, out of the additional revenue generated by this new asset. If we think of the combine and related equipment as the only capital used by a business providing custom harvesting services, the idea is clearer. Ignoring the cost of fuel, labor, and other related expenses, the business's revenue would have to rise in accordance with the annual interest and principal payments if it is to continue to meet its commitments to its banker.

Suppose the business expected to earn a revenue of $30,000 above all other expenses. Although the average annual interest payment at a 20% interest rate could "service its debt," or meet its interest payment, it could not make its total loan payment of $35,782.44.

Suppose a household is considering entering into an agreement to purchase a new residence now under construction. This household needs to borrow $100,000 to finance the purchase of this new home. Assume the household can obtain a 20-year loan calling for monthly payments. Table 13-7 shows how varying interest rates affect the size of the monthly mortgage payments and total interest paid over the entire loan period. If this household could incorporate no more than a $900 monthly mortgage payment into its budget, it could work the monthly mortgage payment into its budget if the interest rate were 8%, but not if the interest rate were 12%. In the latter instance, the monthly mortgage payments would be $266.95 higher than under the lower interest rate. In both instances, higher interest rates may lead to lower levels of investment expenditures—or at least postponed investment expenditures by businesses and households.

Table 13-7
EFFECTS OF A VARYING INTEREST RATE ON A 20-YEAR MORTGAGE

Rate (%)	Monthly Total Payment ($)	Monthly Interest Payment ($)	Total Interest Payment ($)
8	848.78	432.08	103,707.46
12	1,115.73	699.06	167,773.46

THE FEDERAL BUDGET

The annual **federal budget** is one of the largest publications printed each year by the Government Printing Office. The annual federal budget sets forth the President's spending and revenue plans for the coming year. Although some politicians summarily say that the budget is "dead on arrival," the budget often represents a starting point for further negotiations. Often this is due to the overly optimistic assumptions underlying the revenue projections.

Federal Expenditures

Total spending by the federal government, or **federal expenditures**, has grown dramatically over the post–World War II period. Nominal government spending grew from $42.4 billion in 1950 to approximately $4 trillion in fiscal year 2014. A substantial part of this growth can be attributed to inflation. Real government spending in fiscal year 2014 was approximately 70% of the nominal value. Real per capita government spending more than doubled over the 1970–2014 period.

Fiscal policy is carried out at the federal level by Congress and the President. The Federal Reserve, beyond providing advice, has no control over these policies.

The largest single expenditure item in the federal government's fiscal year 2014 budget, presented in Table 13-8, was Social Security benefit payments (24.264%), followed by national defense (17.21%). Medicare and income security programs amounted to 14.59% and 14.65% , respectively. Agricultural programs accounted for less than 1.0% of total federal expenditures, down more than one-half from the 2.6% of the federal budget spent on agriculture in 1987. Total government spending in the United States in fiscal year 2014 amounted to approximately 18.29% of U.S. GDP.[10]

Although many would argue that government spending in the United States is too high and should be cut, it is substantially lower than that in other developed countries. Total government spending is typically 60% of GDP in Sweden, 55% of GDP in the Netherlands, 45% of GDP in the United Kingdom, and 43% of GDP in Germany.

The federal government located in Washington, D.C., is the home of U.S. macroeconomic policy. The Federal Reserve is the home of monetary policy while Congress and the President formulate fiscal policy.   Credit: bbourdages/Fotolia.

---

[10] The U.S. federal government's fiscal year begins on October 1 and ends the following September 30.

Table 13-8
THE FEDERAL BUDGET FOR FISCAL YEAR 2015

Item	Billion dollars	Total Percentage
Receipts:		
Individual income taxes	1,394.6	46.16%
Corporate income taxes	320.7	10.61%
Social insurance and retirement receipts	1,023.5	33.87%
Other	282.7	9.36%
Total receipts	3,021.5	100.00%
Expenditures:		
National defense	603.5	17.21%
International affairs	46.7	1.33%
Health	409.4	11.68%
Medicare	511.7	14.59%
Income security	513.6	14.65%
Social security	850.5	24.26%
Net interest	229.0	6.53%
Other	341.7	9.75%
Total expenditures	3,506.1	100.00%
Surplus or deficit	−484.6	

Source: Economic Report of the President, The Council of Economic Advisors, 2016.

The federal government expenditure totals in Table 13-8 conform with the National Income and Product Account (NIPA) accounting rules followed when measuring U.S. GDP. The complete set of receipt and expenditure transactions of the federal government represents one sector in the set of NIPAs maintained by several government agencies.[11]

Federal Receipts

Believe it or not, no federal income tax existed before 1913. **Federal receipts** from taxes on liquor and tobacco and tariffs on imported goods were enough to offset government expenditures. Before 1913, government spending was less than 10% of U.S. GDP. By comparison, government spending by the federal government in 2015 was 18.29% of GDP.

Figure 13-7*A* shows that the growth in federal revenue was approximately equal to the growth in federal spending until the mid-1970s when the economy experienced the most severe recession since the Great Depression of the 1930s. By the late 1970s, the growth in federal revenue and expenditures converged again until the 1980s, when federal revenues fell because of the 1981–1982 recession and the Economic Recovery Tax Act of 1981, which slashed personal income tax rates. Federal receipts exceeded spending during the 1998–2001 period but have fallen below spending since then. The Tax Reform Act in 1986 during the Clinton administration, the Economic Growth and Tax Relief Act of 2001 and the Jobs and Growth Tax Relief Reconciliation Act of 2003 during the Bush administration, and the American Recovery and Reconciliation Act of 2009 and American Tax Payer Relief Act of 2012

---

[11] The U.S. Department of Commerce maintains the NIPAs, which measure the nation's gross national product (GNP) and national income. The Federal Reserve System maintains a national flow of funds account, which accounts for the flow of funds between agriculture, government, and other sectors of the economy. The U.S. central bank also measures a national wealth or balance sheet account, which captures the assets and liabilities of agriculture and other sectors in the economy.

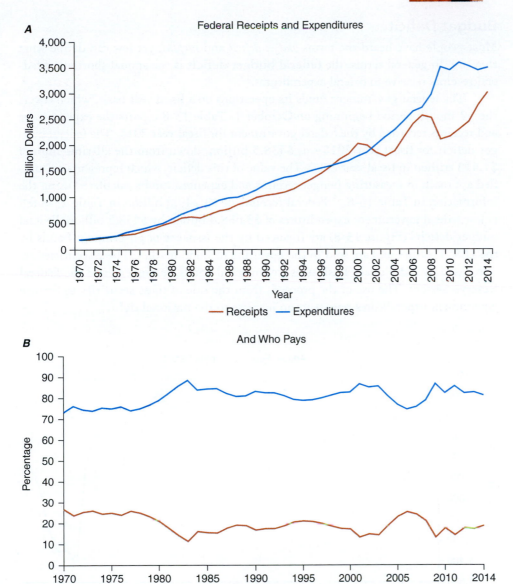

Figure 13-7

(A) A relatively slower growth in nominal federal receipts relative to expenditures occurred over the last five decades. (B) Households are responsible for paying the majority of tax revenue collected by the federal government.

during the Obama administration are examples of enacted fiscal policy legislation that attempted to deal with economic and budgetary issues existing at that time.

The vast majority of federal government revenue comes from tax payments made by households. In fiscal year 2015, 47.41% of federal receipts were in the form of individual income tax payments. Another 32.78% were in the form of Social Security contributions.[12] Corporate income taxes were 10.58%. Indirect business taxes, which are taxes that businesses treat as expenses (i.e., licensing fees, business property taxes, federal excise taxes, customs duties, and sales taxes), and others accounted for 9.23%.

---

[12] Individuals contribute a specific percentage of their wage and salary income through payroll taxes, with their employer matching this amount.

Budget Deficit

The **federal budget deficit** is equal to the amount by which federal expenditures exceed federal receipts. It represents the change in national debt during the year.

Most people have heard the terms *budget deficit* and *surplus*, yet few can define what they are. In general terms, the **federal budget deficit** is the annual shortfall of federal receipts relative to federal expenditures.[13]

The federal government funds its operations on a fiscal year basis, which covers the 12-month period beginning on October 1. Table 13-8 reports the expenditures and receipts recorded by the federal government for fiscal year 2015. The federal budget deficit for fiscal year 2015 was $438.5 billion, down from the all-time high of $1.413 trillion in fiscal year 2009. The value of this deficit, which represents the unified approach to measuring budget receipts and expenses, can be calculated using the information in Table 13-8.[14] Federal receipts of $3,249.9 billion in Table 13-8 fell below federal government expenditures of $3,688.4 billion by $438.5 billion. Federal budget deficits (Figure 13-8) are financed by the issuance of government bonds by the U.S. Treasury and are sold in the government bond market. These securities are purchased by individuals, businesses, commercial banks, foreigners, and the Federal Reserve System. The use of the proceeds from the sale of these securities to finance government expenditures represents an increase to the **national debt**.

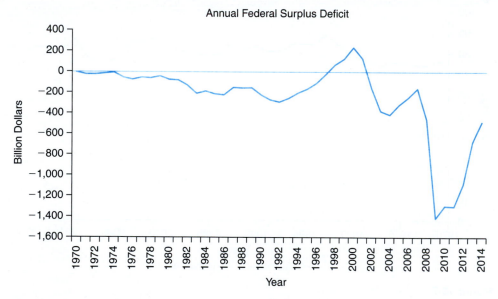

Figure 13-8

This figure illustrates the dramatic growth in the annual federal budget deficit over the past 10 years. The deficit reported here is based on the National Income and Product Account (NIPA) approach. The deficit began to rise in the mid-1970s when the economy suffered a relatively major recession. The 1980s saw the annual budget deficits rise to then unprecedented highs. Reductions in annual deficits in the mid-1990s and surplus in the late 1990s reflected growth in the economy and budget-neutral government spending. The horrific events of September 11, 2001, and subsequent changes in government spending and tax cuts caused deficits to return. The financial crisis that began in 2006 led to the Great Recession during 2007–2009, which, in turn, led to unprecedented levels of federal government spending to avoid what might otherwise have been a depression.

---

[13] The opposite of a federal budget deficit is a federal budget surplus. Because the United States had not experienced a federal budget surplus since 1998, it is commonplace to focus on the size of deficits rather than surpluses.

[14] There are at least four different perspectives on how to define the federal budget deficit. Until recent years, these differences did not lead to substantially different deficit numbers. Today, it is possible to hear one policymaker say the budget deficit is $250 billion and another say the budget deficit is $300 billion when including off-budget items, with both telling the "truth."

THE NATIONAL DEBT

The national debt is (approximately) the sum of the outstanding federal government securities upon which interest and principal payments must be made. The federal budget deficit is the annual addition to the national debt. If the federal budget deficit is $100 billion, the national debt will grow by $100 billion at year end.

The national debt makes headlines. It first rose to a significant level during World War II, reaching about $300 billion. It remained at that level for more than two decades. The national debt rose somewhat during the Vietnam era. Before the 1980s, growth in the national debt was tied largely to either wars or recessions. The growth in national debt in the 1980s, however, reflected the impact of cuts in personal income taxes that included lower marginal tax rates, indexing taxes for inflation, and other changes that contributed heavily to the growth in national debt. Corporate tax rates were also cut.

From 1776 to 1980, a period of 204 years, the United States accumulated its first trillion dollars in national debt. It recorded its second trillion dollar indebtedness by 1984 and its third trillion dollar indebtedness by the end of the decade. By 2005, it had reached $7.7 trillion. As a result of the financial crisis in 2004, the federal government dramatically increased spending. At the end of 2015, the national debt approached $19 trillion. This growth alarms some economists. Should we be concerned? Is the United States going bankrupt?

There are several approaches to evaluating the size of the national debt. We can study the relationship between national debt and GDP, and interest payments and GDP, and judge the ownership of national debt to determine if we are burdening future generations.

National Debt and GDP

The growth in national debt illustrated in Figure 13-9 suggests a federal financing strategy that is out of control. Or does it? Looking at this figure alone, one could certainly find cause for concern. Per capita debt figures tell the same story. In 1950, the ratio of the national debt to the U.S. population would suggest that every man, woman, and child theoretically owed approximately $1,500 to holders of U.S. government bonds. By 2015, this figure had risen to over $60,000 per capita! Whether the public debt is too large can be judged by the same rules one uses to judge private debt—namely, by the size of an individual's existing debt and the individual's earning potential. We can evaluate the size of the national debt by examining the ratio of national debt to gross domestic product or national income.

Figure 13-10 shows that although the trend in the ratio of national debt to GDP has been going in the wrong direction, the levels of these annual percentages are in line with levels experienced in the years just after World War II.

Concern over the size of the national debt should be predicated to the size of U.S. gross domestic product.

Ownership of National Debt

At the end of 2015, the Federal Reserve held 13% of total U.S. Treasury securities outstanding in its portfolio. About 62% of outstanding federal government securities were held by private investors. The more prominent domestic investors were depository institutions (3%), pension funds (3%), mutual funds (6%), and state and local governments (3.4%). About 34% of U.S. federal government securities is held by foreign governments and foreign private investors. This particular percentage is important for one reason: interest payments made to domestic investors as well as state and local governments are payments to ourselves. They are not a drain on the nation's resources, but rather a transfer of income from one group of our society

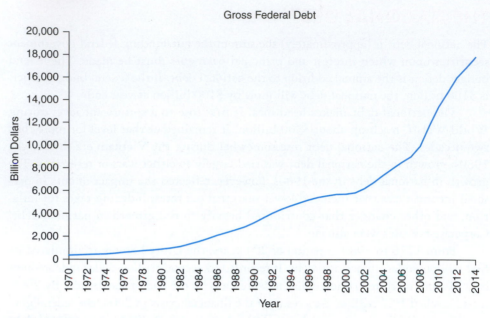

Figure 13-9

This figure illustrates the dramatic growth in national debt beginning in the 1990s. This growth has caused some to question U.S. financial soundness. Are we going broke? Is today's generation overburdening future generations? And who owns this debt?

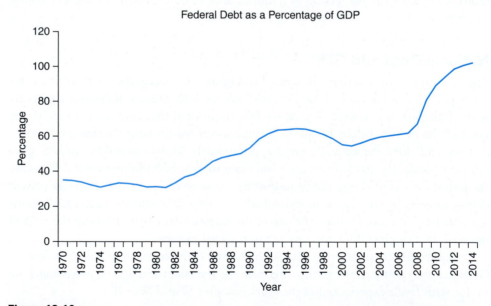

Figure 13-10

National debt has grown dramatically since the early 1980s, but this should not necessarily be alarming if U.S. income and output are growing also. This figure shows that national debt as a percentage of GDP has grown substantially as a result of the 2007–2009 recession and the policies used to address the recession. Federal debt as a percentage of GDP had risen earlier in the 1980s when personal and corporate income tax rates were slashed, without making similar reductions in government spending. The flattening and eventual decline in this ratio in the 1990s reflects the budget-neutrality rules practiced with federal government spending.

(taxpayers) to another (bondholders and those who have ownership claims on firms that hold bonds). Interest payments to international investors, on the other hand, are a drain on the nation's resources.

The fact that 34% of the national debt is held by foreigners, and that portion of interest payments leaves the country, suggests that the drain on the national resources can be large if interest rates rise.

Burdening Future Generations?

Finally, there is the matter of the growth in national debt representing a burden on future generations. There are several perspectives one can take on this issue. Do you expect major corporations to retire all the stock and debt they accumulated when expanding their operations? The answer is, of course, *no*. These firms transcend generations, and they need never retire all capital claims on the business. The real issue is whether these firms can service the claims on the business (dividends to stockholders, interest payments to bondholders, and scheduled interest payments to lenders) from the proceeds of their ongoing operations.

The same thing can be said for the federal government. We should not expect to retire the national debt, but instead keep a healthy relationship between it and the nation's income, which means covering the growth interest costs on the national debt with increased tax revenues. With respect to the principal outstanding, the federal government pays off bondholders when their bonds mature, but it does so by refinancing this debt through selling new securities. This is exactly what large firms such as General Motors do.

One way in which today's growing national debt can represent a burden to future generations is through something called *crowding out*. Crowding out occurs when an increase in government spending either replaces private spending or pushes up interest rates and crowds out some private investment that would have taken place at lower interest rates.

The first type of crowding out is **direct crowding out**, in which the government activity supplants private business activity on a dollar-for-dollar basis. Some examples are government expenditures on rail service, health care, and parks that could be operated by the private sector.

The second type of crowding out is **indirect crowding out** and is of particular interest to us in the present context. When the U.S. Treasury sells new government bonds to finance federal budget deficits, it increases the demand for funds in the nation's money markets and drives up interest rates.[15] These higher rates may mean that some businesses may suddenly find that scheduled investment projects based on a cost of capital of 8% may no longer be feasible from an economic standpoint if the cost of capital jumps to 10%. Figure 13-11 summarizes the chain of events leading to crowding out.

Crowding out, therefore, results in a smaller capital stock of manufacturing assets passed on to future generations and fewer future jobs associated with the operation of these investment goods.[16] This also may translate into lower future productivity growth than otherwise would have been the case if these investments were targeted for modernization of manufacturing operations that capture new technologies.

When the Treasury borrows heavily in the nation's capital markets, it drives up interest rates and "crowds out" businesses that might otherwise borrow capital to finance business expansion.

[15] U.S. government securities are viewed as one of the safest investment outlets in the world. The U.S. Treasury typically has no difficulty in finding buyers for its new security issues, although it may have to offer a higher interest rate to buyers to meet its needs. The issuance of new government securities by the U.S. Treasury does not change U.S. money supply. When the buyers of the bonds write out a check to the U.S. Treasury, the demand deposits at the banks at which these checks are written are reduced. The effect at the national level is neutralized, however, when the government spends these funds and the proceeds from the transaction wind up back in the commercial banking system. The Federal Reserve can purchase additional securities (it is prohibited from purchasing new issues but purchases of securities in the open market have the same effect) to hold down interest rates and avoid serious indirect crowding out in the economy. This, of course, does represent an increase in the money supply, which might later prove to be inflationary.

[16] We can also think of the effects of crowding out on the economy by recalling the production possibilities curve. Instead of boxes of canned fruit and boxes of canned vegetables, we can talk of broad sets of goods and services (e.g., guns and butter). The greater the nation's productive capital stock, the more guns and more butter we can produce and the more the production possibilities curve will lie to the right of the economy's existing curve. Crowding out negates this expansion.

```
┌──────────┐    ┌──────────┐    ┌──────────┐    ┌──────────┐    ┌──────────┐
│Government│    │Government│    │ Treasury │    │          │    │   Some   │
│ Spending │    │  Budget  │    │  Sells   │    │ Interest │    │ Planned  │
│ Exceeds  │───▶│ Deficit  │───▶│   More   │───▶│  Rates   │───▶│Investments│
│  Taxes   │    │  Occurs  │    │  Bonds   │    │   Rise   │    │"Crowded Out"│
│          │    │          │    │          │    │          │    │by High Rates│
└──────────┘    └──────────┘    └──────────┘    └──────────┘    └──────────┘
```

Figure 13-11
Chain of events leading to crowding out.

FISCAL POLICY OPTIONS

Fiscal policy refers to the act of changing government spending and/or taxes to achieve specific policy goals, such as promoting economic growth and stability. To achieve these goals, Congress and the President have certain fiscal policy instruments at their disposal. These fiscal policy instruments can be divided into two distinct categories:

1. **Automatic fiscal policy instruments** are policies that take effect without an explicit action by the president or Congress.
2. **Discretionary fiscal policy instruments** are policies that require the legislative and executive branches of government to take explicit policy actions that would not occur otherwise (e.g., pass a law).

In the discussion to follow, we will assume that the relevant goal for fiscal policy is to eliminate inflationary and recessionary gaps.

Automatic Policy Instruments

Automatic fiscal policy instruments like the income tax code that works 24 hours a day, 7 days a week.

Several types of automatic fiscal policy instruments, which are constantly at work in the U.S. economy, act as built-in stabilizers, automatically watering down the punch bowl just when the party suddenly picks up steam (i.e., the economy suddenly enters into a boom period) and automatically spiking the punch when the party is moving too slowly (i.e., economic activity suddenly starts to slow down).

The progressive income tax system in the United States serves the function of a built-in stabilizer. When taxable income goes up in a boom period, the marginal tax rate also increases. Consider the tax liability of the six different families, each consisting of four people and each having $13,000 in income tax deductions, in Table 13-9. The family of four with an adjusted gross income of $10,000 would owe $1,000 in federal income tax. The family of four with an adjusted gross income of $30,000 has a

Table 13-9
TAX LIABILITY FOR A PROGRESSIVE INCOME TAX SYSTEM[a]

Adjusted Gross Income Over	But Not Over	Tax Base	Marginal Tax Rate (Percentage)	Of Amount Over
0	15,650	0	10.00	0
15,650	63,700	1,565	15.00	15,650
63,700	128,500	8,773	25.00	63,700
128,500	195,850	24,973	28.00	128,500
195,850	349,700	43,831	33.00	195,850
349,700		94,601	35.00	349,700

[a]The tax paid by the family having a taxable income of $77,000 is equal to $6,457 plus 28% of the taxable income exceeding $43,050. The tax paid by the family with $137,000 in taxable income is equal to $23,538 plus 31% over $104,050.

Source: Internal Revenue Service, U.S. Department of the Treasury.

federal tax liability of $3,698. And the family of four with an adjusted gross income of $150,000 would have a tax liability of $30,744. Each of these families fell into a different marginal tax bracket because of its income level. This represents a progressive tax system because, when the families' taxable income rose, the rate at which their income was taxed also rose. This automatic progressive tax rate structure causes aggregate demand in the economy to be less than it would have been if all families of four were taxed at the same flat rate of 20%.

If the economy suddenly slowed down and wages and salaries fell, taxable income would fall and thus be taxed at a lower marginal rate than before. As a result, aggregate demand would not fall by as much as it would have if all families of four were taxed at the same flat rate. In short, our progressive income tax system acts as a built-in stabilizer, buffering the economy from any sudden changes in economic activity.

The program of unemployment compensation in this country also serves automatically to stabilize aggregate demand. When economic activity slows down, workers who have been laid off automatically become eligible to receive unemployment compensation, which keeps their disposable income from dropping to zero until they find other work. When economic activity is strong, however, there is less unemployment, which means that fewer unemployment payments would be necessary.

Both built-in stabilizers tend to dampen the effects of sudden changes in disposable income and, hence, shift in the equilibrium level of national income. Because automatic fiscal policy instruments may remove only part of a recessionary or an inflationary gap in the economy, their use is normally complemented by other policy actions.

Federal government expenditures rose dramatically relative to receipts during the 2007–2009 recession, often referred to as the "Great Recession" due to its severity. As the economy began to slowly recover beginning in 2010, federal government receipts began to rise as personal and business incomes rose, but federal government spending continued to expand due to rising defense and social program spending. It became clear that a policy involving higher income tax rates and fees, combined with cuts in selected government programs, was required to address the abnormally high budget deficits and growth in national debt stemming from the 2007–2009 recession. This could eventually result in higher marginal tax rates, particularly at higher income levels, than those depicted in Table 13-9. It could also be accompanied by policymakers addressing the cost of various federal spending programs, including selected entitlement programs given the rising costs of health care and social services.

Discretionary Policy Instruments

The passage of a law cutting taxes and an increase in government spending authorized by Congress are examples of discretionary fiscal policy instruments or actions. Both involve specific actions for these fiscal policies to take effect.

Discretionary fiscal policy instruments require explicit acts such as passage of laws and regulations to take effect.

Discretionary Tax Policies   There have been 19 significant federal tax cuts passed since the end of World War II according to the U.S. Treasury Department. Let's look at the last three pieces of major discretionary tax policies.

The first major piece of tax legislation that cut taxes was the Economic Recovery Tax Act (ERTA) of 1981 passed during the Reagan administration. This act sought to reduce individual taxes by providing across-the-board rate reductions and indexation of tax brackets, and introduced individual retirement accounts (IRAs) for virtually all workers, in which contributions were deductible from otherwise taxable income. Major changes were also made in the estate and gift tax rules to lessen the tax burden on smaller estates. It also included the modified accelerated cost recovery system, which allowed businesses to recover the cost of (i.e., depreciate) eligible assets more quickly than previously allowed. This act was accompanied by cuts in government spending for specific programs. The aim of this legislation was to lessen the tax burden on savings and investment and to promote recovery in a sluggish economy in the early 1980s. This tax cut amounted to $68.7 billion in real terms and increased the budget deficit by 2.8% as a percentage of national income. This act led to the largest overall tax reduction in U.S. history. The rise in federal budget deficits led to the Tax Equity and Fiscal Responsibility Act of 1982, which boosted federal revenues. The Tax Reform Act of 1984 substantially revised the federal tax system, and the Tax Reform Act of 1986 attempted to simplify the tax code.

The second major piece of tax legislation that cut taxes was the Economic Growth and Tax Reform Reconciliation Act of 2001 passed under the Bush administration shortly after the horrific events of September 11 reduced the marginal tax rates on taxable income, including lowering the top tax rate from 39.6% to 35% over a period of years, and created a 10% income tax bracket applied to the first $6,000 of taxable income for individuals ($12,000 for married couples filing jointly). Exemptions for children were raised from $500 to $700, eventually reaching $1,000 by 2010. This legislation also increased allowable IRA contributions from $2,000 to $5,000 (indexed for inflation) and reduced the top "death" tax rate from 55% to 45% and raised the exemption from taxes from $675,000 to $3.5 million. This tax cut amounted to $75.8 billion in real terms and decreased the budget deficit as a percentage of national income by 2.8%.

The third major tax legislation that cut taxes was the Jobs and Growth Tax Relief and Reconciliation Act of 2003, which was passed under the Bush administration. This was the third largest tax package in U.S. history. Among other things, this act reduced the long-term capital gains tax rate (investments longer than 1 year) from 15% to 10%. It also capped the tax rate on dividend income at 15%. Both features were designed to boost investment expenditures in the economy. This legislation also reduced the top-end tax rate on taxable income from 35% to 33%. Both the 2001 and 2003 acts passed during the Bush administration expired in 2010. Given the growing budget deficits and national debt in recent years, these lowered tax rates may be targets for change in the minds of some policymakers.

The Great Recession during 2007–2009 discussed earlier in this chapter, which stemmed in part from the financial crisis beginning in 2006 that led to the bankruptcy of Lehman Brothers in 2008, was also met with expansionary fiscal policy actions.[17] This included such discretionary government spending actions as the Troubled Asset Relief Program (TARP) auto industry program for the auto industry, the TARP bank investment program that involved capital purchases and asset guarantees, and the AIG stabilization effort, among others. In January 2001, the Congressional Budget Office (CBO) projected cumulative *budget surpluses* of $5.9 trillion through 2011. Instead, these government programs and the concurrent decline in the U.S. economy resulted in cumulative *budget deficits* totaling $19.0 trillion in 2015.

The Budget Control Act of 2011 called for substantial increases in taxes and cuts on government spending if efforts to address the growing federal budget deficit were not passed by December 31, 2012. This was referred to in the media as the fiscal

---

[17] Lehman Brothers investment bank borrowed heavily to fund its investing in housing-related assets. It faced huge losses tied to the subsequent subprime mortgage crisis due to its securitizing the underlying mortgages made by its subprime lender, BNC Mortgage. In the first half of 2008, Lehman Brothers stock lost 73% of its value. It filed for Chapter 11 bankruptcy protection in September 2008. Other banks, money market funds, and hedge funds held significant exposures in Lehman Brothers, which lead to significant contagion on a global scale. This contagion domestically included mortgage financer Freddie Mac (Federal Home Loan Mortgage Corporation) and Farmer Mac (Federal Agricultural Mortgage Corporation); the latter was put in a position of noncompliance with its minimum capital requirements.

cliff. The deadline, which was met at the last moment, called for increased taxes in the form of higher payroll taxes (6.2% for income up to $113, 700), a higher tax rate on individuals making more than $400,000 (couples making more than $450,000), which raises the top marginal tax rate from 35% to 39.5%, and higher taxes on investment income for those in the top income bracket. Little or nothing was agreed to on the government spending side in this legislation. This agreement, from a long-run perspective, reduces the size of the projected federal budget deficits, but did little to address U.S. debt, which currently exceeds $19 trillion.

Discretionary Spending Policies The annual formulation of the federal budget by the President and the spending programs authorized by Congress represent discretionary spending policy actions. Perhaps the most noteworthy discretionary spending policy action, the Balanced Budget and Emergency Deficit Control Act first passed in 1986 and later amended in 1987, has an automatic trigger involved. Otherwise known as the Gramm–Rudman–Hollings Act (or simply Gramm–Rudman), this act was an explicit response to the record levels of deficit spending taking place during the 1980s. The Gramm–Rudman act set forth annual maximum allowable federal budget deficits, with the ultimate goal of balancing the budget by 1993 (a zero deficit). While this target date was missed, the combination of zero-based budgeting, reductions in subsidies, lower interest rates and growth in the economy, and budget surpluses and how to spend them became the topic of discussion in 1998. The automatic trigger referred to here was the act's **sequester** feature for cutting spending whenever deficits were projected to exceed the target by more than $10 billion. These spending cuts were to be across the board with the exception of interest on the debt, Social Security benefits, and certain other entitlement programs. Congress typically enacts an across-the-board sequester by discretionary actions designed to either cut spending or raise (enhance) revenue. Congress in recent years has framed funding proposals by looking for budget offsets in existing programs. A sequester was also employed in 2013 to cut federal spending.

Issues likely to be debated in the coming years include the integrity of the Social Security Trust Fund, projected to be bankrupt in 2037, and specific spending acts targeted to such discretionary expenditures on defense programs versus infrastructure needs.

Fiscal Policy and Aggregate Demand

The application of fiscal policy has a direct effect on U.S. aggregate demand and, hence, on the equilibrium level of GDP and interest rates. This is different from monetary policy, which targeted the rate of interest. These policies can be applied either in an expansionary context or in a contractionary context.

Increasing government spending and cutting taxes are **expansionary fiscal policy actions**, something politicians like to do in election years. Table 13-10 summarizes the effects these actions have upon U.S. product markets. Both of these expansionary actions will shift the aggregate demand curve to the right.

Like monetary policy, fiscal policy can shift the aggregate demand curve to the right (left) to close recessionary (inflationary) gaps in the economy.

Table 13-10
EFFECTS THAT EXPANSIONARY ACTIONS HAVE ON U.S. PRODUCT MARKETS

Expansionary Actions	Effects of Actions
Lower income tax rates	Increases disposable income, which shifts the AD curve to the right
Raise government spending	A component of total spending (G), this shifts the AD curve to the left

> **Table 13-11**
> **EFFECTS THAT CONTRACTIONARY ACTIONS HAVE ON U.S. PRODUCT MARKETS**
>
> | Contractionary Actions | Effects of Actions |
> | --- | --- |
> | Raise income tax rates | Decreases disposable income, which shifts the AD curve to the left |
> | Cut government spending | A component of total spending (G), this shifts the AD curve to the left |

Far less popular with politicians, particularly in an election year, are **contractionary fiscal policy actions**. Instead of increasing spending back in their districts or cutting taxes, they are asked to cut spending and/or raise taxes.

Table 13-11 summarizes the effects of contractionary fiscal policy actions on U.S. product markets. Both of these contractionary actions will shift the aggregate demand curve to the left.

Combating Recessionary Gaps

The fiscal policies of the federal government can be designed to reduce unemployment. Policymakers can adopt expansionary fiscal policies to stimulate aggregate demand in the economy and promote the employment of workers currently seeking jobs. These policies would include reducing income taxes by lowering marginal income tax rates, increasing tax depreciation allowances, adopting other tax credits such as investment tax credits (which stimulate investment), and increasing government expenditures on goods and services.

Reducing income taxes increases the personal disposable income of consumers and retained earnings of businesses, which should lead to an expansion of consumption and investment expenditures. In Table 13-8, we saw that these two components of aggregate demand constitute more than 80% of U.S. GDP.

Increasing government expenditures directly stimulates aggregate demand. Government purchases of goods and services represent approximately 20% of U.S. GDP. To illustrate how these expansionary demand-oriented policies eliminate a recessionary gap, let's examine what happens to the product market equilibrium depicted in Figure 13-12.

Impact of overexpansion
Overexpansion of aggregate demand through expansionary fiscal policy (tax cuts and higher government spending) can lead to hyperinflation just as overexpansion of the money supply by the Federal Reserve.

If aggregate demand shifts from AD_1 to AD_2, aggregate output will increase from Y_1 to Y_2. Additional workers will also be needed to produce this additional output. Because there was plenty of slack in the economy, these policies were not inflationary (i.e., prices remained at P_0 because the AS curve was perfectly elastic over this range of the AS curve).[18] Because of this increase in aggregate demand, the recessionary gap would be reduced from $Y_{FE} - Y_1$ to $Y_{FE} - Y_2$.

If aggregate demand is stimulated to the point at which the economy moves out to AD_3, aggregate output will increase to Y_3. The general price level would begin to rise, reaching the level P_1. The rate of inflation resulting from this action would be the percentage change from P_0 to P_1, and the rate of growth in real GDP would be the percentage change from Y_2 to Y_3. The recessionary gap would be further reduced to $Y_{FE} - Y_3$, but not totally eliminated.

Only if aggregate demand were expanded out to AD_4 would aggregate output achieve full employment Y_{FE}, and the recessionary gap be completely eliminated. The general price level would reach P, which is somewhat inflationary, but does not

---

[18] See Chapter 12 for a further discussion of the complete aggregate supply (AS) curve.

Elimination of Recessionary and Inflationary Gaps

Figure 13-12

Fiscal policies can be used to expand or contract aggregate demand to promote economic growth and price stability.

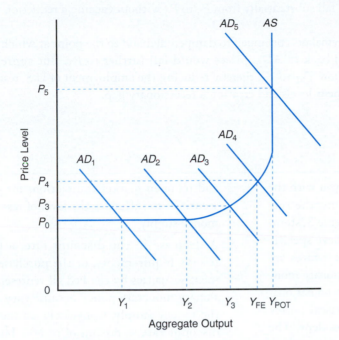

Aggregate Output

swamp the growth in aggregate output from Y_3 to Y_{FE}. The increase in aggregate demand to AD_4 therefore expands the economy to full employment, or to the point where it achieves the lowest unemployment rate and the highest capacity utilization rate without triggering a significant increase in inflation. When consumption, investment, and government expenditures increase in response to the expansionary demand-oriented policies, the aggregate demand curve will shift outward to the right. This leads to greater levels of output in the economy, expands the number of persons employed in the economy, and lowers the unemployment rate.

If aggregate demand were expanded out to AD_5, the economy would reach its capacity to supply goods and services in the current period (Y_{POT}), and while additional output was forthcoming, it would be swamped by the increase in the general price level from P_4 to P_5. This inflation would lower the real wealth of consumers and self-correct aggregate demand back toward AD_4. Congress and the President would also likely pursue contractionary fiscal policies to bring the AD curve back toward AD_4.

Combating Inflationary Gaps

Fiscal policies can also be designed to reduce inflation. Contractionary fiscal policies dampen aggregate demand in the economy and thus reduce existing inflationary pressures. These policies would include increasing income taxes by raising marginal income tax rates, by decreasing tax depreciation allowances, or by reducing or eliminating other tax credits like investment tax credits (which stimulate investment) and decreasing government expenditures on goods and services.

Increasing taxes will reduce the personal disposable income of consumers and retained earnings of businesses and retard the expansion of consumption and investment expenditures. Decreasing government spending on goods and services will also have a dampening effect on aggregate demand.

To illustrate, let's look again at Figure 13-12. Assume that aggregate demand in the economy is represented by the AD_5 curve. If contractionary fiscal policies brought about a shift in this demand curve from AD_5 to AD_4, the aggregate demand curve would intersect the aggregate supply curve at full-employment output Y_{FE}. Prices would fall substantially from P_5 to P_4 without causing a reduction in aggregate output.

If policymakers continued to dampen demand to the point at which the demand curve shifted back to AD_3, prices would fall further to P_3. But aggregate output would fall below Y_{FE} to Y_3, thereby reducing the employment of U.S. resources below full-employment levels and creating a recessionary gap.

SUMMARY

The purpose of this chapter is to acquaint you with the functions of money in a monetary economy, the role of the U.S. central bank (the Federal Reserve System) in regulating the supply of money to achieve specific national economic objectives, how interest rates are influenced by expansionary versus contractionary monetary policies, and the effect of interest rates on national income. The chapter also examines the fiscal policy options available to Congress and the President. The major points made in this chapter are summarized as follows:

1. Money serves as a medium of exchange, a unit of accounting, and a store of value. These functions help us define money. Money has value because it is widely acceptable in exchange for goods and services and because its nominal value is fixed.

2. The Federal Reserve System is comprised of the Board of Governors, the 12 district Federal Reserve Banks, and its member banks. An omnibus banking act in 1980 gave the Fed regulatory powers over the reserves held by all depository institutions in the United States, including savings and loan associations and credit unions.

3. The Fed regulates the supply of money to the private sector, holds the reserves required of depository institutions, provides a system of check collection and clearing, supplies fiduciary currency (Federal Reserve notes) to the economy, acts as the banker and fiscal agent for the Treasury, and supervises the operations of its member banks.

4. Money is created by the banking system through the multiple expansion of deposits. The full effect of an increase in reserves is found by multiplying these new reserves by the money multiplier, which under normal economic conditions is equal to the reciprocal of the reserve requirement ratio. The Federal Reserve has historically influenced the money supply by changing the reserve requirement ratios for deposits. It can also use the discount rate

and its open market transactions for government securities to influence the level of reserves and the money supply.

5. A decrease in the discount rate, a lowering of reserve requirements, or the purchase of government securities by the Fed all represent an expansionary monetary policy because they will expand the money supply. Conversely, an increase in the discount rate, a raising of reserve requirements, or the sale of government securities by the Fed all represent a contractionary monetary policy.

6. Monetary policy affects aggregate demand in the economy during the current period through the market rate of interest. The higher the rate of interest, the lower aggregate demand will be, and vice versa. The effectiveness of monetary policy will be reduced by the presence of currency drains and bank holdings of excess reserves.

7. The federal budget reflects the federal government's revenue expectations and expenditures for a specific calendar or fiscal year. Real government spending has grown steadily over the post–World War II period.

8. The difference between federal spending and expenditures is the federal budget deficit. There are several definitions of the budget deficit, including the deficit based on national income and product account concepts, and the unified budget deficit.

9. The national debt and the annual federal budget deficit are interrelated. The national debt at the end of the year is approximately equal to the national debt at the beginning of the year plus the federal budget deficit.

10. Two measures of the burden of the national debt are (1) the ratio of national debt to U.S. GDP and (2) the ratio of interest payments on the national debt to U.S. GDP.

11. Foreign investors own 34% of the national debt. State and local governments, banks, life insurance

companies, and individuals hold much of the balance of U.S. government securities.

12. The national debt does not necessarily represent a burden on future generations. Just as there is no reason to believe that General Motors should retire its indebtedness, there is no need for the national government to retire its debt. It is important that the relationship between the national debt and national income and output remain healthy.

13. Crowding out because of rising interest rates associated with larger federal budget deficits can retard investment expenditures and slow the growth of the economy's productive capital stock.

14. Automatic fiscal policy instruments are those policies that take effect without explicit action by policymakers. Two examples are the progressive income tax system and unemployment compensation.

Discretionary fiscal policy instruments are those policies that require explicit actions on the part of Congress and the President. Two examples of discretionary fiscal policy are the 2001 and 2003 acts passed during the Bush administration.

15. Expansionary fiscal policy actions involve lowering income tax rates or raising government payments, which would shift the economy's aggregate demand curve to the right.

16. Contractionary fiscal policy actions involve raising income tax rates or lowering government payments, which would shift the economy's aggregate demand curve to the left.

17. A balanced budget requires that federal government receipts equal federal government expenditures. The federal budget deficit in this instance would be equal to zero.

KEY TERMS

Automatic fiscal policy instruments
Contractionary fiscal policy actions
Currency drains
Direct crowding out
Discount rate
Discretionary fiscal policy instruments
Excess reserves
Expansionary fiscal policy actions
Federal budget
Federal budget deficit

Federal expenditures
Federal funds market
Federal Open Market Committee (FOMC)
Federal receipts
Fiduciary monetary system
Indirect crowding out
Legal reserves
Medium of exchange
Money multiplier
National banks
National debt

Near monies
Nominal interest rate
Open market operations
Precautionary demand for money
Required reserves
Sequester
Speculative demand for money
State banks
Store of value
Transaction demand for money
Unit of accounting

TESTING YOUR ECONOMIC QUOTIENT

1. Fill in the blanks in the following table with the appropriate response: "lower" if the variable will decline in value as a result of the policy action or "raise" if the variable will increase in value as a result of the policy action.

Policy Effects on	Expansionary Monetary Policy	Contractionary Monetary Policy
Farm input prices		
Farm crop prices		
Farmland prices		
Net farm income		

2. Suppose the Federal Reserve buys $4,000,000 in government securities from a depositor at the First

National Bank of North Zulch. Let's assume the person deposits this money in this bank. Further assume that the current reserve requirement ratio is 20%.

a. Indicate here what will initially happen to this bank's balance sheet as a result of this transaction.
   Change in reserves _____
   Change in loans _____
   Change in deposits _____

b. Indicate here what will eventually happen to the U.S. *banking system* as a result of this transaction.
   Change in reserves _____
   Change in loans _____
   Change in deposits _____

3. Review the following graph and answer questions (a), (b), and (c).

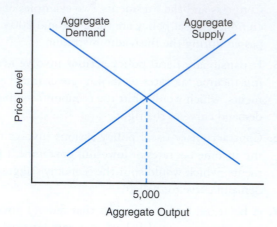

Aggregate Demand

Aggregate Supply

Price Level

5,000

Aggregate Output

a. If the full-employment level of output is $4,000, is the economy depicted in an inflationary or a recessionary gap? What *specific* monetary policies would you recommend, if any, to correct this?

b. If the full-employment level of output is $6,000, is the economy depicted in an inflationary or a recessionary gap? What *specific* monetary policies would you recommend, if any, to correct this?

c. If the full-employment level of output is $5,000, is the economy depicted in an inflationary or a recessionary gap? What *specific* monetary policies would you recommend, if any, to correct this?

4. Place a T or F in the blank appearing next to each statement.

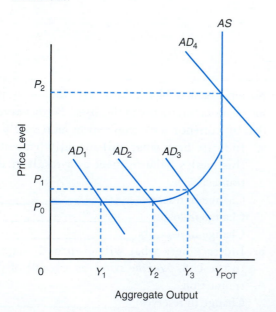

AS

AD_4

P_2

AD_1 AD_2 AD_3

Price Level

P_1

P_0

0   Y_1 Y_2 Y_3 Y_{POT}

Aggregate Output

a. _____ Only a small level of inflation occurs as the economy moves from AD_1 to AD_2.

b. _____ A tax cut represents one approach to reducing prices below P_2.

c. _____ This graph depicts the effects of stagflation.

5. Review this graph and answer questions (a), (b), and (c).

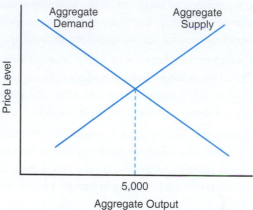

Aggregate Demand

Aggregate Supply

Price Level

5,000

Aggregate Output

a. If the full-employment level of output is $6,000, is the economy in an inflationary or a recessionary gap? What *specific* fiscal policies could be used to correct this?

b. If the full-employment level of output is $4,000, is the economy in an inflationary or a recessionary gap? What *specific* fiscal policies could be used to correct this?

c. If the full-employment level of output is $5,000, is the economy in an inflationary or a recessionary gap? What *specific* fiscal policies could be used to correct this?

6. Given the following figure, explain what type of inflation is taking place and the specific fiscal and monetary policies that may be used to combat this type of inflation if, during a period of years, the economy moves from AD_1 to AD_4.

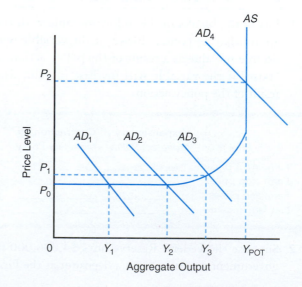

AS

AD_4

P_2

AD_1 AD_2 AD_3

Price Level

P_1

P_0

0   Y_1 Y_2 Y_3 Y_{POT}

Aggregate Output

7. Complete the following table describing the short-run effects of specific macroeconomic policy actions with a "+" denoting an increase in the variable in each column heading and a "−" denoting a decrease (20 points; 1 point each).

Policy Action	GDP Growth Rate	Unemploy-ment Rate	Interest Rate	Inflation Rate
Decrease in federal income tax rate				
Decrease in Federal Reserve discount rate				
Decrease in the level of government spending				
Federal Reserve sells government bonds				

8. Given the following demand and supply equation for a market, answer the following questions:

$$M_S = 1/rr_m\,(\text{TR})$$
$$M_D = 45 - 125\,(i) + 1.0\,(Y)$$
$$M_S \equiv M_D$$

where i represents the rate of interest, Y represents national income, rr_m represents the fractional reserve requirement ratio, and TR represents total reserves.

Assume national income in 2015 was $1,200 and is projected to be 5% higher in 2016. Also, assume the reserve requirement ratio is 0.25 and total reserves are equal to 140.

a. What market clearing interest rate would you project for 2016?

b. What level of the money supply would be needed to achieve an interest rate of 8.5% in 2016 (Hint: using whole percentage rather than decimal equivalent; e.g., using 12 rather than 0.12)?

9. Automatic fiscal policy instruments include
   a. policy that requires specific policy actions.
   b. such instruments as actions to eliminate a recessionary or contractionary gap in the economy.
   c. instruments such as Congressional appropriations in the years when a drought or other natural disasters occur.
   d. none of the above.

10. The rate of inflation in the economy will be highest in the _____ range of the aggregate supply curve.

11. The three major types of monetary policy instruments are _____, _____, and _____.

12. The money multiplier is equal to the _____ of the fractional reserve requirement ratio.

13. A fiduciary monetary system is based on the fact that each dollar of currency in the money supply may be redeemed for an equal amount of gold or silver at the request of the currency holder.   T   F

14. By decreasing the discount rate, the Federal Reserve can increase the demand for money.   T   F

15. Lowering the fractional reserve requirement ratio is an example of
    a. contractionary fiscal policy.
    b. expansionary monetary policy.
    c. expansionary fiscal policy.
    d. none of the above.

16. Expansionary monetary policy would likely
    a. increase farmland prices.
    b. increase export demand for agricultural commodities.
    c. increase net farm income.
    d. all of the above.

REFERENCES

Board of Governors of the Federal Reserve System: *Federal Reserve bulletin*, Washington, D.C., various issues, U.S. Government Printing Office.

Economic Report of the President, Washington, D.C., various issues, U.S. Government Printing Office.

U.S. Department of Commerce: *Business Conditions Digest*, Washington, D.C., various years, U.S. Government Printing Office.

14

Consequences of Business Fluctuations

The goals of macroeconomic policy are to promote employment, price stability, and economic growth. The general economy, as we all know, experiences rising unemployment, inflation, and economic stagnation from time to time. Macroeconomic policymakers must use the monetary and fiscal policy tools available to them to cure these economic ills.

The purpose of this chapter is to examine the nature of fluctuations in business activity in the general economy, outline the macroeconomic policy actions the federal government can take to achieve specific objectives, and illustrate the effects of these actions on macroeconomic variables such as the market rate of interest, the rate of inflation, the unemployment of labor and capital, and the rate of growth in real GDP. Chapter 15 will address what this all means for the farm sector of the economy and implications for the broader U.S. food and fiber industry.

FLUCTUATIONS IN BUSINESS ACTIVITY

The U.S. economy traditionally goes through periods of ups and downs in the level of its business activity. These business fluctuations, which are often referred to as **business cycles**, are typically thought of in terms of movements in the economy's GDP, interest rates, or unemployment rate. We will discuss the nature of business fluctuations in the general economy, the major indicators of this activity, and the policy actions normally taken to address these fluctuations.

Business cycles reflect periods of expansion and recession in an economy. These periods culminate with a peak (expansion) or a trough (recession).

Nature of Business Fluctuations

Figure 14-1 illustrates the general nature of business fluctuations in the economy, which has four distinct phases. Periods of recession and expansion occur between

The price of gasoline back in 1935, as shown in this photo, was 15 cents a gallon, a far cry from the $4.00 seen in 2008. The growth in the price of gasoline can be attributed in part to inflation. Oil-producing nations are well aware of the rate of inflation when they attempt to set the price of crude oil.   Credit: Dorothea Lange/Getty images.

peaks of these cycles. As the cycle reaches a trough, or "bottoms out," and the recovery begins, the economy enters another **expansionary period**, which is often referred to as a "boom" in business activity. Not every expansionary period reaches a new high. In Figure 14-1, for example, the expansionary period can be divided into two parts: (1) economic recovery and (2) economic expansion. The economy is said to be recovering up to the point where aggregate output or GDP equals the previous peak. The remainder of the expansionary phase above the peak represents true expansion of the economy. An expansionary phase may end for one or more reasons, and then a new **recessionary period** will begin. If the level of business activity falls sharply, it may be classified as a depression rather than a recession. Of course, nothing in modern U.S. economic history rivals the depths of the Great Depression of the 1930s.

Four of the last seven periods of expansion in the economy coincided with U.S. participation in war activities. A period of expansion ended in 1929 with the Great Depression, which lasted well into the 1930s. The economy experienced a boom in economic growth associated with World War II manufacturing activity. The U.S. economy also experienced an unprecedented period of expansion during much of the 50-year period beginning in 1950.

There are different definitions of what constitutes a recession in the U.S. economy. The U.S. Department of Commerce takes the position that the economy is in a recession when it experiences two consecutive quarters of negative growth in real gross domestic product.[1] The National Bureau of Economic Research (NBER) has devoted a considerable amount of its resources to defining and measuring business

---

[1]We continue to be concerned with the real, as opposed to nominal, growth of the economy for reasons outlined in Chapter 12. We have deducted the effects of inflation to focus on the growth of the economy's purchasing power (i.e., its real gross national product).

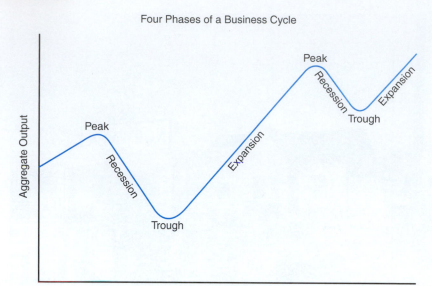

Figure 14-1

The traditional business cycle (of fluctuation) has four phases: (1) an expansionary phase, (2) a peak that marks the end of the expansionary phase, (3) a recessionary phase, and (4) a trough that marks the end of the recessionary phase. Most cycles result in a higher peak in economic activity during a period of prolonged economic growth in the economy; however, this need not always be true. During the 1930s, the recessionary phase became so pronounced that it was labeled the Great Depression.

fluctuations in the economy. After studying business activity "after the fact," the NBER defines the "official" timing of the recession. The NBER's identification of recessions does not necessarily have to contain two consecutive quarters of negative real economic activity. The latest recession occurring during the 2007–2009 period is referred to by many as the "Great Recession" in that it lasted longer than any recession dating back to bank panics in the early 1900s as well as the depth to which negative growth occurred during this period.

Indicators of Economic Activity

Of course, businesses and policymakers cannot wait for the NBER to tell them that the economy experienced a recession 12 months ago. Some follow indicators of business activity that coincide with, lag behind, or lead business fluctuations in the economy.

The U.S. Department of Commerce regularly publishes a series of lagging, coincident, and leading indicators that serve to tell businesses and policymakers what is happening in the economy and what is likely to happen in the near future.

Coincident Indicators   The **coincident indicators** move concurrently with business activity in the economy. Examples of coincident indicators are information on current industrial production, the number of employees on payrolls, personal disposable income, and manufacturing sales. These indicators help explain current business activity.

Lagging Indicators   The **lagging indicators** of business activity usually indicate a change in economic activity about one-quarter (i.e., 3 months) after the fact. Examples of lagging indicators are business inventories, labor cost per unit of output, the average duration of unemployment, and the average interest rate that banks charge their best customers (i.e., the prime rate).

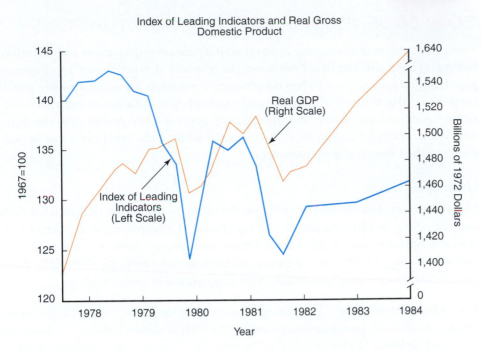

Index of Leading Indicators and Real Gross Domestic Product

Figure 14-2

The index of leading economic indicators, depicted by the solid line, indicates business fluctuations before they actually occur. The 1978–1984 period was chosen because it illustrates how well this index can forecast the direction of fluctuations in real GDP in subsequent quarters. Note in particular the double peak, or M-shaped activity, forecasted during the 1980–1982 period.

Leading Indicators   The **leading indicators** of business activity are defined so as to indicate business fluctuations before they occur; that is, they are expected to peak approximately one- to two-quarters before business activity actually peaks in the economy. Figure 14-2 shows a classical case where the index of leading economic indicators indicated future GDP fluctuations 6 to 9 months later. Examples of leading economic indicators are new orders for consumer goods, new building permits, new investment in plant and equipment, and changes in selected prices and the money supply.[2]

Forecasting Models   As an alternative to using leading indicators to make forecasts of what will happen to the economy in the future, many businesses and policymakers employ the services of sophisticated computer models. These models reflect the relationships between past economic behavior of producers and consumers, and selected prices, interest rates, and other variables thought to explain economic behavior.

There are numerous commercial and government-sponsored models that are capable of projecting events in the economy in general and the farm business sector in particular.[3] Their major advantage is their ability to examine what would happen if conditions were to take on alternative values. This gives businesses and policymakers a feeling for the range of outcomes that are likely to occur under specific sets of conditions.

Economic indicators

Government policymakers and businesses pay close attention to trends in a variety of indicators, including the closely watched index of leading economic indicators. Many of these indicators are published by the U.S. Department of Commerce.

---

[2]The U.S. Department of Commerce regularly publishes a set of indexes capturing these and other indicators in the publication *Business Conditions Digest*. These include an index of 12 leading indicators, an index of 4 coincident indicators, and an index of 6 lagging indicators.

[3]Examples of economic models of the U.S. economy that contain a farm sector are the models developed by Wharton Econometric Forecasting Associates (WEFA).

CONSEQUENCES OF BUSINESS FLUCTUATIONS

Two consequences of fluctuating business activity are **unemployment** and **inflation**. Unemployment will rise in periods when the economy is experiencing a recessionary gap. Inflation generally rises when the economy is experiencing an inflationary gap. It is often said that the goal of monetary and fiscal policy is to eliminate inflation (inflationary gaps) and unemployment (recessionary gaps) in the economy. Yet the terms unemployment and inflation are frequently used without the benefit of a precise definition of their meaning or measurement.

Unemployment

Unemployment, broadly defined, refers to the idling of part of the civilian labor force and the idling of business plants and equipment. Unemployment of part of U.S. scarce resources results in the loss of output and savings, both of which affect the potential future growth of the economy and the economic suffering endured by those workers and businesses whose resources are unemployed or underemployed.

Consequence of recession
One of the consequences of recessionary trends in the economy is rising unemployment of labor and capital.

Unemployment of Labor   The **unemployment rate for labor** is simply equal to the number of civilians unemployed divided by the total of those civilians employed *and* those unemployed. To be included in "unemployed" in this instance, one must be actively seeking but not finding employment. The denominator represents the size of the **civilian labor force**. The civilian labor force specifically excludes (1) individuals in the military; (2) individuals in institutions such as prisons, mental hospitals, or nursing homes; and (3) individuals under age 16.

The **unemployment rate** is then found by dividing the number of unemployed persons by the size of the total civilian labor force, or

$$\text{Annual unemployment rate} = \frac{\text{Number of unemployed persons}}{\text{Size of total civilian labor force}} \quad (14.1)$$

In 2015, for example, the unemployment rate was 5.28% (i.e., $0.0528 = 8.3 \div 157$). Figure 14-3 shows what has happened to the unemployment rate since 1970. This figure shows that the unemployment rate in 2015 of 5.28% was well below the unemployment rate of almost 25% during the Great Depression.[4]

Unemployment of labor falls into several categories. **Frictional unemployment** refers to the continuous flow of people who are changing jobs and, therefore, are currently unemployed. **Cyclical unemployment** refers to unemployment associated with business fluctuations. **Seasonal unemployment** refers to unemployment associated with changes in business conditions that are seasonal in nature. Workers in fruit- and vegetable-processing plants and construction workers in northern states are two examples of workers who may be seasonally unemployed. **Structural unemployment** refers to those workers who are unemployed because of structural changes in the economy brought about by technological change that does away with their jobs. Farm laborers whose jobs have been replaced by tomato pickers are examples of structurally unemployed workers.

Unemployment of Capital   The equivalent concept for capital (i.e., plant and equipment) is the **manufacturing capacity utilization rate**, established by the U.S.

---

[4]Some economists argue that this unemployment rate understates the *true* unemployment rate because it does not account for discouraged workers who are no longer seeking employment and therefore are not included in the total civilian labor force (the denominator in the unemployment rate). This phenomenon is referred to as hidden unemployment. The labor force participation rate reflects the percentage of available people of working age who are actually in the civilian labor force.

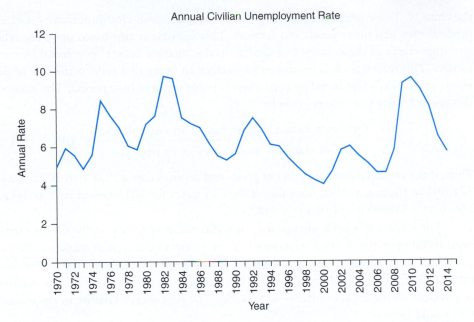

Figure 14-3

Compared with the high unemployment rates observed during the Great Depression in the 1930s when unemployment reached 25%, today's unemployment rate seems rather small. The unemployment rate is determined by dividing the number of unemployed persons by the total civilian labor force. Some economists argue that during a recession hidden unemployment occurs because workers are discouraged from seeking work. Therefore, the true unemployment rate would be understated.

Source: From *Economic Report of the President,* The Council of Economic Advisors.

Department of Commerce, or actual output divided by current manufacturing capacity. The lower the capacity utilization rate, the higher the unemployment of capital. **Full employment** refers to the employment of labor and capital when the economy is producing at its maximum noninflationary level of output.

In recent years, full employment GDP has been thought to occur at an unemployment rate for labor of about 4% and a capacity utilization rate for business capital of 86%.

Inflation

Inflation is generally defined as a sustained rise in the general price level (i.e., weighted average of all prices). Two key terms in this definition require special emphasis.

1. The term *sustained* rules out any temporary rise in prices.
2. The term *general price level* rules out the use of the term *corn price inflation* or *fuel price inflation.*

A temporary rise or spike in the general price level that returns to its initial level is not inflationary. The rise in the price of a single commodity also need not be inflationary, particularly if the prices of substitute goods or services have fallen. Instead, only a sustained rise in the general price level, as measured by the **consumer price index (CPI)**, which accounts for changes in the prices of all goods and services purchased by consumers, represents their inflation.

The CPI is one of the most closely watched statistics coming out of Washington. The CPI reflects changes in the cost of living. The cost-of-living statistic is based upon more than 100,000 individual price quotations obtained monthly from approximately 25,000 retail stores, 20,000 households, and 20,000 tenants in 85 cities across

Consequences of inflation
One of the consequences of inflationary trends in the economy is rising inflation facing both consumers (CPI) and producers (PPI).

the country. These prices include the price of food, housing, clothing, transportation, medical care, and other goods and services. This statistic is also based upon the relative importance of these goods and services in the "market basket" purchased by consumers. This information is combined to produce an index that reflects the cost of the goods and services purchased by consumers relative to some base period. The value of this index for any one year is given by

$$\text{CPI} = \frac{\text{Cost of standard market basket in current year}}{\text{Cost of standard market basket in base year}} \times 100 \qquad (14.2)$$

The consumer price index (CPI) reflects the cost of a "basket" of consumer goods and services today relative to some base period.

Thus, if the standard market basket of goods and services cost $35,000 in 2015 versus $21,084 in the base period, the value of the CPI index for 2015 would be 166.6 (i.e., $166.6 = [35,000 \div 21,084] \times 100$).

The impact of a price change for a specific group of goods on the CPI is calculated based upon the relative importance of the group in the typical market basket of goods and services purchased by consumers. For example, assume that the cost of food eaten away from home has gone up by 10% during the period and that the relative importance of these expenditures is 5.712%. The percentage change in the overall CPI would be

$$\begin{aligned} W_{\text{FAFH}} \times \%\Delta P_{\text{FAFH}} &= \%\Delta \text{CPI} \\ 0.05712 \times 10.0 &= 0.5712\% \end{aligned} \qquad (14.3)$$

or approximately six-tenths of 1%. The variable W_{FAFH} represents a weight reflecting the relative importance of food consumed away from home to all goods and services purchased by consumers. The relative importance weights used for food consumed away from home and other goods and services reflect expenditure patterns observed over a recent period and are typically updated every 2 years. As shown in Table 14-1, housing carries the greatest weight (42.427%), followed by transportation (17.688%) and food (14.914%).

Annual percentage changes in the CPI are of more than idle interest for many consumers. Beyond reflecting what happens to the **purchasing power** of our income, or real disposable personal income, these changes are used to automatically adjust the nominal level of income received by specific groups of consumers as a cost-of-living adjustment (COLA). Steelworkers' wages, for example, are directly tied to the CPI. As a part of their union contract, steelworkers get a raise of $0.01 per hour for every 0.3 points in the CPI index. For example, if the CPI rose from 72.6 to 82.4, reflecting a 13.5% inflation rate faced by consumers, this means that steelworkers' nominal hourly wages rose by almost $0.33 to offset the effects of inflation. Social Security benefits also go up automatically whenever the rate of inflation (i.e., the percentage change in the CPI) exceeds 3% annually. Veterans and retired federal workers receive similar protection. As a result of these COLAs, it is estimated that a 1% increase in the CPI triggers more than $2 billion in additional federal government expenditures. According to Bureau of Labor Statistics estimates, the nominal income of more than one-half of all American households is affected by movements in the CPI.

Different age groups of consumers will have different weights. Young unmarried consumers will spend a different percentage of their consumption dollar on food and nondurable goods than a family of four will. Do the relative importance weights in Table 14-1 describe the relative importance of these goods and services to you?

Measure of inflation The rate of inflation facing consumers is measured by the percentage change in the CPI.

The annual rate of inflation for consumers can be measured by computing the percentage change in the CPI from one period to the next, or

$$\text{Annual inflation rate} = \frac{\text{Current CPI} - \text{Previous CPI}}{\text{Previous CPI}} \qquad (14.4)$$

Table 14-1
CONSUMER PRICE INDEX

Expenditure Category		Relative Importance Weights
Food and beverages		14.914
Food at home	7.660	
Food away from home	6.173	
Alcoholic beverages	1.080	
Housing		42.427
Residential rent	5.765	
Homeowners rent	23.942	
Household furnishings and operation	4.702	
Other housing	2.253	
Apparel and upkeep		3.731
Transportation		17.688
New vehicles	4.632	
Used vehicles	1.773	
Motor fuel	5.482	
Public transportation	1.106	
Other transportation	4.695	
Medical care		6.231
Recreation		5.647
Education		2.944
Communication		3.142
Other goods and services		3.276
All items		100.000

Source: Bureau of Labor Statistics, U.S. Bureau of Labor Statistics.

For example, if the CPI in 2015 was 236.9 and in 2014 was 232.9, the annual rate of inflation for 2015 would be

$$\text{Annual inflation rate} = \frac{236.9 - 232.9}{232.9}$$
$$= 0.0171, \text{ or } 1.71\% \qquad (14.5)$$

Figure 14-4 shows what happened to the rate of inflation from 1970 to 2014. We see the downward trend in inflation that occurred since the late 1970s and early 1980s.

The U.S. Department of Commerce also prepares a **producer price index (PPI)**, which accounts for changes in the prices of goods and services purchased by producers. The rate of inflation for producers, as measured by the percentage change in the PPI from one period to the next, is calculated in a manner similar to the CPI; that is, it uses base year quantities at current prices and compares them with base year quantities at base year prices.

The cost of unemployment is clear—particularly to those who are unemployed. Unlike unemployment, however, inflation affects all of us in a direct and immediate way. It reduces the purchasing power of consumers' disposable income and the purchasing power of producers' profits. Persons particularly hurt by inflation include

1. workers whose salary does not increase enough to offset the effects of inflation (e.g., those on a fixed income),
2. lenders who make loans at fixed interest rates and who are unfavorably surprised by the extent of inflation,
3. individuals or businesses who sign contracts that do not account for inflation, and

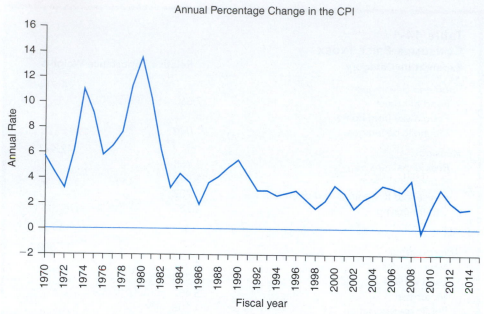

Figure 14-4

Inflation is defined as a sustained rise in the general price level. The annual percentage change in the consumer price index (CPI) measures U.S. consumers' inflation rate. Not all consumers are confronted by the same rate of inflation, however. Young consumers typically purchase a different basket of goods and services than do retired consumers. Inflation also varies between rural and urban areas and across major geographical regions of the country.

Source: From *Economic Report of the President,* The Council of Economic Advisors.

Real GDP an important variable used to describe the growth of the domestic economy, is measured by dividing nominal GDP by the implicit GDP deflator. This deflator is the broadest measure of prices in the economy.

4. individuals or businesses who hold money rather than assets that go up in value with inflation.

Consider the example presented in Table 14-2, where we determined what you would have to earn in 2015 to be able to afford the same basket of goods and services you purchased with $20,000 in 2005, if the annual rate of inflation during this period was either 5% or 10%.

Table 14-2 suggests that your salary would have to be $77,812 in 2015 if you wanted to be able to buy the same bundle of goods and services that $30,000 bought

Table 14-2
SALARY NEEDED TO BREAK EVEN WITH A 5% AND 10% ANNUAL INFLATION RATE

| | Salary Required | |
Year	5%	10%
2005	$30,000	$30,000
2006	31,500	33,000
2007	33,075	36,300
2008	34,729	39,930
2009	36,465	43,923
2010	38,288	48,315
2011	40,203	53,147
2012	42,213	58,462
2013	44,324	64,308
2014	46,540	70,738
2015	48,867	77,812

Table 14-3
NOMINAL AND REAL GROSS DOMESTIC PRODUCT

2000 Nominal GDP	2015 Nominal GDP	2015 Real GDP
100 oranges @ 20¢ = $20	120 oranges @ 27¢ = $32	120 oranges @ 20¢ = $24
80 apples @ 50¢ = $40	90 apples @ 60¢ = $54	90 apples @ 50¢ = $45
Total spending = $60	Total spending = $86	Total spending = $69

in 2005, if the annual rate of inflation were 10%. You would only have to earn $48,867 by 2015 to break even if the annual rate of inflation were only 5%. This underscores the problems faced by those on fixed incomes or those who receive annual pay raises that are less than the rate of inflation.

Real GDP is found by dividing nominal GDP by the implicit GDP price deflator. The implicit GDP price deflator is nothing more than the broadest price index one can use when accounting for inflation when assessing the magnitude of the nation's output. This price index includes all the goods and services consumers purchase and purchases of other final goods by businesses, governments, and foreign buyers. Real GDP is thus a measure of output that attempts to highlight changes in levels of production at the economy level between two different periods in time.

Let's examine an example of an economy producing two products: apples and oranges (Table 14-3). We cannot assess the aggregate output of this economy in a single statistic that includes market values, because we cannot add apples and oranges. The implicit GDP price deflator for this two-product economy in 2015 would be 124.63 (i.e., [$86/$69]100). This example shows that real GDP in 2015 expressed in constant or 2000 prices would be $69, which reflects a 15% real growth in this hypothetical economy since 2000. A comparison of nominal GDP over this period would have suggested a 24.6% growth rate, due in part to the growth in physical consumption of apples and oranges and in part to the increase in the prices of these two commodities. The latter does not represent a real expansion of the economy.

Types of Inflation Like unemployment, there are also several types of inflation: demand-pull inflation, cost-push inflation, and stagflation.

When the aggregate demand for goods and services is rising and the economy is approaching full employment, **demand-pull inflation** will occur. The result is a rise in the general price level.[5] This phenomenon is illustrated in Figure 14-5. In this figure, we assume that the AS curve represents the economy's current aggregate supply curve and AD_1, AD_2, and so on represent a series of alternative aggregate demand curves.[6]

Figure 14-5 suggests that if the aggregate demand curve were to shift from AD_1 to AD_2, there would be virtually no change in the general price level. If the aggregate demand curve instead shifted from AD_1 to AD_3, the general price index would increase from P_0 to P_1. This increase in demand, therefore, is inflationary, although mildly so.[7] The inflation rate in this instance would be equal to

Types of inflation There are several types of inflation, including demand-pull and cost-push inflation, which are named after their effect on demand and supply curves resulting in higher prices in the economy.

---

[5]This also corresponds to the inflationary gap, in which desired spending is greater than the economy's full employment output, which was introduced in Chapter 12 and will be discussed in considerable detail in Chapter 15.

[6]This aggregate supply curve contains the full range of potential outcomes in the economy. The portion of the AS curve associated with outputs between 0 and Y_2 is called the Keynesian or depression range. The segment associated with Y_2 through Y_{POT} was illustrated earlier in Figure 12-11 and is referred to as the normal range. The perfectly inelastic segment of AS is known as the classical range.

[7]Demand-pull inflation begins to occur before full employment of the nation's resources is achieved. Some economists refer to this as **premature inflation**.

Figure 14-5

Demand-pull inflation occurs when changes in demand result in a sustained rise in the general price level. The increase in aggregate demand from AD_1 to AD_2 was not inflationary. The increase in aggregate demand from AD_2 to AD_3, however, was mildly inflationary because the general price level rose from P_0 to P_1. Finally, the increase in aggregate demand from AD_3 to AD_4 would be highly inflationary. The general price level rose from P_1 to P_2.

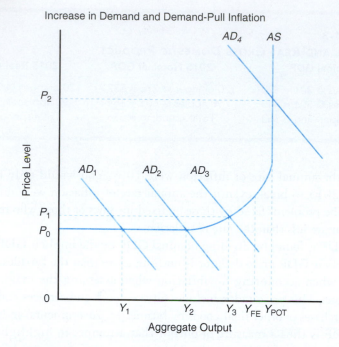

$(P0 - P1) \div P1$. Finally, if the aggregate demand curve were to shift from AD_1 to AD_4, the economy would reach its capacity to supply additional goods and services in the current period, Y_{POT}. This desired level of expenditures would be highly inflationary because the general price index would increase from P_0 to P_2. In the last two situations, increases in demand would be "pulling up" the general price level.

A second form of inflation, which occurs when the economy is not at full employment, is **cost-push inflation**. This form of inflation was prevalent during the 1970s. In both instances, prices were rising even though the economy was nowhere near full employment. Cost-push inflation, which is sometimes referred to as market-power inflation, can arise for at least two reasons: (1) union monopoly power and (2) business monopoly power.

Some unions may have enough bargaining power in the labor market to impose wage increases on employers, who in turn raise their prices to maintain current profit margins. This wage–price spiral results in a rise in the general price level. Businesses with monopoly powers can also raise their prices if they desire higher profits. Workers will then demand higher wages to compensate for losses in their standard of living (they no longer can buy the same bundle of goods as a result of this price increase). This, in turn, causes the monopolist to raise his or her prices further to secure his or her desired profit margin. This wage–price spiral also results in a rise in general price levels.

A phenomenon first experienced in the late 1970s and early 1980s is **stagflation**, which refers to the existence of increasing inflation during a period when the economy is experiencing rising unemployment. Rising prices and a slowing economy occurred again in 2007/2008 as a result, in part, of rising fuel prices. This term originates from the combination of economic *stag*nation and in*flation*. Obviously, different macroeconomic policies are needed in each of the foregoing situations if policymakers desire both to promote the growth of the economy and to stabilize prices (i.e., eliminate inflation).

Short-Run Phillips Curve

The Phillips curve was developed to describe the trade-off between inflation and unemployment in the short run. Studies have shown that it is less well suited to this task in the longer run.

We would all agree that rising unemployment and rising inflation are bad. Rising unemployment hurts those who are unemployed *and* those who would have sold

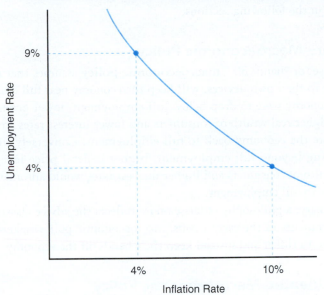

Figure 14-6

The short-run relationship between the rate of inflation and rate of unemployment is captured by its Phillips curve.

goods and services to them. Inflation hurts practically everyone by reducing the purchasing power of income and wealth. Are these two problems interrelated? Yes, says British economist A. W. Phillips, who first noted the interrelationship between these two consequences of business fluctuations. He observed that wages increased more rapidly when England's unemployment rate was low than when the unemployment rate was high. This suggested that inflation of wages and other prices does not wait until full employment GDP is reached, which explains the normal range of the aggregate supply curve. The Phillips curve seems to hold for specific periods of time, but structural changes and other factors shift the aggregate Phillips curve to the right or to the left from time to time. Few would question the short-run nature of this curve.

Figure 14-6 presents the short-run Phillips curve for the hypothetical country of Lower Slobovia. It suggests that an inflation rate of 10% is associated with an unemployment rate of 4%, and an inflation rate of 4% is associated with an unemployment rate of 9%. Thus, policies that fight inflation will—at least in the short run—lead to higher unemployment. And policies to stimulate employment will lead to higher inflation in the short run. This has obvious implications for the U.S. choice of macroeconomic policies.

MACROECONOMIC POLICY OPTIONS

Policymakers adopt several approaches toward trends in market activity in the economy. Classical economists such as Adam Smith long argued that macroeconomic policymakers should do nothing, and that the economy will somehow self-correct for undesirable trends in unemployment, inflation, and recessionary GDP gaps.

During the Great Depression of the 1930s, British economist John Maynard Keynes advanced the notion that federal governments should take an activist position in dealing with undesirable trends in the economy. He recommended using demand-oriented monetary and fiscal policies either to stimulate or to retard economic activity, depending on whether it was experiencing a recessionary or an inflationary gap.

The Reagan era of the 1980s saw the advancement of supply-side arguments designed to stimulate productivity as a means of expanding aggregate output while lowering inflation. The general features of each of these macroeconomic policy options are discussed in the following sections.

Laissez-faire Macroeconomic Policy

A **laissez-faire**, or "hands off," **macroeconomic policy** assumes that markets in the economy, left to their own devices, will keep the economy near full employment output. If the economy were to drop below full employment, lower general price levels and, hence, higher real wealth of consumers and lower interest rates would automatically stimulate the economy back to full employment. Conversely, if the economy were expanding beyond full employment, higher general price levels and, hence, lower real wealth of consumers and higher interest rates would automatically pull the economy back to full employment.

In summary, a philosophy of laissez-faire reflects the advice classical economists gave to governments in the early 1900s: macroeconomic policymakers should resist the temptation to tinker and instead keep their hands off the economy.

Demand-Oriented Macroeconomic Policy

The Great Depression caused real GDP in the United States to fall 30% from 1929 to 1933, forcing both economists and policymakers to reconsider the wisdom of a laissez-faire macroeconomic policy. Keynes' book, titled *A General Theory of Employment, Interest and Money*, discussed the importance of consumer disposable income as the engine for macroeconomic growth along with interest rates. Keynes prescribed that an economy should pursue aggressive macroeconomic policies in its money and product markets to boost aggregate demand when recessionary trends exist and slow aggregate demand when inflationary pressures exist.

> **Demand-oriented policies**
> The macroeconomic policies described in Chapter 13 can be thought of as demand-oriented policies. They accomplish their goal by shifting the aggregate demand curve to achieve the desired policy objective.

In 1934, Keynes actually published an open letter to President Franklin D. Roosevelt in the *New York Times* that outlined how the United States could lift the economy out of the Great Depression by following aggressive **demand-oriented macroeconomic policies**, shifting the aggregate demand curve in Figure 14-7 to the right. The use of demand-oriented macroeconomic policies became so popular in the post–World War II period that President John F. Kennedy announced in an economic policy speech that he was a "Keynesian."

The role of demand-oriented macroeconomic policy can be likened to being the host of a party; that is, policymakers should know when to remove the punch bowl before the party gets out of hand and when to bring the punch bowl back before the party gets too dull. In economic jargon, policymakers attempt to promote the economic growth of the economy without stimulating inflation. If inflation begins to rise, demand-oriented policy actions can be taken to dampen aggregate demand for goods and services, which reduces inflationary pressures but slows economic growth if the economy is not at full employment. If unemployment begins to rise, demand-oriented policy actions can be taken to stimulate aggregate demand. These actions will reduce unemployment but may be inflationary if the economy is approaching full employment.

Figure 14-7 shows that expansionary demand-oriented macroeconomic policies will increase aggregate output from Y_1 to Y_2. The general price, however, would rise from P_1 to P_2, signaling an increase in inflation. The extent to which general price levels will rise depends on the elasticity of that segment of the *AS* curve intersecting the *AD* curve. The more inelastic or steeper the *AS* curve, the greater the increase in the general price level for a given increase in demand.

The reverse is also true. Contractionary demand-oriented policies will decrease the general price level, but when aggregate output declines, the unemployment rate

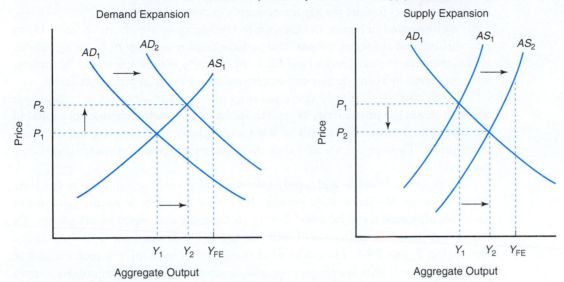

Figure 14-7

The implementation of demand-oriented and supply-oriented macroeconomic policies can have different impacts on the macroeconomy. An expansionary demand-oriented policy will shift the aggregate demand curve from AD_1 to AD_2, causing aggregate output to rise from Y_1 to Y_2 and the general price level to rise from P_1 to P_2. An expansionary supply-oriented policy will shift the aggregate supply curve from AS_1 to AS_2, causing aggregate output to rise from Y_1 to Y_2 and the general price level to fall from P_1 to P_2.

will rise.[8] And the less inelastic or flatter the *AS* curve, the smaller the increase in the general price level for a given increase in demand.

Keynes' proposal to actively use monetary and fiscal policies to stabilize the economy at full employment was "the law of the land" until the 1960s, when monetarists such as Milton Friedman began to question attempts to fine-tune the economy. Monetarists and other critics of aggressive monetary and fiscal policies suggested that a steady rate of growth in the money supply alone, using the monetary policy instruments discussed in Chapter 12, was a more attractive macroeconomic policy alternative to the prescriptions of Keynes.

Many economists are critical of the demand-oriented, activist macroeconomic policies because they allow politicians to use fiscal policy to manipulate the economy for political short-term goals that may not be desirable in the long run. They argue that lags in the adoption and full implementation of macroeconomic policies may begin to take effect after the economy has largely corrected itself, and thus cause either an unnecessary recession or inflationary pressures. The economy, they argue, is more like a supertanker than a row boat; a row boat can be reversed almost instantaneously, but a supertanker requires more than 10 miles for its course to be reversed.

Supply-Oriented Macroeconomic Policy

Stagflation is a relatively recent phenomenon. Stagflation refers to the presence of a stagnant economy accompanied by rising inflation. **Supply-side economics**, which

---

[8] This inverse relationship between inflation and unemployment was first popularized by British economist A. W. Phillips. He advanced a graphical relationship between these two important macroeconomic variables that came to be known as the Phillips curve. More recent research has shown that, although this relationship may hold in the current period, changes in technology and productivity distort this relationship over the long run.

gained popularity when President Ronald Reagan entered office in early 1981, involved attempts to shift the aggregate supply curve to the right.

As illustrated in Figure 14-7, a shift in the aggregate supply curve from AS_1 to AS_2 will increase aggregate output (and perhaps employment) from Y_1 to Y_2. Importantly, the general price level would fall from P_1 to P_2. Stagflation would be successfully countered by rising output and employment and lower general price levels.

To be successful, supply-side economics must lead to increases in technology, which increases the productivity of capital, increases the productive stock of capital in the economy, increases the supply of labor and/or labor productivity, and decreases input prices. These policies would shift the economy's production possibilities curve outward.

In practice, demand- and supply-oriented policies are both in force at any time in the economy. Macroeconomic policies designed to promote economic growth and lower unemployment can be found during recessionary and expansionary phases of a business cycle. These policies will shift the aggregate demand curve to the right as depicted in Figure 14-5. Policies to fund research that leads to new technologies or fund development that lowers cost per unit of input will shift the aggregate supply curve to the right over time. More will be said about these policies in Chapter 15.

SUMMARY

The purpose of this chapter was to examine the economic consequences of fluctuations in business activity and the potential macroeconomic policy responses to these fluctuations. The major points made in this chapter are summarized as follows:

1. Expansionary monetary and fiscal policies can be employed to combat unemployment, and contractionary monetary and fiscal policies can be used to combat inflation.

2. In recent years, supply-side economists have proposed that demand management policies cannot attack the problems of stagflation. These economists have proposed tax cuts to shift the supply curve to the right to lower unemployment and inflation.

3. There are four phases to a business cycle: (1) expansion, (2) peak, (3) recession, and (4) trough. Farmers generally fare well when the economy enters a recovery period and inflation is low. When anti-inflationary policies are instituted, however, and interest rates rise, farmers fare rather poorly.

4. Businesses and policymakers often study published indicators of economic activity. Leading indicators, for example, are expected to indicate business fluctuations *before* they occur. Computer forecasting models are also often used for this purpose.

5. Two consequences of fluctuating business activity are unemployment of labor and inflation. Unemployment rises during periods of recession, and inflation rises in periods of expansion.

6. The unemployment rate reflects the percentage of the total civilian labor force that is unemployed. The equivalent concept for capital is the manufacturing capacity utilization rate.

7. *Inflation* is defined as a sustained rise in the general price level. The inflation rate for consumers is measured by the percentage change in the consumer price index (CPI). The inflation rate for producers is measured by the percentage change in the producer price index (PPI).

8. When the aggregate demand for goods and services is rising and the economy is either approaching or beyond full employment, demand-pull inflation will occur. Cost-push inflation, which is also referred to as *market-power inflation*, occurs when unions or imperfectly competitive firms impose an increase on the prices they charge for their goods or services.

9. The term *stagflation* refers to the existence of increasing inflation during a period when the economy is stagnant or experiencing little or no economic growth.

10. The short-run Phillips curve for an economy shows the trade-off between the rate of inflation and the rate of unemployment. Structural changes in the economy will shift this curve over time.

KEY TERMS

Business cycles

Civilian labor force

Coincident indicators

Consumer price index

Cost-push inflation

Cyclical unemployment

Demand-oriented macroeconomic
   policies

Demand-pull inflation

Expansionary period

Frictional unemployment

Full employment

Inflation

Lagging indicators

Laissez-faire macroeconomic policy

Leading indicators

Manufacturing capacity utilization
   rate

Premature inflation

Producer price index (PPI)

Purchasing power

Recessionary period

Seasonal unemployment

Stagflation

Structural unemployment

Supply-side economics

Unemployment

Unemployment rate

Unemployment rate for labor

TESTING YOUR ECONOMIC QUOTIENT

1. The nation, over a period of time, experiences cycles in business activity.

   a. In the graph template given in this question, draw and label the four phases of a business cycle.

   Real GNP | _____ Time

   b. What are the two major economic consequences of business fluctuations in the nation's economy?

   c. In the graph template, illustrate the relationship between these two consequences. Label all parts of this graph. What is the name given to this graph?

2. If the total labor force is comprised of 250 million people, there are 6 million people in the armed services and institutions, and there are 225 million civilians employed in the economy, calculate the nation's unemployment rate.

3. If the consumer price index was 166.6 in 2014 and 163.0 in 2013, calculate the annual rate of inflation that occurred in 2014.

4. Given the following hypothetical information about aggregate consumer price levels in an economy, answer the following questions.

	Weight	2014	2015
Food and beverages	20.0	210.0	210.0
Housing and furnishings	40.0	160.0	190.0
Transportation	10.0	200.0	240.0
Entertainment	10.0	250.0	290.0
Other goods and services	20.0	210.0	230.0

   a. What was the rate of inflation facing consumers in this economy in 2015?

   b. What likely happened to aggregate demand in this economy during 2015? Why?

   c. What can we say about the quantity clearing the food and beverage market in this economy during 2015? Why?

5. Define or graphically illustrate each of the following terms:

   Recessionary gap

   Potential GDP

   Demand-pull (excess demand) inflation

   Inflationary gap

Cost-push inflation

Stagflation

Phillips curve

6. The term *laissez-faire macroeconomic policy* refers to
   a. government regulation to maintain fair or equal opportunity in the economy.
   b. Federal Reserve policy to support fair or consistent commercial bank interest rates.
   c. fiscal policy to maintain full employment.
   d. the assumption that without intervention the economy will converge toward full employment.

7. The short-run Phillips curve shows
   a. the relationship between aggregate output and aggregate expenditures.
   b. the relationship between consumption and investment.
   c. the relationship between consumption and disposable income.
   d. none of the above.

8. Expansionary "supply-side macroeconomic policies" would include
   a. increasing government spending to shift the aggregate demand curve to the right.
   b. increasing government spending to shift the aggregate supply curve to the left.
   c. policies to shift the aggregate supply curve to the right.
   d. none of the above.

9. Inflation erodes the purchasing power of wages earned by consumers. T F

10. Demand-pull inflation may occur when union and business monopoly power is high, resulting in a wage–cost spiral. T F

11. The inflation rate in 2015 can be calculated by subtracting the CPI in 2014 from the CPI in 2015 and dividing this difference by the CPI in 2015. T F

12. A sustained rise in the general price level is known as _____.

13. The phase of the business cycle where aggregate output is declining is known as the _____.

14. The flow of people who are changing jobs and are thus currently unemployed is known as _____.

REFERENCES

Bureau of Labor Statistics: *Consumer Price Index*, Washington, D.C., various issues, U.S. Government Printing Office.

Economic Report of the President, Washington, D.C., various issues, U.S. Government Printing Office.

🍎15

Macroeconomic Policy and Agriculture

Chapter Outline

Macroeconomic policy can have a large impact on U.S. farmers and ranchers, particularly for those who borrow to finance the purchase of land and other fixed assets as well as variable production inputs. It also has an impact on commodity prices where a strong export market is vital to maintain producer profitability. This chapter focuses on what happens to the farm sector in the nation's economy over the short run when policymakers respond to unwanted macroeconomic trends, and the implications for other sectors in the U.S. food and fiber industry. Emphasis is placed on key relationships at the macro, market and micro, or firm level. The farm sector in most countries closely satisfies the conditions of perfect competition discussed earlier in Chapter 8. Other sectors in the nation's food and fiber industry, such as manufacturers of farm equipment, may more closely resemble one or more of the forms of imperfect competition discussed in Chapter 9.

    The discussion of specific macroeconomic policies on the general economy and the farm sector in this chapter will be presented in the following context: the impact of policies designed to reduce/eliminate a **recessionary gap** and the impact of policies designed to reduce/eliminate an **inflationary gap**. The mythical country of Lower Slobovia provides the backdrop for this discussion.

A HISTORICAL U.S. PERSPECTIVE

The relationship between macroeconomic policy and the farm sector in the United States over the last 50 years can be described by a passage from Charles Dickens's *A Tale of Two Cities*: "It was the best of times, it was the worst of times." The U.S. farm sector has experienced periods of strong economic performance; it has also experienced periods of weak economic performance. If the general U.S. economy is experiencing hardships, this does not mean that the farm sector is negatively affected in a major way. Likewise, a strong general economy does not necessarily guarantee a healthy farm sector.

The 1970s represented an era of rising real net farm incomes. Furthermore, the combination of rising real net farm incomes and declining **interest rates** made many farm owners "paper" millionaires as farmland values rose sharply. This decade was characterized by cheap credit. Some years saw the **rate of inflation** exceeding interest rates, resulting in negative **real interest rates**. The value of the dollar was also extremely low, reflecting low real interest rates. The annual federal budget deficit varied considerably more in the 1960s. The 1974–1975 recession and corresponding fiscal stimulus explained part of this variability. Another key characteristic of the 1970s economy was the rising price of fuel orchestrated by the Organization of Petroleum Exporting Countries (OPEC). By the end of the decade, the real price of fuel was double what it was at the beginning of the decade.

The 1980s saw a reversal of fortunes for many farmers and ranchers when interest rates rose dramatically, real net farm incomes fell, and farmland prices tumbled. Historical discussion of the 1980s is often referred to as the "farm financial crisis." Nominal interest rates reached the 20% range in the early 1980s as the Federal Reserve sought to gain control over inflation and inflationary expectations. Many rural commercial banks failed as a result of the farm financial crisis. The value of the dollar rose sharply as well. For the farm sector, which is highly capital intensive and dependent on export demand for much of its production, the effects of contractionary monetary policy by the Federal Reserve were particularly harsh. These policies led to a double-dip recession in 1980 and 1981–1982. Expansionary fiscal policy in the form of tax cuts caused federal budget deficits to rise sharply, which also put upward pressure on interest rates. The nominal prime interest rate reached 21% at one point during 1983. This year also saw real net income in the farm sector fall to levels not seen since the Great Depression as shown in Figure 15-1. As inflation came down to single-digit levels in the mid-1980s, the Federal Reserve began to adopt expansionary monetary policies, interest rates fell, and the general economy expanded. Congressional appropriations to support net farm income were required to help support a recovery in the farm sector. Finally, the federal budget deficits began to rise to levels not seen since the turn of the century.

The U.S. economy entered the 1990s with a then unprecedented level of public and private debt. Consumer spending slowed and a brief recession occurred. This decade saw rising federal budget deficits, a dramatic growth in a negative trade balance, a continuing crisis in the savings and loan industry, and more commercial bank failures early in the decade. The Federal Reserve walked a tightrope in pursuing a monetary policy that kept interest rates low to promote economic growth and yet high enough to avoid stimulating inflationary expectations. This decade also saw a major shift in farm policy, referred to as "freedom to farm," in an effort to return the farm sector to more free market–based decisions. A dramatic decline in crop commodity prices late in the decade, resulting in part from the Asian financial crisis, required a substantial infusion of government payments. By 1999, government payments accounted for almost one-half of net farm income. The bottom line is that the farm sector experienced greater price variability and downside risk associated with net farm income.

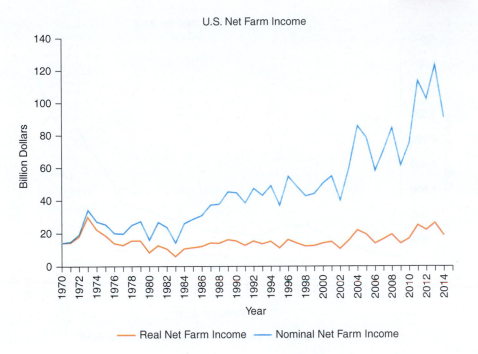

U.S. Net Farm Income

— Real Net Farm Income — Nominal Net Farm Income

Figure 15-1
Trends in nominal and real net farm income.

Source: Economic Research Service, U.S. Department of Agriculture

The U.S. economy entered the 2000s on a strong note but slowed in 2001 as the economy recorded a minor (two-quarter) recession stemming in part from the terrorist actions on September 11, 2001. The Federal Reserve lowered its short-term interest rate 12 times leading up to 2004 in an effort to stimulate the economy, causing interest rates to fall to 40-year lows. The federal budget surpluses incurred in the late 1990s gave way to substantial budget deficits beginning in 2004 with more red ink to follow. The price of crude oil on the New York Mercantile Exchange rose to $147 a barrel in 2008. The Federal Reserve was faced with the dilemma of rising inflation and slower economic growth. Compounding the problems at that time was the subprime lending problem that led to home mortgage foreclosures and the major recession (which some call the "Great Recession" because of how long it lasted and its depth) during the 2007–2009 period. Net farm income continued to vary annually in the 2000s as the farm sector competed in an increasingly global economy. Renewable fuel mandates expanded the industrial demand for corn, which had indirect effects on other crop commodities like soybeans as well as livestock commodities. Rising fuel and energy-related input costs raised farm operating expenses but gave increased impetus to increased corn-based fuel mandates that increased specific crop commodity prices that were capitalized into farmland values.

The decade of the 2010s thus far has seen a much slower macroeconomic recovery than seen after recent recessions. Slow growth in both domestic and global demand, coupled with rising productivity, had kept the civilian **unemployment rate** high by historical standards until it finally fell to 5 percent in late 2015. While the general economy has been slow to expand after the 2007–2009 recession, net farm income reached record levels in 2011. A severe drought in the midwestern states led to lower net income levels in 2012, but one that was still almost 50% above the 2002–2008 average. Low interest rates, coupled with high cash rental rates, contributed to double-digit growth rates in farmland values in most years until low commodity prices in 2015 caused the growth in farmland values to slow nationally, as shown in Figure 15-2. This has led some to speculate that farmland in regions like the Corn Belt and the Northern Plains might be a dark-horse candidate for an asset bubble.

Nominal net farm income has grown substantially since the mid-1980s. After accounting for inflation, however, we see that real net farm income has grown much more slowly.

Figure 15-2

Trends in nominal and real farmland values.

Source: Economic Research Service, U.S. Department of Agriculture.

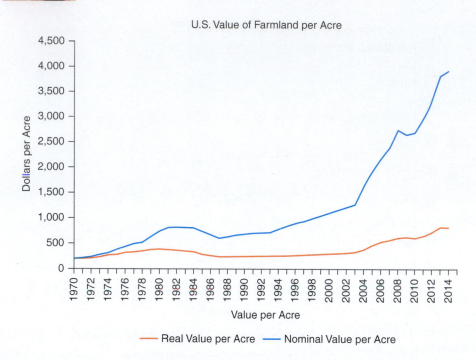

THE BIG FIVE

There are five key macroeconomic variables that can have an impact on the economic performance and financial strength of the farm sector. This is not to say that other events external to the farm sector cannot also have an impact. These five variables, however, have a consistent track record in affecting the farm sector and farm operator families over time. These variables in no particular order are given as follows:

1. Rate of inflation
2. Rate of interest
3. Rate of unemployment
4. Rate of growth in real GDP
5. Rate of foreign exchange

We will refer to these variables as the "Big Five." How do these variables affect the farm sector of our economy? Each variable is discussed here. As you will see, many of these five variables are interrelated.

Rate of Inflation

The rate of inflation can affect the price farmers pay for variable production inputs. Remember the farm sector comes as close as any sector of the economy to satisfying the conditions for perfect competition. Thousands of farmers produce a homogeneous product. For example, there are hundreds of thousands of farmers producing No.2 yellow corn. They are largely price takers in variable production input markets such as seed, fertilizer, and chemicals. This means that variable farm input manufacturers as well as manufacturers of farm machinery and equipment can pass along their rising costs directly to farmers and ranchers. As the Federal Reserve responds to rising inflation, it will begin to follow more **contractionary monetary policy** than before, which will raise interest rates on farm operating as well as farm mortgage loans.

Rate of Interest

Rising interest rates affect the level of farm interest expenses. Many farmers borrow hundreds of thousands of dollars or more each year to finance operating expenses

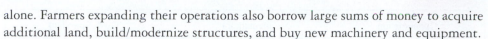

alone. Farmers expanding their operations also borrow large sums of money to acquire additional land, build/modernize structures, and buy new machinery and equipment.

The simple capitalization formula for asset values can be used to underscore the impact that interest rates have on farmland values. Equal to cash rent divided by a long-term interest rate like a 10-year constant maturity bond rate, the capitalized value of farmland will rise when interest rates fall, *ceteris paribus*. As interest rates fell sharply during the 2005–2014 period, farmland prices rose dramatically. Conversely, rising interest rates would have resulted in a correction in farmland values like that seen in the early 1980s (see Figure 15-2).

Rate of Unemployment

The rate of unemployment, aside from its effect on the aggregate demand for goods and services in the economy, affects rural employment opportunities particularly near urban areas. Off-farm income is an important source of income and internal finance for many farm operator families. This can be as a result of the crop farmer working off the farm during months where no production activity is taking place. More likely it comes from the farmer's spouse and other family members having a job in town. In many years, off-farm income is greater than net farm income.

Rate of Growth in Real GDP

As the economy expands, consumer incomes expand, although not necessarily at the same rate. A rise in consumer disposable income translates into an increase in the demand for all goods considered a normal as opposed to an inferior good. As you will recall from the elasticity discussion in Chapter 5, most agricultural products are normal goods, although associated with a low income elasticity of demand. Thus, a growing economy and rising disposable income does not mean that the demand for agricultural products will grow at the same rate. For example, if consumer incomes grow at a 10% rate in a given year, this does not mean that the demand for carrots will rise by 10%. Why? The income elasticity for carrots is closer to zero than one.

The impact of a change in real national income is greater on other sectors in the economy than it is on the farm sector due to the relatively low income elasticity for raw farm products.

Rate of Foreign Exchange

More often referred to as the value of the dollar, the **exchange rate** indicates how many units of foreign currency (e.g., Japanese yen) are required to "buy" a U.S. dollar to complete a transaction denominated in dollars. A weaker dollar against a foreign currency like the Japanese yen means that the yen "goes farther" in buying U.S. products. This exchange rate can have a significant impact on the export demand for U.S. agricultural products. A strong dollar can depress the export demand for U.S. agricultural products, while a weak dollar can enhance export demand. During the late 1990s when the financial crisis known as the "Asian Flu" weakened the strength of many Asian currencies relative to the U.S. dollar, the price of wheat per bushel fell almost in one-half. This had a significant impact on the net farm income of U.S. wheat producers.

IMPACTS OF MACROECONOMIC POLICY ACTIONS ON THE GENERAL ECONOMY

Several trends in the general economy that all economists, including agricultural economists, watch closely are trends in what we referred to above as the "Big Five." Adverse trends in the rate of inflation, rate of unemployment, rate of growth in **real GDP**, rate of interest, and the rate of foreign exchange or exchange rate from the farm sector's perspective can lead to financial stress in the farm sector. Over the last 5 years leading up to 2016, the farm sector benefited from low inflation, low interest rates,

Figure 15-3

Existing aggregate product and labor market in Lower Slobovia.

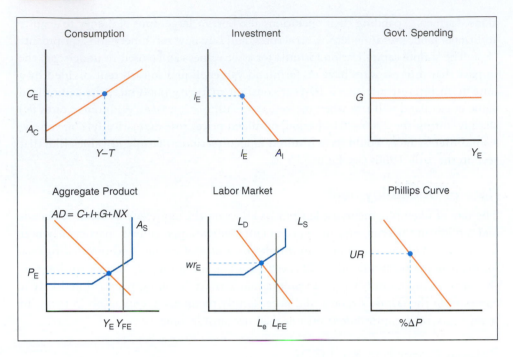

and a relatively weak dollar. It is important to understand how events in the economy and changes in macroeconomic policy can affect the farm sector.

The Real Economy

Chapters 12 through 14 discussed a variety of macroeconomic policy applications and their implications for the general economy. This included how monetary and fiscal policy can be used to close either a recessionary gap or an inflationary gap in the economy under normal circumstances.

Let's consolidate the discussion of topics covered in previous chapters by examining two new figures. Figure 15-3 captures these initial conditions in the aggregate product and labor market in the mythical country of Lower Slobovia before a macroeconomic policy shock is invoked. Here we see the initial equilibrium levels of consumption, investment, and government spending as well as the level of employment and the trade-off between inflation and unemployment in the economy illustrated by the Phillips curve.

The six individual graphs in Figure 15-3 relate to many of the Big Five variables. The Phillips curve, for example, illustrates the current trade-off between the rate of inflation and the unemployment rate associated with activity in Lower Slobovia's labor and aggregate product markets. Net exports, reflecting the rate of foreign exchange (value of the dollar relative to foreign currency), are captured in the product market as a component of GDP. These six graphs tell us a great deal about the Lower Slobovian general economy, including the general price level P_E (think in terms of the CPI for now) in the economy, the average wage rate in the labor market, and the potential for demand-pull and cost-push inflation by examining the Y_E/Y_{FE} ratio in the aggregate product market and the L_E/L_{FE} ratio in the labor market. Y_{FE} represents the **full employment output** of the economy, while L_{FE} represents the full employment labor force. As these two ratios approach 1.0, inflationary pressures increase in the economy from both a demand-pull and cost-push perspective. These six graphs in a highly aggregative way summarize the activity in this nation's **real economy**. What is missing is the Lower Slobovian economy's financial market, where the interest rate, the fifth and last of the Big Five variables, is determined.

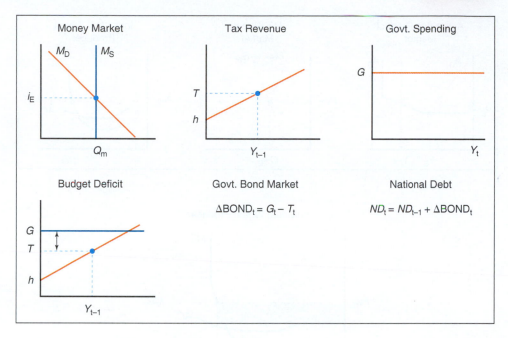

Figure 15-4
Existing conditions in the money market and the federal budget in Lower Slobovia.

The Monetary Economy

The discussion of monetary and fiscal policy and the relationship between these policies and the nation's money market and government bond market in the two previous chapters can be summarized as shown in Figure 15-4. This graph collectively represents the monetary segment of the nation's economy.[1] We see the equilibrium interest rate in the money market is influenced by monetary policy actions which shift the money supply curve to the right (expansionary) or the left (contractionary).

We also see the level of federal tax revenue and federal government spending in Lower Slobovia is influenced by fiscal policy and the implications its policy has for Lower Slobovia's federal budget deficit, the need for the nation's treasury to borrow funds in the government bond market to pay its bills, and what all this means for the level of the national debt in Lower Slobovia at the end of the year.

Figures 15-3 and 15-4 together summarize the existing events in Lower Slobovia's general economy. The economy's markets reflect a series of equilibrium associated with the Big Five variables. As addressed in Chapter 14, if these two figures depicted a recessionary gap in the aggregate product market, we would expect to see either expansionary monetary or fiscal policy, or both. If these two graphs instead depicted an inflationary gap in the aggregate product market, we would expect to see either contractionary monetary or fiscal policy, or both.[2]

MACRO–MARKET–MICRO LINKAGE

The final link needed to examine the effects of macroeconomic policy on the farm sector in Lower Slobovia involves understanding the various factors that shift the demand and supply curves for a particular commodity, including changes in disposable income and the role played by the income elasticity of demand discussed earlier in this book.

Figure 15-5 shows the market clearing price and quantity in the Lower Slobovian wheat market associated with aggregate demand and national income. Assuming that wheat producers in Lower Slobovia are perfect competitors, and thus price takers

---

[1] We can essentially ignore the stock market for the purposes of this discussion since stocks represent an asset on an owner's balance sheet and a liability on other balance sheets, thereby canceling each other at the aggregate or macro level.
[2] Full employment or YFE is less than potential output or YPOT, which represents the maximum output possible in the economy. It can be argued that full employment output represents the goal of macroeconomic policy.

Figure 15-5

Existing conditions at individual market and firm level in Lower Slobovia.

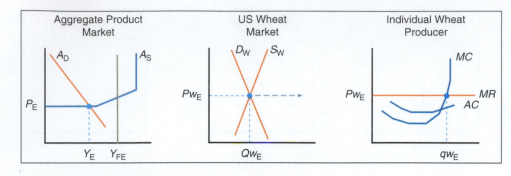

Figure 15-6

Profit at the firm level in Lower Slobovia. The shaded area is equal to average profit times the quantity produced.

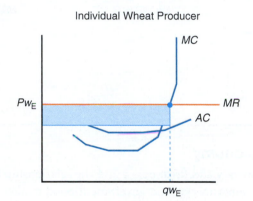

in the wheat market, we see the level of production planned by the individual wheat farmer is determined by the price taken from the wheat market.[3]

Before ending this discussion, we must examine whether the individual producer in Figure 15-5 is making an economic profit. If we study the last panel in Figure 15-5 with the help of Figure 15-6, we see that the individual producer in this case is making a profit. First we see that the producer's average profit, or average revenue (market price) minus average cost, is positive. Total economic profit in Figure 15-6 is equal to the height of the shaded rectangle (average profit) times the length of the shaded rectangle (quantity).

In summary, Figures 15-3 through 15-6 enable us to graphically examine the impacts that a change in macroeconomic policy at the macro level can have under conditions of perfect competition at the micro level.

IMPACTS OF MACROECONOMIC POLICY ACTIONS ON AGRICULTURE

The macro–market–micro equilibrium conditions presented in Figures 15-3 and 15-5 provide a basis for examining the effects of changes in monetary and fiscal policy on a perfectly competitive producer, namely, our wheat farm in Figures 15-5 and 15-6. In the case of imperfect competition, the last two graphs in the bottom row of Figure 15-5 would collapse into one. We will discuss the case of imperfect competition in the following section.

---

[3] We have initially assumed conditions of perfect competition, which are nearly satisfied in the U.S. wheat market. Wheat is produced by hundreds of thousands of wheat producers, and the demand curve reflects both the domestic and export demand for wheat. We will discuss how this discussion differs under imperfect competition later in this chapter. In addition, we are obviously ignoring the effects of government programs designed to address issues associated with program crops like wheat discussed earlier in Chapter 11.

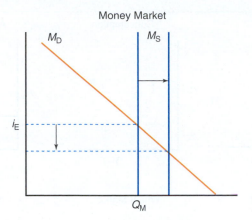

Money Market

Figure 15-7
Expanding money supply tends to reduce interest rates in the money market.

Effect of Expansionary Monetary Policy

Traditional implementation of **expansionary monetary policy** involves the use of one or more policy actions: (1) buying government bonds, (2) lowering the discount rate, and (3) lowering the fractional reserve requirement ratios. Each of these actions was discussed in previous chapters. Figure 15-7 shows the effect on interest rates as the money supply curve shifts to the right.

Lower interest rates stimulate investment not only because interest rates falling makes borrowing more attractive but also because businesses expect future growth in demand, which encourages businesses to enhance the firm's current capacity. Remember, a change in the money supply is a leading economic indicator, suggesting economic expansion (contraction) if the money supply increases (decreases). Lower interest rates and expanding national income stimulate purchases of consumer goods and services by consumers in addition to spurring activity in the nation's housing market. This set of events would change 180 degrees if the Federal Reserve were instead to implement contractionary monetary policy.

A general equilibrium in both the money market and the aggregate product market would produce the direction of activity in the economy shown in Figure 15-8. We see in the top row in this figure that investment expenditures expanded as interest rates declined, and consumer expenditures increased as incomes increased. The general price level would increase as aggregate demand rises, resulting in some demand-pull inflation. Additional labor was needed in the short run to produce the level of output needed to meet demand. This increases wage rates in the labor market and, if sustained, would also lead to cost-push inflation in the economy.[4] If we assume that the rise in wage rates in general does not affect wheat producers at least in the short run, the marginal cost curve of the individual wheat producer and thus the supply curve in the wheat market would not shift. Given the activity in the general economy, which leads to higher income levels, we would expect the price of wheat in the wheat market would rise since the income elasticity of demand for wheat is positive (i.e., it is a normal good). While not shown in Figure 15-8, we know the increased demand for labor will lead to a lower unemployment. This, in combination with rising inflationary pressures, means that the economy is moving down the Phillips curve. This macroeconomic policy action may lower the unemployment rate, but it will increase the general price level as well.

Just where the economy is on the aggregate supply curve shown in Figure 15-9 will influence the rate of inflation incurred in the economy. If aggregate demand was

Expansionary monetary policy which lowers interest rates typically helps the farm sector by increasing net farm income and farmland values. Contractionary monetary policy typically has the opposite impact on the farm sector.

[4] The argument can be made that the rising marginal costs of production will be offset to a degree by lower interest expenses since interest rates are falling.

Figure 15-8

Macro–market–micro effects of expansionary monetary policy in Lower Slobovia.

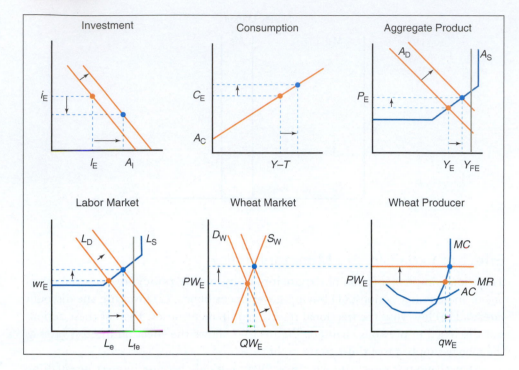

in the depression or Keynesian range, there obviously would be no inflation. If aggregate demand was instead in the classical range, we could see double-digit inflation or even hyperinflation. Figure 15-8 reflected an economy at the lower end of the normal range, so inflation was relatively small given the expansion in aggregate demand and subsequent increase in wage rates paid to labor.

The last two graphs in the bottom row of Figure 15-8 show that expansionary monetary policy will increase demand for wheat (domestically since incomes have increased, and globally because lower interest rates will make the dollar cheaper and expand the export demand for wheat). This increases the price of wheat. The individual wheat producer will respond by expanding production. How does this affect the net income of the individual wheat producer? Figure 15-10 suggests the price of wheat would increase from PW_{E1} to PW_{E2}. The wheat producer responds by increasing production from qw_{E1} to qw_{E2}. As a result, the wheat producer's revenue from the market would increase from the shaded area in the lower left-hand corner graph to the larger shaded area in the lower right-hand graph. The inset text box in Figure 15-10 further suggests that the producer's net farm income would therefore increase. While the producer's expenses increased due to the increased level of production, as would

Figure 15-9

Ranges of aggregate supply curve. The U.S. economy typically operates somewhere in the normal range.

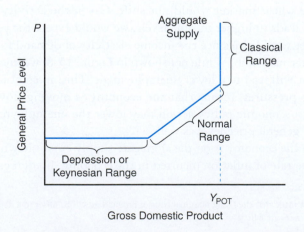

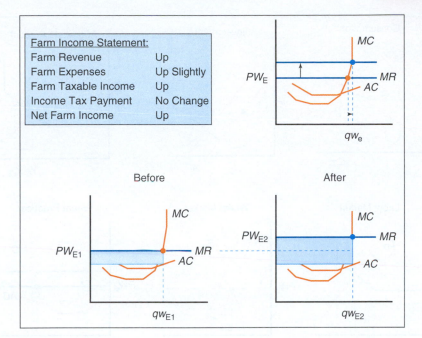

Figure 15-10

Implications for the farm sector in Lower Slobovia at the micro level.

farm taxable income, income tax payments in the current year are assumed to be unchanged in the current period since these payments are not due until early next year. Finally, since net farm income in the current period is higher and interest rates are lower under this expansionary monetary policy, the capitalized value of farmland would rise, as would the producer's end-of-year equity or net worth. Thus, we can conclude that expansionary monetary policy is good for this wheat producer and for others producing a normal product (demand increases with an increase in income).

The entire process covered in Figures 15-8 through 15-10 for expansionary monetary policy actions would be reversed if a contractionary monetary policy action was adopted. A shift of the money supply curve in Figure 15-7 would increase interest rates, lower GDP and national income, make exports more expensive, and lead to a decline in net farm income. This decline, coupled with a higher interest rate, would lower the capitalized value of farmland and lower the producer's equity.

Effect of Contractionary Fiscal Policy

Contractionary fiscal policy involves increasing the marginal income tax rate, cutting government spending, or both. Figure 15-11 assumes that policymakers have chosen to increase taxes by increasing the marginal income tax rate and leave government spending at current levels. This action would increase taxes on the previous

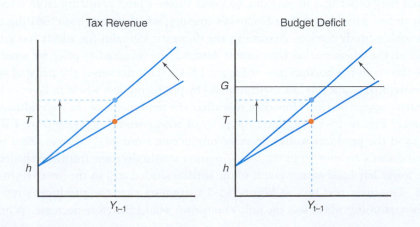

Figure 15-11

Increasing the marginal income tax rate increases the slope of the federal revenue curve. This has the effect of reducing the gap between G and T or the budget deficit in the short run.

Figure 15-12
Macro–market–micro effects of contractionary fiscal policy in Lower Slobovia.

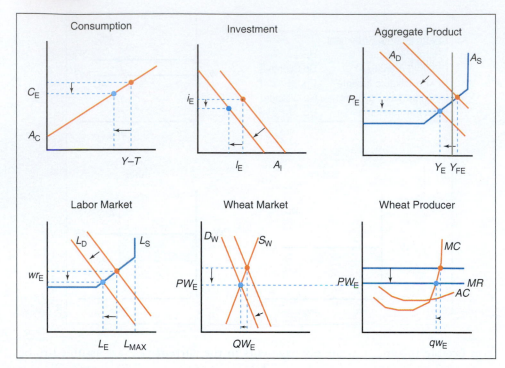

period's income as shown in the left-hand graph, which implies a reduction in the federal budget deficit. This is seen as the narrowing of the distance between T (taxes) and G (government spending) in the right-hand graph. An **expansionary fiscal policy** action would produce the opposite results.

A general equilibrium in both the money market and aggregate product market as a result of this contractionary fiscal policy action would produce the direction of activity in the economy shown in Figure 15-12. We see that consumption expenditures would decline as a result of the decrease in disposable income. Investment expenditures would also decline as profit expectations decline based on the expectation of declining consumer disposable income from an increase in marginal tax rates or reduction in government spending. Both would reduce aggregate demand. Declining wage rates in the labor market as the demand for labor falls would result in less cost-push inflation. This, along with less demand-pull inflationary pressures in the aggregate product market, lowers the general price level in the aggregate product market, or disinflation.

The last two graphs in the bottom row of Figure 15-12 show that contractionary fiscal policy will decrease the demand for wheat as domestic income falls. Global demand may offset this in part due to lower interest rates resulting from a reduced deficit to be financed by Lower Slobovia's treasury, which means less crowding out in the loanable funds market. Assuming the domestic demand for wheat has a larger impact on the wheat market than export demand, the commodity price for wheat will fall as shown in the bottom row of Figure 15-12. This decreases the price of wheat. The individual wheat producer will respond by producing less wheat in favor of better alternative opportunities. How does this affect the net income of the individual wheat producer? Figure 15-13 suggests the price of wheat would decrease from PW_{E1} to PW_{E2} and the producer would decrease production from qw_{E1} to qw_{E2}. As a result, this producer's revenue from the wheat market would decrease from the shaded area in the lower left-hand corner graph to the smaller shaded area in the lower right-hand graph. The inset text box in Figure 15-13 suggests that the producer's net farm income (assuming wheat was the only enterprise) would therefore decrease. While the

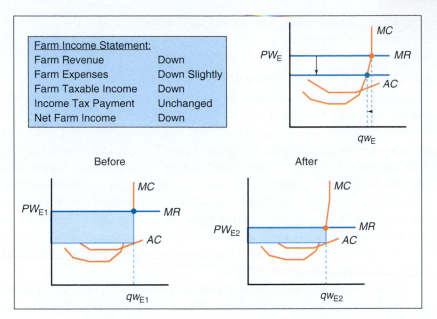

Figure 15-13
Contractionary fiscal policy implications for the farm sector in Lower Slobovia at the micro level.

wheat producer's expenses decreased due to the decreased level of production, as would farm taxable income, the income tax payment is not payable until the following year in Lower Slobovia. Finally, since net farm income is lower and interest rates are lower under this contractionary fiscal policy, the capitalized value of farmland would fall, assuming lower net farm income offsets the effect of lower interest rates, and the producer's end-of-year net worth or equity would decline. This would be compounded if government payments to the wheat producer were reduced by a cut in discretionary programs. Thus, we can conclude that contractionary fiscal policy is, on balance, not good for this wheat producer and for others producing a normal good.

The entire process covered in Figures 15-8 through 15-13 for contractionary fiscal policy actions would be reversed if an expansionary fiscal policy action was adopted. A downward shift of marginal income tax rate function as shown in Figure 15-14 in Lower Slobovia would lower the government's tax revenue and increase the country's federal budget deficit in the short run. This action would be adopted by macroeconomic policymakers if it was felt that the economy needed a fiscal stimulus, or boost in private spending. The tax cut would increase consumer disposable income and business expected profits, thereby leading to greater spending by consumers and businesses. The major difference between expansionary monetary policy actions and expansionary fiscal policy actions is what happens to interest rates. Unlike expansionary monetary policy, increasing deficits in the short run by cutting taxes would mean that the Lower Slobovian treasury would have to borrow more to cover government

Contractionary fiscal policy which increases marginal tax rates or cuts government spending typically lowers net farm income and farmland values, particularly if government payments to farmers are decreased.

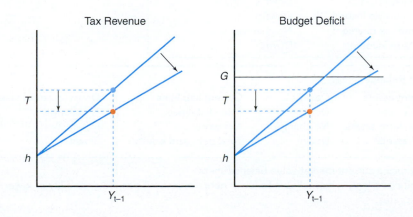

Figure 15-14
Cutting the marginal income tax rate reduces the slope of the federal revenue curve. This increases the gap between G and T or the budget deficit in the short run.

spending which, in turn, would lead to increased crowding out in the loanable funds markets, driving up interest rates.

Microeconomic Performance Implications

We can summarize the effects of the policy actions on farmers and ranchers in Lower Slobovia by examining Figures 15-15 and 15-16. In Figure 15-15, we concluded that expansionary monetary policy is good for Lower Slobovian farmers and ranchers since it leads to higher net farm income in the short run. The rise in net farm income will enable farmers and ranchers to retain more profits, which leads to higher retained earnings and hence current assets. The combination of higher net farm income and lower interest rates will lead to higher capitalized farm values, which will increase their net worth or equity in their farming operations. Thus, expansionary monetary policy is seen as a win-win proposition for the Lower Slobovian farm sector. Conversely, contractionary monetary policy that increases interest rates will likely reverse many of the impacts summarized in Figure 15-15.

In Figure 15-16, we concluded that contractionary fiscal policy is not beneficial to Lower Slobovian farmers and ranchers. While interest rates will decline as less crowding out takes place in the loanable funds market, net farm income will decline as consumer disposable income falls or government spending is cut. Thus, while export demand will benefit from a lower value of the dollar, domestic demand will

Figure 15-15
Summary of short-run impact that expansionary monetary policy has on farmers and ranchers in Lower Slobovia.

Farm Income Statement:

Farm revenue	Up
Farm expenses	Up slightly
Farm taxable income	Up
Income tax payment	No change
Net farm income	(Up)

Farm Balance Sheet: 1/

Current assets	Up	Current liabilities	Up (accrued taxes)
Land values 2/	Up	Fixed liabilities	No change
Other farm assets	No change	Net worth	(Up)
Total assets	Up	Total debt and equity	Up

1/ Assumes a current market value balance sheet.
2/ Capitalized land values rise due to higher net farm income and lower interest rates.

Figure 15-16
Summary of short-run impact that contractionary fiscal policy has on farmers and ranchers.

Farm Income Statement:

Farm revenue	Down
Farm expenses	Down slightly
Farm taxable income	Down
Income tax payment	No change
Net farm income	(Down)

Farm Balance Sheet: 1/

Current assets	Down	Current liabilities	Down (accrued taxes)
Land values 2/	Down	Fixed liabilities	No change
Other farm assets	No change	Net worth	(Down)
Total assets	Down	Total debt and equity	Down

1/ Assumes a current market value balance sheet.
2/ Capitalized land values fall due to lower net farm income despite possibility of lower interest rates.

fall with a decline in consumer disposable incomes. Given the relatively low income elasticity of the demand for farm products, however, contractionary fiscal policy is likely to have less negative implications for farmers and ranchers than contractionary monetary policy. That is due largely to the more favorable direction of interest rates. Thus, while Figure 15-16 suggests that net farm income and land values may decline, the impact is considerably less harmful than that of contractionary monetary policy.

Implications for Imperfect Competition

Much of this chapter has focused on those who come close to meeting the assumptions of perfect competition, like an individual wheat producer. To remedy this, let's discuss what is unique about the link between aggregate demand and imperfect competitors. You will recall when discussing the implications of expansionary monetary policy from a macro–market–micro perspective in Figure 15-8 we explicitly captured the relationship between the price-taking wheat producer and the wheat market.

What is unique about imperfect competition is that the middle commodity market is actually incorporated in the micro-level graph. Let's consider a monopoly farm equipment manufacturer in Lower Slobovia. The demand for farm equipment is a derived demand; as net farm income in Lower Slobovia increases, the demand curve for farm machinery to either replace existing worn-out or obsolete equipment or expand capacity shifts to the right. This would shift the demand curve from D_1 to D_2 in Figure 15-17. Marginal revenue is now associated with the downward-sloping demand curve rather than a perfectly elastic relationship the wheat producer faces in the more perfectly competitive wheat market.

The firm still produces where marginal cost equals marginal revenue. The difference is that it sets the price based upon consumers' willingness to pay given by the demand curve. For the D_1 curve, the monopolist would produce quantity Q_1 but price its product at P_1. Given the increase in demand, the monopolist would produce Q_2 and attempt to charge price P_2.

Implications for Other Sectors in the Food and Fiber Industry

The food and fiber industry in any economy is a grouping of input supply firms like John Deere Company, marketing intermediaries like a local elevator, processors who add value to raw agricultural products, and wholesale and retail firms that move the product to its ultimate consumer. There are a number of interdependencies involved. We discussed the derived demand relationship between an input supply firm on the one hand and farmers and ranchers on the other in the previous section assuming the case of imperfect competition. Forward-linked firms from the local elevator to the

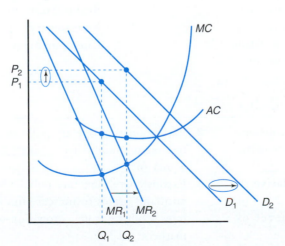

Figure 15-17
Implications for imperfect competition.

retail grocery store are affected by unique events like droughts, hard freezes, and other externalities originating in the farm sector. These externalities show up in forward-linked sectors when surpluses or shortages occur in the farm sector.

SUMMARY

The purpose of this chapter is to examine the manner in which macroeconomic policy actions can affect the economic performance and financial strength of the farm sector of the economy. While the discussion was cast in the context of the mythical country of Lower Slobovia, the conclusions drawn for the short run are largely applicable to the U.S. economy as well. The major points in this chapter are summarized as follows:

1. Expansionary monetary policy which leads to rising disposable income, moderate inflation, a lower unemployment rate, lower interest rates, and a weaker U.S. dollar is generally good for the economic performance of the farm sector at the aggregate level. These policies have historically led to higher net farm incomes and rising farmland values. Contractionary monetary policy typically has the opposite effect.

2. Expansionary fiscal policy which leads to rising disposable income, a lower unemployment rate, and moderate inflation is also generally good for the economic performance of the farm sector at the aggregate level. The exception is the potential for "crowding out" in the loanable funds market as the government borrows to finance deficit spending, which can lead to higher interest rates and a stronger dollar. Contractionary fiscal policy actions, which can include cuts in federal farm programs

and higher tax payments, typically have the opposite effect.

3. The major channels through which macroeconomic policy is transmitted to the farm sector are referred to as the "Big Five" variables. These are interest rates, the *rate of inflation*, the unemployment rate, the rate of growth in real GDP, and the foreign exchange rate.

4. The farm sector realizes these effects through the prices farmers receive for their production, the input prices they pay for production inputs (variable inputs as well as machinery and equipment), the interest rates paid on loans, the value of the dollar's effect on export demand, the availability of off-farm employment for farm operator family members, and the potential appreciation of farmland values.

5. Given the derived demand relationship between the farm sector and the other sectors in the nation's food and fiber industry (input supply, processing, and wholesale/retail trade), the effects that macroeconomic policy has on the farm sector have implications for the others. For example, input supply firms like John Deere will experience declining demand for their product line when the farm sector is weakened by *contractionary monetary policy* actions, and a growing demand when the farm sector is strengthened by expansionary macroeconomic policy actions.

KEY TERMS

Contractionary fiscal policy	Full employment output	Real GDP
Contractionary monetary policy	Inflationary gap	Real interest rate
Exchange rate	Interest rate	Recessionary gap
Expansionary fiscal policy	Rate of inflation	Unemployment rate
Expansionary monetary policy	Real economy	

TESTING YOUR ECONOMIC QUOTIENT

1. The distinguishing difference between expansionary monetary and expansionary fiscal policy is the direction of movement in the _____.

2. A decrease in the value of the dollar relative to foreign currencies is likely to cause the level of imports to _____ and the level of exports to _____.

3. Expansionary monetary policies are likely to cause net farm income to _____ and farm equity to _____.

4. Expansionary monetary policies would likely cause farmland prices to increase due to rising interest rates resulting from a crowding-out effect in credit markets.   T   F

5. Real net farm income increased dramatically over the 1970–2012 period.   T   F

6. A major factor influencing net farm income and farmland values in the 2000s and 2010s was the renewable fuel mandate for producing ethanol. T   F

7. Expansionary monetary policies to promote an economic recovery from the 2007–2009 recession were hampered by existing historically low interest rates.   T   F

8. The nation's food and fiber industry
   a. includes the health care sector.
   b. represents a group in independent sectors including the farm sector.
   c. is affected by changes in farm policy and environmental regulations.
   d. none of the above.

9. Expansionary fiscal policy
   a. leads to growth in aggregate demand resulting from lower interest rates.
   b. causes an increase in national debt.
   c. can be inflationary if the economy is in the Keynesian range of the aggregate supply curve.
   d. none of the above.

10. The national debt in the current year is
    a. equal to the national debt at the beginning of the year minus the annual budget deficit.
    b. equal to the national debt at the end of the year plus the annual budget deficit.
    c. equal to the national debt at the beginning of the year plus the annual budget deficit.
    d. none of the above.

11. The "Big Five" variables
    a. include the rate of interest, the rate of employment, the rate of inflation, the rate of growth in real GDP, and the exchange rate.
    b. represent key linkages between the sectors in the food and fiber industry.
    c. include the rate of interest, the rate of unemployment, the rate of inflation, the rate of growth in real GDP, and the exchange rate.
    d. none of the above.

12. The economy of Lower Slobovia is currently experiencing an interest rate of 12% and a real GDP of $5,000. The level of aggregate demand puts it in the Keynesian range of the country's aggregate supply curve and perfectly elastic range of the country's labor supply curve. Suppose the government has announced its intent to achieve a target interest rate of 6% and a real GDP level of $10,000. Further assume this action would still leave the economy in the perfectly elastic ranges of its aggregate supply and labor market supply curves.
    a. Identify the specific macroeconomic policy action you would recommend the government to take to achieve these twin targets.
    b. What impact will the change in the general economy have on the market equilibrium for wheat in Lower Slobovia?
    c. What impact would you expect the macroeconomic policy you recommended in *part a* on an individual wheat producer's average profit? Total profit?

13. Please complete the blanks in the following table for the Lower Slobovian economy operating in the normal range of the aggregate supply curve:

Policy Effects on	Expansionary Monetary Policy	Expansionary Fiscal Policy	Contractionary Monetary Policy	Contractionary Fiscal Policy
Interest rate				
Real GDP				
Unemployment rate				
Inflation rate				
Exchange rate				
Net farm income				
Farmland value/acre				

REFERENCE

Economic Research Service, U.S. Department of Agriculture: Farm Income and Finances database, www.ers.usda.gov/topics/farm-economy/farm-sector-income-finances.aspx.

🍎16

Agricultural Trade and Exchange Rates

This chapter examines recent trends in U.S. agricultural trade and the importance of exchange rates. The changing product composition of U.S. agricultural trade is examined, and an overview of the changing direction of trade, the factors affecting market potential, and international competition is discussed. Exchange rate determinants are discussed along with the role of currency values influence of international trade.

GROWTH AND INSTABILITY IN AGRICULTURAL TRADE

The **value** of U.S. agricultural **exports** experienced unprecedented growth during the 1970s. This growth was surpassed since then by rapid spurts in the 1990s and 2000s (Figure 16-1). Agricultural export volume growth has been less rapid due to price swings and economic decline from time to time.

Value is the dollar amount of exports, imports or the trade surplus.

Export Boom and Bust

U.S. agricultural exports have been characterized by periods of rapid growth and steep decline. Trade growth has resulted in higher farm prices and incomes but increased exposure to uncontrollable world market forces and greater market instability. Export decline leads to falling farm prices and lower farm incomes.

Since the mid-1990s, U.S. farm exports have expanded to a record $150 billion in 2014. **A weaker dollar** and global economic recovery that led to stronger demand for U.S. agricultural exports led to this growth. In addition, lower trade barriers and trade liberalization stimulated export expansion and opened markets previously closed to U.S. products. Together, these factors made U.S. agricultural exports more competitive on the world market and increased foreign demand.

U.S. agricultural trade has experienced more instability during the past three decades. In contrast to the stable markets of the 1950s and 1960s, the 1970s, 1980s, and 1990s were a virtual rollercoaster for most U.S. farm and agribusiness industries. Figure 16-1 shows that the destabilizing effects of world market forces on U.S. agriculture began in the early 1970s and have continued to the present. Greater

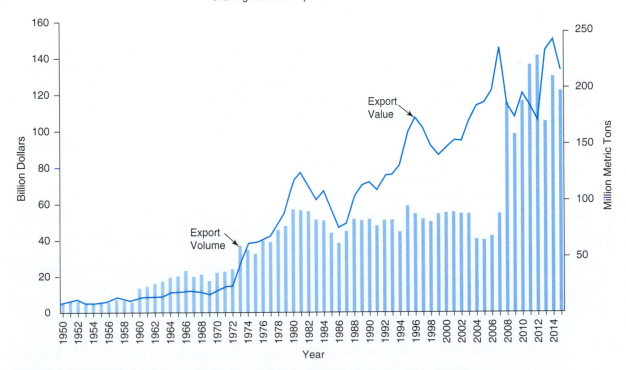

Figure 16-1

U.S. agricultural exports reached record levels in 2012 due to high levels of economic growth in many countries. Record high prices led to record levels of the value of exports. Exports are expected to remain volatile, creating additional uncertainty for farm and agribusiness managers and rural communities.

Source: U.S. Department of Agriculture, 1970, and U.S. Department of Agriculture, FATUS, 2012.

Grain shipped to former Soviet Union from Charleston, South Carolina.   Credit: Volodymyr Kyrylyuk/Fotolia.

uncertainty for agriculture has also meant increasing instability for input supply businesses, marketing firms, and related sectors of the agricultural and nonagricultural economies. Rural communities have also shared in the boom or bust fortunes of U.S. agriculture.

Moves toward Trade Liberalization

The Uruguay Round Agreements (URA), which involved 125 nations, were completed on December 15, 1993, ratified by the U.S. Congress, and signed into law on April 15, 1994. The URA was implemented on January 1, 1995. The URA represents the most substantial reform of agricultural trade undertaken by GATT parties and an important first step to additional trade liberalization in subsequent rounds of negotiations. Major provisions of the URA included the following:

Creation of the WTO The Uruguay Round of GATT created the World Trade Organization, lowered import tariffs, reduced export subsidies, and curtailed spending on domestic farm support.

Tonnage is the volume of weight in metric tons.

1. Created the World Trade Organization (WTO) to implement the URA provisions, provided a dispute settlement body, and provided a trade policy review mechanism.
2. Provided market access by converting quotas to **tariff** equivalents, then phased in a reduction of tariffs by an average of 36% over 6 years, with a minimum of 15% reduction for each tariff and minimum market access growing from 3% to 5% of domestic consumption.
3. Reduced export subsidies by 36% in expenditure and 21% in **tonnage** over 6 years.
4. Lowered trade-distorting domestic agricultural support by 20% over 6 years (the United States and the European Union are already in compliance, so no additional reductions in farm support programs are required by these countries).
5. Uses accepted international standards for food safety and animal and plant health regulations, harmonizes those standards in order to prevent their use as nontariff barriers to trade, and establishes disease/pathogen-free zones within geographic boundaries of countries affected by certain diseases and pests. Countries may maintain standards more stringent than the international standards if they are based upon sound scientific evidence.

In addition to multilateral trade liberalization, the United States has negotiated free trade agreements with many other countries. The most recent trade negotiations are the Trans-Atlantic Trade and Investment Partnership (TTIP) with the 28-nation European Union and the Trans-Pacific Partnership (TPP) agreement with Australia, Brunei, Canada, Chile, Japan, Malaysia, Mexico, New Zealand, Peru, Singapore, and Vietnam. The TPP negotiations have concluded, while the TTIP negotiations are still in progress.

A new round of multilateral trade negotiations (MTNs) in agriculture began in 2000 under the auspices of the World Trade Organization. Under the URA of GATT, agriculture was obligated to begin new negotiations in 2000 or revert to more restrictive trade measures agreed to before the end of the URA. While attempts were made in November 1999 to launch a comprehensive MTN in Seattle, Washington, involving all sectors, these efforts failed because of disagreement between the United States and the Cairns group in one camp and the European Union, Japan, and some developing countries in the other. In addition, many environmental groups staged disruptive demonstrations that precluded delegation members from attending some key sessions and further hindered progress at the meetings. A new round was started in Doha, Qatar, in November 2001.

The Doha Development Agenda, as the current round of multilateral trade negotiations is called, involves 162 WTO members and has economic development of poorer countries as one of the focal points. Other major issues include negotiating reductions in government support of farm programs that distort trade, which will affect U.S. policy, eliminating export subsidies, and increasing access to all markets for agriculture, services, nonagricultural products, and intellectual property.

> **MTN and the Doha Round** A new round of multilateral trade negotiations, the Doha Development Agenda, started in 2001. Agricultural trade liberalization is an important part of these talks.

THE IMPORTANCE OF AGRICULTURAL TRADE

U.S. agriculture's increased dependence on foreign markets has made U.S. farmers and rural communities more sensitive to forces and events beyond U.S. borders. Weather in other countries; foreign, domestic, and trade policies; and political events in importing nations all have a direct and profound impact on the well-being of U.S. agriculture. Increased uncertainty associated with greater dependence on international trade means that U.S. producers and agribusinesses need to become better informed about global events affecting their operations. Improved understanding of global forces will be crucial for those managers trying to make better decisions about the impacts of U.S. macroeconomic policies, export competitor policies, or changing price trends on their businesses.

Increased Export Dependence

U.S. agriculture is relatively **export dependent**. More than 20% of all U.S. agricultural output was exported in 2013, accounting for 28% of farm income. About 10% of all animal products are exported. This compares to 24% of plant products.

Greater Dependence on Imports

Since the 1980s, a growing share of domestic food consumption has reflected greater **import dependence**. In 2014, about 20% of all U.S. food consumption expenditures would be attributed to imported foods and beverages. About 12% was attributed to animal products, while 27% was attributed to plant products.

In some cases, however, **imports** play a vital role in supplying the processing needs and capacity of U.S. food processing plants. While imported tomatoes and peas may compete with U.S. production during some parts of the year, without imports

> **Import dependence** The share of agricultural production attributed to imported goods.

> **Imports** represent the amount of goods entering a country's borders.

many U.S. food processors could be forced to reduce plant capacity, relocate processing facilities, or close the plant. These alternatives would limit the market for U.S.-grown products.

Imported foods satisfy consumer demand for a diverse market bundle of goods. Imports represent a lower-cost food product. By purchasing imports at relatively lower prices, consumers are actually working to increase their disposable incomes. Imports also represent more variety of choice to the consumer, thereby increasing the competition for U.S.-produced goods and effectively lowering the cost to the consumer.

Noncompetitive imports such as tropical oils, coffee, tea, and cocoa are major sources of essential food products. These imports represent additional product variety for consumers at a lower cost. In many cases, limiting imports would reduce or eliminate available supplies, resulting in substantially higher food costs and a lower standard of living.

U.S. import shares for animal products ranged from less than 1% for poultry to 95% for fish and shellfish. Import shares of tropical products are about 97% for coffee, cocoa, tea, and species.

USDA economists estimate that a minimum of $1.29 in additional business activity is generated for every dollar of agricultural exports. Agricultural exports generate nearly one million jobs for the U.S. economy annually. One new job is created for every $147,000 in agricultural exports. Of these jobs, 286,000 are in the farm sector, while 637,000 are in nonfarm sectors. Food processing, transportation, manufacturing, and services account for the majority of nonfarm jobs attributed to U.S. agricultural exports.

Indirect economic impacts of export activity are also important. With additional export earnings, farm households can purchase durable goods, equipment, building supplies, and other capital and consumer products; thus, additional purchasing power is spread throughout the total economy. These spinoff effects should always be considered when estimating economic impacts of agricultural trade. Indirect economic benefits to services, such as transportation and wholesale and retail trade, lead to increased investment, greater operating capacity, and heightened competitiveness for U.S. agriculture.

THE COMPOSITION OF AGRICULTURAL TRADE

The composition of agricultural exports and imports has changed dramatically over the last two decades. U.S. food and fiber exports have become more value-added, and imports have become essential high-quality food products. The role of both exports and imports is discussed in the following section. Important changes affecting the traded product mix are emphasized.

The Role of Agricultural Exports

Commodity mix of U.S. exports The majority of U.S. agricultural exports are now products such as frozen broilers, fresh produce, flour, vegetable oils, and hides and skins.

Bulk products, such as wheat, corn, and soybeans, account for 40% of U.S. agricultural exports. Consumer-oriented goods such as meats, fruits, vegetables, dairy products, and other processed goods represent about 40% of exports, while the balance is intermediate goods such as hides, skins, and vegetable oils.

The Role of Agricultural Imports

U.S. agricultural imports have increased since 2000 from $39 billion to $113 billion in 2015. Agricultural imports have become an important source of low-cost, high-quality foods for consumers, and they are a major source of foreign competition for some U.S. producers. Import growth marked the emergence of a U.S. market for a large variety of high-quality food products at reasonable prices, regardless of their source or national origin. Growth in imported food products also heightened consumer awareness of food safety issues and international environmental concerns, and focused attention on the important role other countries play in supplying critical U.S. food needs.

About half of all U.S. agricultural imports are horticultural products. Raw products and semi-processed goods, such as flour, meals, and oil, are the fastest-growing components of trade.

DIRECTION OF U.S. AGRICULTURAL TRADE

The direction of U.S. agricultural trade has shifted dramatically over the last two decades. Developed nations in the EU have declined in importance as markets. Developing countries like China have increased in their importance as markets for U.S. agricultural exports.

Major Export Markets

Trade patterns can be viewed on a regional or country basis. Canada, China, Mexico, Japan, and South Korea are consistently the top five markets for U.S. agricultural exports. They account for more than half of all U.S. exports. These five are followed by Taiwan, Hong Kong, Indonesia, the Netherlands, and the Philippines. Two-thirds of all U.S. agricultural exports are destined for one of these top 10 markets.

Much of the growth in exports to Mexico and Canada is attributed to the North American Free Trade Agreement (NAFTA). Canada, Mexico, India, China, and Italy are the top five suppliers of imports to the U.S. agricultural market, accounting for nearly half of the total. These countries are followed by Brazil, Indonesia, Ireland, Chile, and New Zealand. All but two, Ireland and New Zealand, are developing countries. These top 10 suppliers account for about two-thirds of all U.S. food imports. Again, the prominence of developing countries is important.

Major Import Suppliers

Major suppliers of food and fiber products to the United States have changed significantly.

Developing countries in Asia and Latin America are supplying an increasing amount of food and fiber to the U.S. market. Brazil is the largest supplier of frozen concentrated orange juice and a major supplier of coffee. Most developing country suppliers of food to the United States tend to specialize in those products that require hand labor to produce and harvest, thereby capitalizing on their natural advantage, low-cost labor.

U.S. AGRICULTURAL TRADE PERFORMANCE

The importance of international trade to agriculture and other sectors of the economy has been well documented in previous sections of this chapter. What may not be apparent, though, is the importance of trade to the overall economy and recent trends in trade performance.

U.S. merchandise trade as a share of gross domestic product has been rising steadily since the mid-1960s. In recent years, trade has consistently accounted for 25% of U.S. gross domestic product (GDP). Recent efforts to open the U.S. economy

Top U.S. export markets   The top markets for U.S. agricultural exports are Canada, Mexico, Japan, the European Union, and Asian countries such as South Korea and China. As a group, developing countries are larger markets than developed countries.

Top import suppliers   Canada is the top import supplier of food products to the U.S. market. Mexico, the European Union, and Australia follow.

and increased U.S. dependence on trade make it critical to understand the performance of trade and what it means for agriculture.

The Balance of Trade

The **balance of trade** is commonly used to measure overall trade performance. The merchandise trade balance measures the difference between total revenues earned from exports and total expenditures on imports.[1] If exports exceed imports, the nation is said to have a **trade surplus**. If more is imported than exported, the nation has a **trade deficit**. A trade deficit may be portrayed as a bad thing, but the merchandise trade account is only one component of a nation's total balance of trade. Placing emphasis on monthly or even quarterly reports of merchandise trade statistics may mislead policymakers, resulting in misdirected macroeconomic policy, overcorrection, and futile attempts to reduce short-term imbalances in trade.

Several factors were responsible for large U.S. trade deficits in recent history. First, the United States underwent its longest sustained period of economic growth since World War II, stimulating strong demand for imported goods. Nearly two-thirds of the U.S. merchandise trade deficit was accounted for by oil/petroleum products and automobiles. Second, the U.S. dollar rose to record levels against other currencies, leading to less expensive imports and more expensive exports. Thus, the balance of trade declined. Third, a proliferation of nontariff barriers to trade such as import licenses, quotas, and export subsidies were used by key trading partners, such as Japan and the EU, to limit imports to protect domestic employment from foreign competition. Finally, because of a large domestic market and inward orientation, many U.S. firms lacked the necessary incentives and expertise to aggressively pursue international marketing opportunities.

U.S. agriculture has consistently generated a trade surplus every year since 1960. Although this surplus was relatively small during the 1950s and 1960s, it grew rapidly during the 1970s and 1980s and reached $36 billion in 2012, falling by one-half since then. Further, agriculture exhibited strong trade performance despite rapid growth in imports.

The balance of trade has two important implications for the economy as a whole and for agriculture in particular. First, a trade surplus provides capital to a nation because the nation is selling more than it is buying. This capital surplus may be used to satisfy foreign debt or to increase both domestic and foreign investment, enhance consumption by increasing imports of foreign goods, or save for future consumption and investment. For agriculture, the trade surplus has been an important means of partially offsetting the large U.S. trade deficit and stimulating the import of key food and fiber products at a lower cost to consumers. Second, trade deficits, particularly if chronic, may reduce a nation's capital reserves, leading to policies to restrict or limit imports or necessitating the need to borrow foreign capital. The trade balance also has some important impacts on the exchange value of currencies, which is discussed in the following section.

Remember that a trade deficit is not necessarily bad and that a trade surplus is not necessarily good. When trade is in surplus, a country is giving up more than it receives. A trade deficit indicates that a country is receiving more goods than it is giving up. Neither condition can be sustained indefinitely. Each case of surplus or deficit must be analyzed based on the particular circumstances causing the underlying trade position. Neither chronic surpluses nor chronic deficits are in the best interests of any

---

[1] Merchandise trade is normally reported as the nation's trade balance, in contrast to trade in services, such as shipping, insurance, banking, and tourism, which are referred to as trade in invisibles.

Figure 16-2
Exchange rates influence trade.

Credit: semisatch/Fotolia.

nation. Care should be taken not to overemphasize any short-term condition or report of a partial measure of trade or performance.

The decoupling of international capital movements and trade flows has major implications for industries that are dependent on exports, compete with foreign imports, and tend to be capital intensive (Figure 16-2). U.S. agriculture is one such industry.

EXCHANGE RATES AND THE FOREIGN EXCHANGE MARKET

In this chapter, we have learned that many factors influence the level and direction of agricultural trade. The **exchange rate**, the number of units of foreign currency that can be exchanged for one unit of domestic currency, is one of the most important factors affecting the level and the destination of U.S. agricultural exports. Relative prices determine which goods are traded and where they are shipped. For the businessperson making the decision of where to buy and where to sell, the process is one of converting one currency to another at the prevailing rate of exchange and comparing the prices. The difference in relative prices determines the flow of goods and services and, therefore, the patterns of international trade.

The exchange rate is the number of units of foreign currency that can be exchanged for one unit of domestic currency.

Exchange Rates Defined

Trade is affected by exchange rates through the price linkage. Exchange rates are used to convert international market prices to domestic currency equivalents. If the exporter's currency declines in value, for example, the importer's cost of foreign exchange will decrease, thereby lowering the commodity price in the import market and increasing the quantity demanded. Consequently, world prices will rise, inducing exporters to increase sales on the world market. If the exporter's currency increases in value, the foreign currency cost to the importer will rise, resulting in higher prices and a decline in the quantity demanded of imported goods.

Today, the currencies of most developed nations are fully convertible to one another at market-determined rates. As such, the exchange rate may be thought of as the value, or price, of one currency in terms of another currency. For example, assume that the euro (€) is valued at $0.69 and the British pound sterling (£) at $1.56. In this case, the exchange rate is equivalent to the number of U.S. cents, or dollars, one euro or one British pound sterling will buy. The reciprocal of this relationship reflects the number of foreign currency units per U.S. dollar. For instance, if €1 = $0.69, then $1 = €1.44, which can be calculated as 1 ÷ $0.69. The U.S. dollar, in effect, will buy €1.44. The relationship between the euro and the pound can also be expressed as the **cross rate**, which is the exchange rate between two currencies as calculated from the value of a third currency. For example, the cross rate between the euro and pound is calculated by multiplying the number of dollars per one pound by one euro per number of U.S. cents as

$$(\$1.56/£) \times (€/\$0.69) = (€1.56/£0.69) = €2.26/£ \qquad (16.1)$$

The Foreign Exchange Market

Currency values are determined on the foreign exchange market where they are traded. Global currency trading exceeds $2.0 trillion each business day, far surpassing world trade in goods and services. In fact, international financial transactions account for about 90% of the foreign exchange market activity around the world. Currencies are traded in major international centers in London, New York, and Tokyo.

In reality, there is no "Big Board" on which currencies are traded or any one foreign exchange market. There are, however, a large number of private and institutional users of foreign currency who have foreign exchange departments and are in continuous communication regarding currency values and quantities traded. The foreign exchange market resembles a network of telephone lines, cables, facsimile machines, and online computer services, linking all the major money centers of the world. Important participants in foreign exchange markets are (1) traditional users that include importers, exporters, tourists, speculators, traders, individual firms, and investors who buy and sell currencies to settle international accounts; (2) commercial banks that buy and sell currencies to traditional users and that maintain

Loading coastal barge with grain in Rotterdam, the Netherlands.   Credit: Dave/Fotolia.

correspondent relationships with banks in other countries; (3) foreign exchange brokers who act as market intermediaries between banks for the purpose of settling commercial accounts by buying or selling currency; and (4) central banks, which act as the lenders of last resort by intervening in foreign exchange markets to balance national foreign exchange earnings and expenditures.

The primary function of foreign exchange markets is the transfer of funds among countries and currencies. Much of the activity on foreign exchange markets is related to the exchange of assets such as stocks, bonds, or bank accounts. Currencies are traded to maximize the expected returns on interest rates or dividends in various countries. Many institutions and individuals now maintain their wealth in both domestic and foreign assets to reduce the risk of large swings in asset values.

The volume of foreign exchange transactions has increased in recent years. Changes in the world economy are now much more important to the United States than ever before. Much of the growth in global financial flows has been aided by the development of a fully integrated capital market in which financial assets can be transferred among countries almost instantaneously. Improvements in telecommunications, such as electronic funds transfer and facsimile machines, along with the existence of a 24-hour foreign exchange market, have facilitated the growth in financial transactions. Also related to the growth of global capital flows has been the development of the **Eurodollar** market, which refers to U.S. dollars on deposit outside the United States.

Since 1973, exchange rates of most major currencies have been allowed to fluctuate in response to changes in supply and demand conditions on foreign exchange markets. This type of market mechanism is referred to as a flexible exchange rate system, which is synonymous with a floating exchange rate system. The current flexible rate system is in contrast to the fixed exchange rate system, which was established in 1944 at Bretton Woods, New Hampshire, and existed until 1973. The Bretton Woods Agreement required government central banks to intervene in foreign exchange markets in order to maintain the relative values of their currencies at fixed, agreed-upon rates.

Exchange rates alone give little or no indication about the relative strength of currencies or the economies they represent. For example, the fact that the euro is trading at $0.69 and the Japanese yen (¥) at $0.0115 does not indicate that the euro is 80 times stronger than the yen or that the EU economy is stronger than that of Japan. What is important, however, is the relative change in currency value over time. Exchange rates are highly sensitive, fluctuating in many cases on a daily or even an hourly basis. Currency variability is important because exchange rates can increase and decrease in value.

Although exchange rate indices are useful in understanding the relationship between the dollar and a group of other currencies, nothing is revealed about the movements of individual currencies relative to the dollar. Determining the actual exchange rate change between two currencies requires specific information about actual currency values. The Japanese yen was quoted at ¥109.37 on August 18, 2004. When this rate of exchange is compared to ¥118.66 (the 12-month low), it can be determined that the U.S. dollar actually declined in value by 7.8%. This decline in the value of the U.S. dollar is called a currency depreciation and is calculated as

$$(\yen118.66 - \yen109.37) = (\yen9.29/\yen118.66) = (0.0783 \times 100) = 7.83\% \quad (16.2)$$

Currency depreciation refers to a decrease in the foreign price of the domestic currency. For instance, the U.S. dollar would buy ¥109.37 on August 18, 2004, and 1 year earlier one dollar would buy only 1.48 Swiss franc (SF). While the U.S. dollar depreciated in value, the Swiss franc increased in value.

Currency depreciation is a decrease in the foreign price of domestic currency.

Currency appreciation is an increase in the foreign price of domestic currency.

Currency appreciation refers to an increase in the foreign price of the domestic currency. Suppose that 1 year ago the exchange rate between the U.S. dollar and Mexican peso (P) was P3.35/$ and that today the rate is P5.92/$. To determine the appreciation in the value of the dollar, calculate the percentage change as

$$(P5.92 - P3.35) \div P3.35 = 76.7 \tag{16.3}$$

The U.S. dollar buys 76.7% more Mexican pesos today than it did 1 year ago.

THE INTERNATIONAL MONETARY SYSTEM

International trade and exchange rates are heavily influenced by the international monetary system, designed to allow for the maximum flow of international trade and investments and provide an equitable distribution of the gains from trade among nations. The main purpose of the international monetary system is to provide for the orderly and timely settlement of international financial obligations. Operations of the current system are controlled by rules and regulations, customs, and institutions agreed to and supported monetarily by the member nations. Understanding of the international monetary system is important because it sets rules and guidelines for exchange rate operations related to international agricultural trade.

The Gold Standard and the Interwar Years

From 1880 to 1914, world trade operated on the gold standard under which each nation's currency was defined relative to a predetermined value of gold. Exchange rates between nations were essentially fixed. For example, 1 oz. gold = $20.67 = £4.25; therefore, £1 = $4.87. The importance of gold as a medium of exchange should not be overemphasized, because gold had only a reserve currency role and most commercial transactions were conducted in paper currency. However, this period was characterized by economic prosperity and stability due mainly to free trade and income growth, so there is a tendency to give undue credit to the role that gold had during this period. The gold standard ended with the beginning of World War I in 1914.

The interwar years (1914 to 1939) were characterized by flexible exchange rates and a high degree of exchange rate instability, competitive currency devaluations in an attempt to export unemployment, and high tariff protection. This was one of the most chaotic periods in world monetary history. During this period, the United States passed the Smoot–Hawley Tariff Act (1930), which effectively raised duties of goods entering the United States to 59%, one of the highest levels in history. Other countries passed similar protective measures, leading to a precipitous decline in world trade. Many analysts attribute the onset and worsening of the Great Depression to the restrictive trade policies of this period.

The Bretton Woods System

Contemporary international monetary policy had its beginnings at Bretton Woods, New Hampshire, in July 1944, when the United States, the United Kingdom, and 42 other countries met to determine the structure of the new international monetary system, which would emerge from the chaos of World War II. The primary purpose of Bretton Woods was to devise a plan for postwar reconstruction and economic stability. Several important lessons had been learned from the time the gold standard ended in 1914 until the beginning of World War II. First, according to some analysts, when nations pursue their own interests without regard for the interests of other nations, all may suffer. Second, a stable monetary system, coupled with international financial liquidity, is critical for achieving economic growth and development worldwide.

Finally, currency instability, excessive use of exchange controls, and policies that restricted international trade would undermine economic growth, the achievement of full employment, and the hopes for political stability following the war. Institutions, mechanisms, and policies established by the Bretton Woods System provide the basis for today's international monetary system.

The Bretton Woods System was essentially a return to a gold standard, but with some currency flexibility. From 1945 to 1971, exchange rates operated under a crawling peg system in which currencies were pegged to each other, but allowed to adjust within a limited range. The United States played a pivotal role under Bretton Woods. The exchange value of the dollar was set relative to gold at $35 per ounce. The United States agreed to exchange dollars for gold with other monetary authorities at the fixed price, but was not obligated to sell gold on the private market. For purposes of commercial trade and investment, foreign currency convertibility directly into dollars was limited. Other nations fixed the value of their currencies relative to the dollar and therefore implicitly to the price of gold. Because of this strong linkage to gold, the Bretton Woods System is sometimes referred to as the gold-exchange standard. Further, other countries agreed to intervene in foreign exchange markets to prevent the exchange rate from varying by more than 1% above or below its fixed price. In essence, currencies were fixed relative to each other and the dollar and were free to fluctuate only within a limited range.

The United States was firmly established as the world's central banker under the Bretton Woods System. Consequently, tremendous financial and political responsibility was placed on the United States. Most other currencies did not become fully convertible into dollars until the late 1950s and early 1960s. During this time, the U.S. dollar was the only **intervention currency**, creating, in effect, a dollar gold standard maintained by the United States. The role of intervention currency meant that the U.S. dollar was the only currency used by other nations' monetary authorities to keep currencies within their predefined ranges. The expanded role for the dollar resulted in many private international transactions being financed in dollars. The emergence of the Eurodollar market in the 1960s was a direct result of these events. New York became a major international financial center. Both of these events would have major consequences for global monetary stability by the early 1960s.

In the early 1960s, the United States began running a relatively large balance of payments deficits, which were financed in dollars. High capital outflows for investment in other countries, coupled with excessive increases in the money to fund the Vietnam War, acted to increase U.S. inflation and worsen the trade deficit. By 1970, foreign dollar holdings exceeded $40 billion, representing potential claims against U.S. gold reserves, which had declined by 50% to only $11 billion. **Currency devaluation**, or an action taken by a monetary authority to deliberately decrease the value of the dollar from a fixed level relative to other currencies, was not possible because of the reserve role of the dollar.

When U.S. balance of payments deficits persisted in the face of sharply lower gold reserves, it became apparent that in order for the United States to stimulate exports in an attempt to reduce its mounting trade deficit, currencies would have to be realigned. Attempts to convince West Germany and Japan to revalue their currencies failed because neither nation was willing to sacrifice its export competitiveness. Expectations that a U.S. dollar devaluation was imminent led to massive capital flight when foreign investors and residents alike sought to relocate their funds in markets with higher potential returns and more stability. On August 15, 1971, President Richard Nixon suspended convertibility of the dollar and imposed a 10% import duty. The Bretton Woods era had ended. Attempts to revive the system failed. The dollar was devalued to $38 per ounce of gold in 1971 and to $42 per ounce in 1973,

Currency devaluation is an action by a government to reduce the value of a nation's currency.

but more and more nations severed their ties to the fixed exchange rate system. Ironically, however, the dollar remained an international currency, even increasing in importance after losing value relative to other currencies.

The Bretton Woods System offered several advantages and disadvantages. Exchange rate stability was established, in that currency values could be changed only after consultation with the International Monetary Fund (IMF) and then only by a maximum of 10% of the initial fixed value. The United States began to accumulate dollar liabilities to other nations, while foreign countries acquired dollar assets. Dollar reserves became preferable to gold. For the United States, large balance of payments deficits could be financed with a reserve currency accepted worldwide. However, foreign individuals and institutions held dollar balances in interest-bearing accounts, which had to be settled with dollars paid by the United States. The primary drawback of this system, however, was the inability of the U.S. government to change the value of the dollar. The government's ability to pursue domestic economic policy objectives was severely limited because the value of the dollar was, in effect, determined by foreign monetary authorities as a residual of their own currency values.

The Present International Monetary System

Since March 1973, world trade and investment have operated under a managed float exchange rate system. Under a **managed float**, or "dirty" float as it is sometimes called, the government intervenes periodically to change currency values. This is in contrast to a **floating exchange rate** system in which currency values are determined by market supply and demand conditions, with minimal government intervention. The managed float is especially important for U.S. agriculture because many countries manage exchange rates to give themselves a competitive advantage in the export of food and fiber products, thus eroding the U.S. competitive advantage on the international market.

Exchange Rate Arrangements   Current exchange rate arrangements may be classified into three basic categories: pegged, managed float, and independent float. Figure 16-3 shows the major exchange rate arrangements and the number of countries subscribing to each arrangement. In recent years, the International Monetary Fund reported that 75 currencies, mostly those of developing nations, were pegged (fixed) to one other major currency; 43 of these were pegged to the U.S. dollar. Therefore, when the U.S. dollar appreciates, each of these 43 currencies appreciates also. The currencies of 56 countries were defined to have limited flexibility. Of the 187 members of the International Monetary Fund, only 31 allowed their currencies to operate under an independent float, with minimal government intervention. Among the most important were the United States, United Kingdom, Switzerland, Japan, Korea, Canada, Mexico, Australia, Brazil, and New Zealand.

International Monetary Institutions   Two important institutions were created as a result of Bretton Woods: the IMF and the International Bank for Reconstruction and Development, also known as the World Bank. The IMF was created to maintain exchange rate stability and facilitate settlement of temporary balance of payments difficulties. Today, IMF member nations finance temporary balance of payment deficits by borrowing from the international reserve fund, which was established in 1947. Each nation member was allocated a quota based on volume of trade, level of income, and international reserves. This quota was paid to the IMF, 25% in gold or dollars and 75% in domestic currency. Countries could then borrow against this stock of reserve currency. By 1967, demand for international reserves far exceeded available supply.

The International Monetary Fund   works to settle temporary balance of payments problems. Special drawing rights (SDR) are used among IMF members to settle official debt.

Exchange rate value determined   The value of a floating exchange rate is determined by market supply and demand conditions for currencies.

Floating exchange rates   Only 31 nations allow the value of their currencies to be determined by market forces. Among these are the United States, Japan, Canada, Australia, the United Kingdom, Switzerland, and New Zealand.

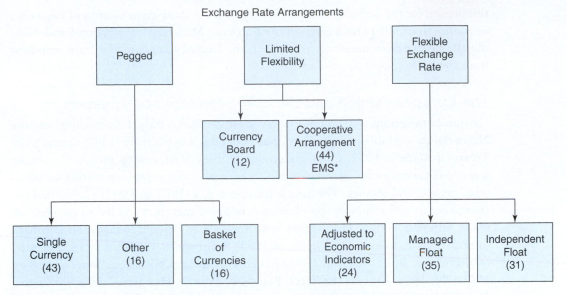

Figure 16-3

Only 31 countries, including the United States, Japan, Canada, Mexico, and Australia, allow their currencies to float independently on the foreign exchange market. Most countries manage, peg, or fix, their currencies relative to another currency, a basket of currencies, or the special drawing rights.

Source: John B. Penson, Oral Jr. Capps, Rosson III C. Parr, and Richard T. Woodward. *EMS (Exchange Rate Mechanism II replaced EMS on January 1, 1999).

To aid with the shortage of international reserve currency, the IMF created **special drawing rights (SDR)**, an artificial currency in the form of accounting entries to supplement reserve assets of the IMF. SDRs are not backed by gold or any other international reserve currency, but represent authorized reserve assets of the IMF. Further, SDRs cannot be used in private transactions, but are reserved for use only by central banks to settle official balance of payments deficits or surpluses. All funds borrowed from the fund must be repaid within 3 to 5 years to avoid tying up reserves for long periods. The IMF remains the cornerstone of the present international monetary system.

In contrast to the IMF, the World Bank was established to provide long-term development assistance. After post–World War II reconstruction was complete, the World Bank expanded its role to include 20 to 30 development projects in developing countries. In addition, the International Development Association was created to finance projects in the poorest countries for the development of infrastructure and agriculture, and the improvement of literacy. Later, the International Finance Corporation was established to stimulate private investment in developing countries.

The European Monetary System

One of the most significant developments in recent monetary history was the creation of the European Monetary System (EMS). In March 1979, eight members of the EC— France, Italy, West Germany, Spain, Belgium, Denmark, the Netherlands, and Ireland—formed the EMS to promote exchange rate stability, economic growth, and employment. Proponents claim that the EMS has been effective in fostering low inflation, low interest rates, and high levels of investment among members, and has reduced the gap between economic growth of member nations.

Major features of the EMS included more complete monetary integration and the creation of a common currency among members. To accomplish these goals, the EMS provided for (1) the creation of a common currency, the euro, (2) a range of fluctuation for the currency of each country, and (3) short-term balance of payments assistance to member states through the European Monetary Cooperation Fund. During the European monetary crisis in 1992, the United Kingdom and Italy withdrew from the EMS.

The European Union and the European Monetary System

Six European nations—Belgium, France, West Germany, Italy, Luxembourg, and the Netherlands—established the European Economic Community (EEC) through the Treaty of Rome in 1957. Complete economic integration among member countries was one of the major objectives of the EEC, with plans for a common market for products, people, and monies. The Single European Act (1987 to 1992) established the criteria for formal and more complete integration of members and led to the creation of the European Union. The EU now has 28 members. Britain, Ireland, and Denmark joined in 1973, followed by Greece in 1981, Spain and Portugal in 1986, and Austria, Finland, and Sweden in 1995. EU enlargement negotiations were started on March 30, 1998. Cyprus, the Czech Republic, Estonia, Hungary, Latvia, Lithuania, Malta, Poland, Slovakia, and Slovenia joined on May 1, 2004. Croatia became a member in 2013.

The EMS was based on a system of fixed but flexible exchange rates, with each member country pegging its currency to every other member's currency. EMS currencies were allowed to fluctuate relative to the U.S. dollar by 2.25% until 1993 when large depreciations forced member governments to widen the band to 30%. This type of exchange rate mechanism was called an *adjustable peg*, allowing central banks to intervene on foreign exchange markets to affect exchange rates directly, or governments could alter monetary or fiscal policies.

The Treaty of Maastricht (1992) created the basis for the European Monetary Union (EMU). Five convergence criteria were used to allow member countries to adopt the **euro**: government budget balance, public debt, inflation, interest rates, and exchange rate stability. The treaty was ratified by all EU members and all joined the EMU except Britain, Denmark, Greece, and Sweden. The European Central Bank was also created to implement and coordinate the EMU and its policies. The euro replaced the ECU on January 1, 1999, and euro coins and notes replaced all national currencies of the 11 EMU members in 2002. The euro is presently used by consumers, retailers, businesses, and public authorities as the only legal tender.

The Treaty of Maastricht created the European Monetary System, with the euro as its common currency.

EXCHANGE RATE DETERMINATION

Under the current flexible exchange rate system, the foreign exchange value of currencies is determined by the interaction of the supply and demand for currencies. Market forces combine to determine the equilibrium exchange rate, which clears the foreign exchange market. In the short run, interest rates, fiscal and monetary policies, and expectations about important economic variables affect movement in exchange rates. Over the long run, however, price changes and the balance of trade are the key determinants of change in currency values. Although both short-run and long-run factors are clearly important, a comprehensive exchange rate theory has not yet been fully developed. Therefore, an explanation of each main factor is necessary to gain a complete understanding of the forces affecting currency values.

Demand and Supply of Foreign Currencies

To simplify the analysis of exchange rate determination, it will be assumed that only two currencies are relevant, the U.S. dollar and the euro. In reality, there are more

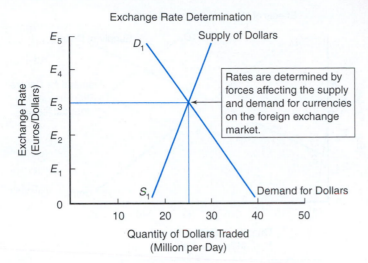

Figure 16-4

Exchange rates are determined by the interaction of forces affecting the supply and demand for currencies on the foreign exchange market. Exchange rates influence how competitive industries such as agriculture are on the international market. If the exchange rate rises, the price of U.S. goods to European buyers increases.

than 100 currencies traded worldwide, giving rise to numerous rates of exchange between any two.

Figure 16-4 illustrates the relationship between the dollar and the euro (€) on the foreign exchange market. The foreign demand for U.S. dollars is negatively sloped, indicating that the higher the exchange rate, the lower the quantity of dollars demanded. The lower the exchange rate (the fewer number of euros required to buy dollars), the cheaper it is for Europeans to import goods and services and to invest in the United States. When this occurs, the quantity of euros demanded by European residents will increase.

The supply of U.S. dollars is positively sloped, indicating that the higher the exchange rates, the greater the quantity of dollars supplied to the market. At higher rates of exchange, U.S. residents receive more euros for their dollars, making European goods, services, and investments cheaper and more attractive. The supply of euros to the United States therefore increases. The initial equilibrium conditions denoted by the supply and demand conditions in Figure 16-4 reflect a market exchange rate of €3=$1, and $25 million traded each day.

Relative Interest Rates

One of the major factors affecting the value of exchange rates in the short run is interest rate differentials between countries. Assume that U.S. interest rates increase relative to prevailing interest rates in Europe. This can be viewed as an increase in demand for U.S. assets. Foreign investors, speculators, businesses, and institutions would transfer funds to the United States to be placed in interest-yielding securities and other financial assets. The net result of higher U.S. interest rates is an increase in demand for dollars.

The effects of rising U.S. interest rates on the exchange rate and quantity of dollars traded are shown in Figure 16-5. Rising U.S. interest rates cause the demand for U.S. dollars to increase from D_1 to D_2. The exchange rate increases from €3/$ to €4/$ and the quantity of dollars traded expands from $25 million to $40 million per day. Because the exchange value of the dollar has increased, the value of the euro has depreciated. By calculating the inverse of the €/$ exchange rate, it can be determined that the euro depreciated from $0.33/€ to $0.25/€ as a result of higher U.S. interest rates. It is important to remember that an increase in European interest rates relative to interest rates in the United States will increase the demand for European assets, driving up the value of the euro and driving down the exchange value of the U.S. dollar. Interest rate declines have the opposite effects.

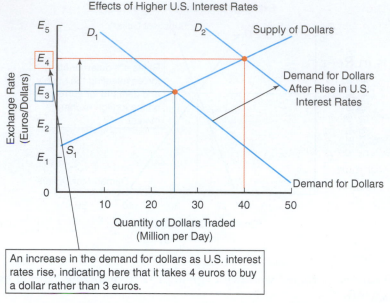

An increase in the demand for dollars as U.S. interest rates rise, indicating here that it takes 4 euros to buy a dollar rather than 3 euros.

Figure 16-5

Any force or factor affecting the supply or demand for either the euro or the U.S. dollar will influence the rate of exchange between the two currencies. For instance, higher U.S. interest rates increase the demand for dollars in Europe, leading to an upward shift in the demand curve to D_2. The price of the dollar in euro, its exchange rate, increases from €3/$ to €4/$.

Forces affecting exchange rates Exchange rates are determined by many factors. Interest rates, inflation, economic growth rates, the balance of trade, expectations, and government policy are among the most important forces.

Aside from market forces, exchange rates can be influenced by government intervention for economic or political reasons. For example, if government authorities believe that the exchange rate is too high, thereby damaging the competitiveness of exports, foreign exchange can be sold to increase its supply and force the value of the currency to decline. Similarly, if government thinks the exchange rate is too low, undermining the ability to import foreign goods and threatening to cause inflation, it can intervene in the market by purchasing foreign exchange and causing the exchange rate to appreciate. However, government actions usually have only short-term impacts on currency values, with market forces dominating over the longer term.

Governments have several options for obtaining the currency required for market intervention. One option is to draw upon the stocks of foreign currency held strictly for intervention purposes, called international reserves. In addition, domestic currency can be printed and used to purchase foreign exchange. Finally, government bonds can be sold to raise the necessary cash to purchase needed amounts of foreign currency. Which option(s) is used will depend upon the overall financial strength of the country and its underlying policy objectives.

Monetary and Fiscal Policies Monetary and fiscal policies affect interest rates and therefore have major implications for subsequent impacts on exchange rates. For example, contractionary monetary policy, other things constant, raises interest rates, thereby causing the exchange rate to appreciate. Monetary contraction can be accomplished either directly by raising the discount rate or indirectly by decreasing the money supply in the economy. Expansionary monetary policy has the opposite effects of lowering the interest rate and causing the exchange rate to depreciate. It is important to note that even expectations of changes in the discount rate or money supply can have the same effects on changes in the exchange rate. Fiscal contraction, all else constant, decreases aggregate demand in the economy, leading to reduced demand for money and lower interest rates. The exchange rate subsequently declines in value relative to other currencies. Fiscal expansion has the opposite effects of increasing the demand for money and raising interest rates, resulting in exchange rate appreciation.

Changes in Relative Prices

Differing rates of inflation between countries are a major cause of exchange rate changes. In fact, any factor that changes the general level of price within a nation can have major effects on the exchange rate. For example, if the level of inflation in the United States rose relative to inflation in other countries, U.S. goods and services would become more expensive. In response to higher U.S. prices, other countries would export more to the United States. As U.S. foreign purchases increased, the supply of dollars on the world market would expand. Other countries would also import less from the United States, thereby reducing the demand for dollars. When the supply of dollars increased and the demand for dollars declined, the value of the dollar, or its exchange rate, would fall.

Inflation rates can also affect exchange rates through monetary policy. For example, expansionary monetary policy would increase the general level of prices in the economy. When inflation increases, the exchange rate would depreciate. Contractionary monetary policy would have the opposite effect, such as currency appreciation due to lower inflation. Therefore, monetary policy affects exchange rates through mutually reinforcing interest rate and inflation differentials.

Balance of Trade Impacts

The balance of trade is one of the most important long-run factors affecting the level of exchange rates. However, since moving to a flexible exchange rate system in the early 1970s, it has become less clear whether exchange rates cause changes in the balance of trade, respond to the trade balance, or both. The consensus is that large, chronic trade deficits and surpluses *do* have a major impact on the value of currencies, but there appears to be a significant lag time for exchange rate changes to occur.

Based on the balance of trade approach to exchange rate determination, the exchange rate balances the value of a country's exports and imports. If a nation experiences a trade deficit, the value of the exchange rate in domestic currency will depreciate. Exports will become cheaper in other countries, while imports will become more expensive domestically. The net result is that exports rise and imports decline to balance trade. Figure 16-6 illustrates that a U.S. trade deficit with the EU would increase dollars available on the world market, shifting the supply curve for dollars to the right to S_2. The supply of dollars increases because as goods are imported from Europe, they are paid for in U.S. currency, thereby raising the number of dollars available in foreign exchange markets. As a result, the exchange value of the dollar depreciates relative to the euro. Following this analysis, we may conclude that when world demand for U.S. goods and services increases, the exchange value of the dollar would appreciate. Conversely, a decrease in demand for U.S. goods and services leads to dollar depreciation.

The rate of adjustment in the trade balance depends on how responsive, or elastic, imports and exports are to exchange rate changes. The more elastic the import demand and export supply are, the faster the rate of adjustment in the balance of trade. Although important, the balance of trade approach to exchange rate determination fails to explain how the U.S. dollar appreciated during the early 1980s when the U.S. trade deficit reached record levels. Neither is it useful in explaining why the U.S. trade deficit did not fall when the dollar depreciated in the mid-1980s. It is obvious that other more important short-run determinants of exchange rates, such as interest rates, were offsetting the effects of the trade balance during these periods.

Figure 16-6

When U.S. imports exceed U.S. exports, U.S. dollars in the foreign exchange market exceed other foreign currencies. As a result, the supply of dollars increases to S_2, causing the exchange value of the dollar to decline from €3/$ to €2/$. At the new exchange rate, exports are less expensive to foreign buyers, and foreign imports are more expensive to U.S. consumers. Consequently, U.S. imports will fall and U.S. exports will increase.

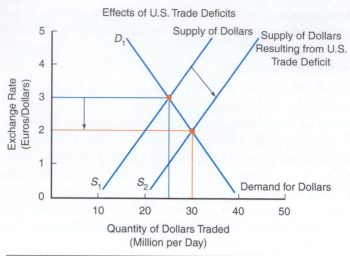

A trade deficit causes an increase in the supply of dollars on currency markets, thereby weakening the dollar and lowering the exchange rate from 3 euros per dollar to 2 euros

The Role of Expectations

With the advent of information technology such as cable and facsimile, the instantaneous electronic transfer of capital, and almost 24-hour money center operation, private and public decision-makers are constantly aware of global events affecting their returns on assets invested in other countries. Consequently, the role that expectations play in determining exchange rates has taken on added importance in recent years. For example, if Japanese investors learn that U.S. monetary policy has tightened, their expectations will be for higher U.S. interest rates and lower inflation, both of which reinforce an appreciation of the U.S. dollar in foreign exchange markets. Their reaction may be to convert large amounts of yen to dollars, with the *expectation* that the dollar will appreciate. By purchasing dollars, the Japanese actually increase the demand for dollars, thereby causing the exchange value of the dollar to appreciate even before the monetary policy is implemented.

Market psychology also is important in determining relative currency values. For example, if the dollar has been appreciating in value relative to the currencies of other major developed nations, investor speculation about continued appreciation can have an important impact. If fundamental factors, such as a large U.S. trade deficit, suggest that the dollar should be declining, then investors may refrain from investing in dollar assets until after the dollar falls. Inaction on the part of market participants may lead to a shortage of dollars on foreign exchange markets, causing the value of the dollar to actually increase when it was expected to decline. Conversely, if investor confidence is eroded by market factors or political action, a dollar sell-off could follow, resulting in even further depreciation of the dollar relative to other currencies.

EXCHANGE RATES AND U.S. AGRICULTURAL TRADE

Exchange rates are key in determining the international prices for agricultural products, thereby determining how much other countries purchase. Although two-thirds of total world trade is conducted in U.S. dollars, every international sale involves two basic transactions: (1) an exchange of currency and (2) an exchange of goods or services. Therefore, even if U.S. wheat is sold to Mexico and the sale is in U.S. dollars, someone, usually the buyer, must convert Mexican pesos to U.S. dollars to make the purchase.

Exchange rates are especially important to agriculture in the United States, where exports account for a large share of agricultural production. In addition, exchange rates have become crucial variables in transmitting macroeconomic policies to the trade sector and in influencing the final outcome of U.S. farm policies.

The exchange rate is one of the most important prices in the economy due to its influence on exports and imports. When the dollar increases in value relative to other currencies, the price of U.S. goods valued in those currencies rises. It takes more foreign currency to buy the same quantity of goods, even if U.S. commodity prices have remained constant. When the dollar declines in value, the prices of U.S. goods decline in terms of foreign currency.

As shown in Figure 16-7, changes in the exchange value of the U.S. dollar have been very closely correlated with changes in U.S. agricultural exports over the last two decades. The exchange value of the dollar relative to other currencies has important effects on how competitive U.S. agriculture actually is at any particular time. For example, the real value of the U.S. dollar decreased by one-third from 1970 to 1979, and U.S. agricultural exports increased in real terms more than eightfold. From 1981 through 1985, the dollar increased by 42% when agricultural exports fell by 31%. Recent USDA estimates indicate that more than 25% of the increase in U.S. agricultural exports since 1985 can be accounted for by changes in the exchange rate. During this period, the dollar declined in value by 21%, and exports increased by 15%. The exchange rate has become one of the critical links between U.S. agriculture and international trade.

Important relationship between the U.S. dollar and exports U.S. agricultural exports and the exchange value of the dollar are inversely correlated, but many other factors also influence the level and direction of trade.

Exchange Rate Indices

The exchange rate index referred to in Figure 16-7 is a real trade-weighted exchange rate index. **Real exchange rate** indices show the change in the value of one currency relative to other currencies, accounting for different inflation rates between countries. This type of index is useful to assess the effects of currency changes on trade flows. A trade-weighted index is necessary because a currency may appreciate against

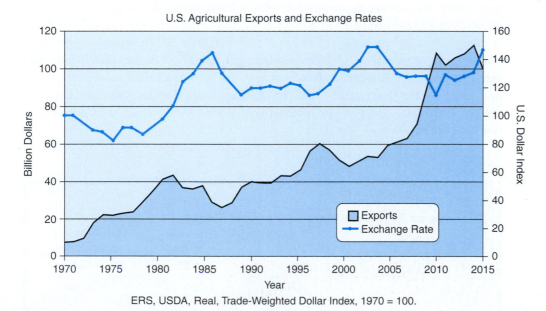

Figure 16-7

The real value of the U.S. dollar and agricultural exports have a strong negative correlation. When the dollar rises, exports fall, and when the dollar declines, exports rise. Strong export growth is typically associated with a weaker U.S. dollar exchange rate.

Source: John B. Penson, Jr., Oral Capps, Jr., C. Parr Rosson, III, and Richard T. Woodward.

some currencies and depreciate against others during the same time period. Further, average indices may not be reflective of trade with any particular group of countries. For example, to account for agricultural trade patterns, a realistic index would need to take into account the relative importance of Japan, the EU, and developing countries as trading partners. The index used is a **bilateral trade-weighted exchange rate index** and is determined by using U.S. agricultural trade volumes with major trading partners to calculate weights, and then multiplying those weights by the real exchange rate of each country. This is sometimes referred to as an effective exchange rate index.

Exchange Rate Impacts on Prices

The exchange rate link between domestic market prices and international prices has become especially important. For example, the hypothetical case shown in Table 16-1 illustrates how exchange rate changes can affect the price of domestic goods on the international market. The initial situation is as follows: the U.S. price of wheat is $100 per ton, and the exchange rate between the U.S. dollar and the Japanese yen is ¥120 = $1.00. The international price of U.S. wheat is then ¥12,000, calculated as ¥120 × $100/ton.

Case 1   If the United States experiences a shortfall in wheat production, for example, the domestic wheat price could rise to $110 per ton. If the exchange rate remains the same at ¥120/$, then the international price of U.S. wheat becomes ¥13,200. U.S. wheat has become more expensive to Japanese buyers due solely to a reduction in supply and the subsequent rise in prices.

Case 2   Now assume that the U.S. wheat price is again $100 per ton, but the exchange rate increases to ¥132 = $1. The cost to the Japanese buyer increases to ¥13,200 due solely to the increase in the value of the dollar relative to the yen. In this case, the U.S. dollar has appreciated relative to the yen and therefore buys more yen. Conversely, the yen has depreciated in value relative to the dollar and will buy only $0.0076 at ¥132/$ compared to $.0083 at ¥120/$. The dollar-per-yen exchange rate is found by taking the inverse of 132 or 1/132 = 0.0076. When yen are converted to dollars by Japanese buyers, each yen is able to purchase fewer dollars than before. As a result, the cost of U.S. wheat in Japan has risen due to the increased value of the dollar or, conversely, the lower value of the yen. Exchange rate changes, therefore, have had the same ultimate effect on the international price of U.S. wheat as an increase in domestic wheat prices.

Table 16-1
EXAMPLE OF EXCHANGE RATE EFFECTS ON PRICES

Case	U.S. Wheat Price $/ton	Exchange Rate ¥/$[a]	International Wheat Price ¥/ton
Initial situation	$100	¥120 = $1	¥12,000
Case 1			
U.S. price increase	$110	¥120 = $1	¥13,200
Case 2			
Exchange rate increase	$100	¥132 = $1	¥13,200

[a]¥ measured as number of Japanese yen per one U.S. dollar.

CONSIDERATIONS FOR POLICY COORDINATION

Macroeconomic and agricultural policies combine to influence the performance of agricultural trade. Monetary and fiscal policies represent the major elements of macroeconomic policy, which affects interest rates, inflation rates, and the exchange value of the dollar. Macroeconomic policies also influence the overall level of economic activity and incentives to import and export. U.S. agriculture, the broader U.S. economy, and the rest of the world are linked through the macroeconomic policies of the United States.

Macroeconomic Policy Coordination

Monetary policy influences the supply of money in the economy by changing the amount of money or credit available to the banking system. Expansionary monetary policy lowers the prevailing interest rate, causing the exchange rate to depreciate. Consequently, net exports increase due to the lower exchange rate, while restrictive monetary policy, a smaller money supply, tends to strengthen the value of the dollar by raising interest rates. As a result, exports decline and imports rise.

Fiscal policy adjusts government spending through government expenditure and tax regulations. Expansionary fiscal policy increases interest rates and exchange rates and, therefore, lowers exports. Contractionary fiscal policy has the opposite effects.

Although the short-run, direct effects of monetary and fiscal policies are fairly clear, their interaction is less clear but critical to agricultural exports. Expansionary fiscal policy puts upward pressure on the exchange value of the dollar because real interest rates rise. If, however, the money supply is expanded by purchasing bonds, the demand for money also increases, putting upward pressure on inflation and downward pressure on the value of the dollar. Although interest rates rise, the appreciation of the dollar would be partially offset. This policy mix would benefit agriculture in the short run because it would hold down interest rates and the exchange value of the dollar while stimulating exports.

Restrictive monetary policy accompanied by expansionary fiscal policy would cause real interest rates and the value of the dollar to rise. The failure to coordinate these two key elements of macroeconomic policy almost certainly contributed to the downturn in U.S. agricultural exports in the early 1980s. Fears of rampant inflation prompted the Federal Reserve Board to adopt a restrictive policy of money supply growth in the early 1980s. As a result, inflation was reduced from the double-digit levels of the late 1970s to less than 4% in 1983.

Expansionary fiscal policy by the federal government pushed the budget deficit over $100 billion in the early 1980s, and it exceeded $200 billion by 1983. Credit markets were squeezed by federal borrowing, causing interest rates to increase dramatically. The combination of high interest rates and low inflation resulted in a record high level of real (inflation-adjusted) interest rates. When foreign investors responded to record rates of return on dollar-valued assets, the value of the U.S. dollar, measured in its exchange rate relative to other currencies, increased 40% in real terms. The 40% dollar appreciation almost certainly contributed to a 32% drop in farm exports between 1981 and 1983.

To summarize, the soaring U.S. federal budget deficit has been financed by borrowing liquid funds from all over the world. Insofar as the trade deficit is a loan from the sellers of goods and services to buyers of goods and services, the federal deficit is, in effect, a fundamental cause of the U.S. trade imbalance. To avoid policy conflict in the future will require careful coordination of both monetary and fiscal policies.

Policies affect exchange rates Macroeconomic and agricultural policies are important factors influencing exchange rates and, therefore, trade.

Domestic Agricultural Policy Coordination

U.S. agricultural policy can play a major role in eroding or enhancing the competitive position of farm exports

High loan rates enacted by earlier farm bills acted as a price umbrella under which competitors could increase production. Instead of purchases being made by foreign customers, the U.S. government became the buyer of last resort. Government stocks of agricultural commodities rose when exports fell; the U.S. share of the world export market for food and fiber products declined. Subsequent farm bills changed this policy, allowing for more price flexibility to ensure U.S. export competitiveness.

In summary, for U.S. agriculture to remain competitive in the world market will require careful coordination of agriculture and macroeconomic policies. In fact, it has been argued by Schuh (1990) that macroeconomic policies now have more influence on agriculture than do farm policies. This has occurred primarily because of the increased interdependence between agriculture and the international market. Exchange rates have become one of the most important variables linking agriculture to the world economy. Currency values influence not only agricultural exports but also international competition and investment decisions. Exchange rates are especially important to U.S. agriculture because such a large share of production is exported and a large share of consumption is imported. For U.S. agriculture to maintain its competitive position on the world market will require agricultural policies that ensure flexible commodity prices and well-coordinated macroeconomic policies so that large swings in exchange rates do not undermine international competitiveness or discourage the importation of U.S. food and fiber products.

The importance of macro policy   Macroeconomic policies are likely more important determinants of U.S. agricultural trade than farm policy. This is due mainly to the development of a well-integrated global capital market.

SUMMARY

This chapter reviews recent trends in agricultural trade, focusing on the importance of those trends for U.S. farms and agribusinesses. The importance of trade to the overall economy and the changing product mix for U.S. agricultural trade were discussed. Changing markets for U.S. agricultural exports were highlighted, followed by a discussion of trade performance and its importance to agriculture. The purpose of this chapter is also to provide an overview of the international monetary environment affecting U.S. agricultural trade, emphasizing exchange rates, the foreign exchange market, and the international monetary system. Exchange rate determination is explored, with emphasis on important links to agriculture, including interest rates, prices, the trade balance, and expectations. The importance of exchange rate impacts on U.S. agricultural trade is examined, with special attention focused on prices, international competition, and agricultural policy. Finally, some considerations for the coordination of macroeconomic and agricultural policy are discussed, emphasizing the importance of maintaining the competitive position of

U.S. agriculture. The major points in this chapter are summarized as follows:

1. Increasingly unstable trade has led to greater uncertainty for farm and agribusiness managers.

2. Distortions to agricultural trade worldwide prompted the major trading nations to complete an eighth round of negotiations in the General Agreement on Tariffs and Trade (GATT). These negotiations liberalized trade in agriculture by reducing some elements of government support that distort trade. The Doha Development Agenda began in 2001 as the first trade negotiating round under the WTO.

3. U.S. agriculture has become increasingly export dependent, and large shares of major crops are currently sold to other countries. Greater export dependence means that farmers and agribusinesses now have to rely more on transnational factors and events for price determination. Effective decision-making will require improved understanding of global forces and better management of international risks.

4. Imported food products are supplying a growing share of domestic food consumption. Food imports are either competitive or noncompetitive. Imports can be a source of competition for some producers. Consumers depend on food imports for variety, quality, and as a source of low-cost products. The growth in imported foods has heightened consumer awareness of food safety issues and international environmental concerns.

5. Agricultural exports represent important employment opportunities and economic activity for both the farm and nonfarm sectors of the U.S. economy. Value-added exports generate more jobs and have a larger economic impact than bulk commodity exports. Value-added exports are becoming more important to the economy in terms of value, but bulk commodity exports are still important in terms of tonnage.

6. Major markets for U.S. agricultural exports are changing. Canada is the single largest country market for U.S. food and fiber exports. However, trends are changing and customers in the developing countries of Asia and Latin America are becoming more important as customers in Western Europe are becoming less important. Developing countries represent the important growth markets for the future.

7. Developed countries are supplying a major share of food imports to the U.S. market. Competitive imports account for three-fourths of all U.S. agricultural imports. The growth rate of competitive imports far exceeds the growth of noncompetitive imports.

8. U.S. agricultural trade has exhibited strong performance over the last 40 years. Although agricultural trade has declined as a share of total trade, agriculture has generated a trade surplus each year, partially offsetting the total trade deficit. The agricultural trade surplus is important to the nation because it provides additional funds for consumption, investment, or savings.

9. Exchange rates, the number of foreign currency units per unit of another currency, have become one of the most important economic variables affecting U.S. agricultural trade. International financial flows—capital and foreign exchange—have grown to exceed world trade by 40 times, reaching more than $1.25 trillion per day. The value of foreign currencies is determined in the foreign exchange market where supply and demand interact to determine the exchange rate. Currency values can appreciate (increase in value) or depreciate (decrease in value), determining the level and direction of U.S. agricultural trade.

10. The international monetary system has evolved over the last 100 years from a system of exchange rates fixed to gold into a system called managed float, whereby exchange rates fluctuate, but within limits. These limits are controlled by the monetary authorities of the 186 members of the International Monetary Fund (IMF). Some countries peg their currencies to those of other nations, mainly the United States and France. Other nations have formed coordinated exchange rate arrangements, such as the European Monetary System of the European Union, for the purpose of stabilizing exchange rate movements.

11. Exchange rates are determined by the interaction of supply and demand for foreign currencies. In the short run, real interest rates, fiscal and monetary policies, and expectations about key economic variables are important in determining the value of currencies. The balance of trade, differentials in rates of economic growth between countries, and price levels are important in determining exchange rate movements over the long run. A comprehensive theory explaining all of the aspects of exchange rate determination has not been developed.

12. Exchange rates are especially important to U.S. agriculture because exports account for a large share of production. When the U.S. dollar appreciates in value, the price of goods in foreign currency increases, leading to a lower demand for U.S. goods, reduced prices, declining exports, and a loss of market share. U.S. agricultural policies that prevent U.S. prices from declining when the exchange rate appreciates can result in a loss of competitive position and market share.

13. Macroeconomic and agricultural policies combine to influence the performance of U.S. agricultural trade. Restrictive monetary policy raises interest rates and the value of the dollar. Expansionary fiscal policy also raises interest rates and causes dollar appreciation. This policy combination, combined with inflexible farm prices, can cause U.S. exports to decline. Macroeconomic policies that lead to higher real U.S. interest rates also place U.S. agriculture at a competitive disadvantage. For U.S. agriculture to maintain its competitiveness, both macroeconomic and agricultural policies must work together.

KEY TERMS

Balance of payments (BOP)
Balance of trade
Bilateral trade-weighted exchange
   rate index
Cross rate
Currency appreciation
Currency depreciation
Currency devaluation
Euro

Eurodollar
Exchange rate
Export dependent
Exports
Floating exchange rate
Import dependence
Imports
Intervention currency
Managed float

Real exchange rate
Special drawing rights (SDR)
Tariff
Tonnage
Trade deficit
Trade surplus
Treaty of Maastricht
Value
Weaker dollar

TESTING YOUR ECONOMIC QUOTIENT

1. During the past 50 years, agriculture has generated a trade surplus that provides additional funds for investment, consumption, or savings.   T   F

2. Expanded U.S. involvement in world trade resulted in higher farm prices, greater market stability, and increased farm incomes.   T   F

3. The Uruguay Round of the GATT negotiations is the only round that attempts major reform of agricultural trade.   T   F

4. World trade in value-added products has declined relative to world trade in bulk commodities.   T   F

5. Consumers depend on imported foods for quality, variety, and as a source of low-cost products.   T   F

6. Which group of countries most likely represents future markets for U.S. agricultural products?
   a. Developing countries
   b. Developed countries
   c. Centrally planned countries
   d. All of the above

7. Competitive imports substitute for U.S. domestic production causing
   a. higher prices.
   b. increased competition for U.S. farmers.
   c. increased grower returns.
   d. all of the above.

8. U.S. agricultural exports generate _____, _____, and _____ throughout the economy.

9. What is the difference between competitive and noncompetitive imports? Give two examples of each type.

10. Define *balance of trade* and discuss the two important implications for the economy as a whole and for agriculture in particular.

11. An exchange rate can be thought of as the value, or price, of one currency in terms of another currency.   T   F

12. The balance of payments is a record of all international transactions by U.S. private and public entities over a specific period of time, usually 1 year. T   F

13. Currency devaluation is used to deliberately decrease the value of a currency from a fixed level relative to other currencies.   T   F

14. An increase in U.S. interest rates would cause a decrease in demand for U.S. dollars.   T   F

15. An expansionary monetary policy will cause the interest rate to decline and the exchange rate to depreciate.   T   F

16. Monetary contraction can be accomplished directly in the United States by raising the _____ or indirectly by decreasing the _____ in the economy.

17. If the level of inflation in the United States _____ relative to other countries, U.S. goods and services would become more _____.

18. If a nation experiences a trade deficit, the value of the exchange rate in domestic currency will _____.

19. Expansionary fiscal policy
   a. decreases exchange rates.
   b. increases interest rates.
   c. increases exports.
   d. decreases imports.

20. A restrictive monetary policy accompanied by an expansionary fiscal policy would cause
   a. real interest rates to decline and the value of the dollar to rise.
   b. real interest rates to rise and the value of the dollar to decline.
   c. real interest rates to rise and the value of the dollar to rise.
   d. real interest rates to decline and the value of the dollar to decline.

21. As the U.S. dollar depreciates in value, the price of goods in foreign currency declines, resulting in
    a. the decline in demand for U.S. goods.
    b. the decline of prices.
    c. loss of market share.
    d. all of the above.

22. Identify and describe the important participants in foreign exchange markets.

23. The U.S. price of corn is $100 per ton, and the exchange rate between the U.S. dollar and the Japanese yen is ¥120 = $1.00.
    a. Calculate the international price of U.S. wheat for Japan.
    b. If the exchange rate changes to ¥145 = $1.00, calculate the new international wheat price for Japan.

24. What are the two major components of the balance of payments?

REFERENCE

Schuh GE: The evolution of the global economy, *Agrichemical Age*, February 1990.

17

Why Nations Trade

Trade occurs because individuals, firms, governments, or nations anticipate economic gains from the exchange of goods or services. Simply stated, people make money in international commerce by purchasing goods or services in the country where they can be produced the cheapest and selling those goods or services in the country where they are worth the most. Each nation is endowed with a unique set of human, institutional, and natural resources that give rise to different prices and trade relationships among countries.

    The purpose of this chapter is to examine the underlying factors affecting trade patterns and to focus on why nations trade. The concepts of absolute and comparative advantage will be explored, along with an example of each. Gains from trade will be examined, with explanations of the importance of exchange, and specialization, and a discussion of the distribution of the gains from trade.

WHY TRADE?

Trade theories The four major theories of international trade are mercantilism, absolute advantage, comparative advantage, and competitive advantage.

Mercantilism encourages exports but discourages imports, leading to a trade surplus and the accumulation of national wealth.

The question of why nations trade has been at the crux of economic theory and analysis for nearly four centuries. One of the earliest attempts to explain why trade occurs was made by Thomas Munn (1571–1641) in his treatise on *England's Treasure by Foreign Trade*. Munn argued that a nation could become rich and powerful by exporting more than it imported. This philosophy, commonly referred to as **mercantilism**, became the accepted practice among trading nations (Munn, 1928). Highly protectionist trade policies, which were designed to restrict imports, were implemented by major traders such as England. The ensuing trade surplus with other countries could then be settled by an inflow of precious metals, namely, gold and silver. In response to this policy, governments attempted to stimulate exports and restrict imports. When stocks of gold and silver were accumulated, the government became richer and more powerful. Governments realized that restricting imports and expanding exports would lead to additional economic output and employment. Pursuits of mercantilism led to government control of most major economic activity and strong movements toward economic nationalism.

The primary flaw in Munn's argument, and in the practice of mercantilism, was that all nations could not have a trade surplus at the same time. Mercantilism implies a zero-sum game. The only way for a nation to practice mercantilist principles was for it to gain and maintain a trade surplus at the expense of other nations. Recall that these were times of rule by monarchs in most parts of the world. The obvious result of attempting to pursue these mercantilist policies was almost constant conflict among nations. With more gold, rulers could maintain larger armies and navies, leading to tremendous military power. Greater military strength was necessary for consolidation of power at home and to the acquisition of additional colonies abroad.

Although these events occurred in the 17th century, they are highly relevant to any discussion of international trade today. First, the body of knowledge brought forward by classical economists, such as Adam Smith and David Ricardo, was in response to, and an attack on, the mercantilist policies and the role of government in international commerce. Second, except for a brief period during the 19th century, no Western nation has ever been entirely free of mercantilist philosophy and policies. In fact, since the mid-1950s, there has been a resurgence of mercantilist philosophy in the United States, known as neo-mercantilism. Neo-mercantilism led nations with high unemployment in import-sensitive industries to restrict imports to recover lost jobs and increase sagging domestic production.

ABSOLUTE ADVANTAGE

One of the most convincing attacks on mercantilism was offered by Adam Smith in his classic book *An Inquiry into the Nature and Causes of the Wealth of Nations*, published in 1776. The key premise underlying Smith's attack on the mercantilist view was

Unloading wheat in Korea. Credit: Time & Life Pictures/Getty Images.

Absolute advantage
states that a nation will
specialize in the production
and export of goods that it
can produce most cheaply
and import those goods
that other nations can
produce more cheaply.

that for nations to trade, each had to gain something in the exchange of goods. If there were no gains, there was no incentive to trade. Smith's premise was based on the concept of **absolute advantage**. The concept of absolute advantage states that a nation will export those goods that it can produce more cheaply than others and import those goods that other nations can produce more cheaply (Smith, 1937). A nation is said to have an absolute advantage in the production of a good if its costs of production are lower than other nations at prevailing prices and exchange rates. Nations can gain economically by specializing in the production of the goods for which they are most efficient and by trading the excess production with other nations. In this way, resources are used most efficiently and the output of nations will rise, leading to an increase in global economic welfare.

The basis for international trade is the efficiency with which nations combine productive resources to produce goods and services. For example, a farmer may trade grain for dollars and dollars for medical care. A doctor trades medical skills for dollars and dollars for food. Each individual is better off than before: the farmer is healthy and the doctor is well fed. Both individuals are endowed with, or have developed, a unique set of skills that allow them to specialize and practice what they do best, or what they do most efficiently.

Trade among nations is similar to trade between individuals. For example, because of its favorable climate, the United States is an efficient producer of wheat but an inefficient, or high-cost, producer of coffee. Conversely, Mexico is an efficient producer of coffee, but an inefficient producer of wheat. The United States imports coffee from Mexico, which has been produced and marketed at prices below those in the United States. Mexico imports wheat from the United States at less cost than if it were produced in Mexico. With a given amount of income, consumers in both nations can purchase a greater variety of goods, in larger amounts, than if everything were produced domestically. In this way, nations can achieve a higher level of satisfaction from a given income with trade than without it.

The contrasts between the arguments of Munn and Smith are important to understand. The mercantilist doctrine held that one nation could gain from trade only at the expense of other nations and that strict government control of all economic activity and trade was necessary. This was most often accomplished by restricting imports and stimulating exports. Adam Smith, on the other hand, wrote that all nations would gain from free trade and advocated the **laissez-faire** philosophy, which emphasized little or no government control of a nation's economy. Smith further offered that free trade would lead to efficient resource use, thereby maximizing world welfare.

A numerical example of Adam Smith's absolute advantage should help solidify the concept and will serve to develop a conceptual framework for the discussion of comparative advantage. To simplify the analysis, the nation will be assumed to be the relevant economic unit. The use of nations as the economic unit facilitates the analysis of trade. Economic resources can normally be reallocated more easily within nations than among nations. Languages, laws, customs, culture, and institutions are often more alike within a nation. Policies and trade restrictions are generally not as important for domestic transactions as they are for international trade. The nation will be assumed to employ all its resources fully in the production of two goods, wheat and coffee. Further, resources are assumed to adjust efficiently and completely to changing economic conditions within countries. However, resources such as labor are assumed to have differing degrees of mobility among countries. Otherwise the major incentive for trade, different costs of production among countries, would disappear.

Table 17-1 shows that 1 hour of labor is required to produce 10 tons of wheat in the United States and 2 tons of wheat in Mexico. On the other hand, 1 hour of labor produces 6 tons of coffee in Mexico but only 4 tons of coffee in the United

Table 17-1
EXAMPLE OF ABSOLUTE ADVANTAGE MEASURED
AS UNITS OF OUTPUT PER UNIT OF LABOR

	United States	Mexico
	Tons/Man-Hour	
Wheat	10	2
Coffee	4	6

States. Based on the results of this analysis, the United States is the most efficient, or least-cost, producer of wheat, and Mexico is the most efficient producer of coffee. With trade, the United States would specialize in wheat production and exchange part of the surplus for coffee. Mexico would specialize in coffee production and trade part of the surplus for wheat.

The United States could trade 10 tons of wheat (10W) for 10 tons of coffee (10C) and gain 6C, saving 1.5 man-hours of labor. The 10W that Mexico receives from the United States would require 5 man-hours to produce in Mexico. With the 5 man-hours saved, Mexico can produce 30 tons of coffee, or 30C (5 hours times 6 tons per man-hour). By trading 10C for 10W, Mexico has gained an additional 20C, thereby saving 5 man-hours.

Although the concept of absolute advantage has intuitive appeal, it is not very useful in explaining the real-world relationships of trade that are prevalent today. For example, what if a nation has an absolute advantage in the production of all goods due to low-cost labor, abundant natural resources, advanced technology, or superior management skills? Absolute advantage would dictate that this nation would export goods but import nothing. In fact, it can be shown that absolute advantage is only a very special and limited case of the more general principle of comparative advantage. In the following section, we will learn that nations do not need an absolute advantage to gain from trade.

Comparative Advantage

The limitations of absolute advantage to explain trade among nations accurately and realistically led David Ricardo to develop more fully the principle of **comparative advantage**. The law of comparative advantage was first presented by Ricardo in his book *On the Principles of Political Economy and Tax*ation, published in 1817. One of the most important and still unchallenged laws of economics, comparative advantage states that even if one nation is less efficient in the production of a good than another nation, there is still a basis for gains from trade (Ricardo, 1963). A nation should export the good for which its relative, or comparative, advantage is greatest and import the good for which its relative, or comparative, advantage is least. Conversely stated, a nation should specialize in the production and export of the good for which it has the least relative disadvantage and import the good for which its relative disadvantage is greatest. The key point is that trade is based on relative or comparative cost relationships rather than absolute costs.

A number of simplifying, yet important, assumptions were made by Ricardo in his explanation of comparative advantage: (1) a two-nation, two-good world, (2) completely open and free trade, (3) free movement of labor within nations, but no movement of labor between nations, (4) constant production costs, (5) absence of costs of transport and transfer of goods, (6) constant technology, and (7) that labor is the only factor of production or that it is used in the same fixed proportion in the production

Comparative advantage states that a nation will produce and export those goods for which its relative, or comparative, cost is lowest.

of both goods and that labor is homogeneous. This final assumption is called the labor theory of value and is often not used to explain the law of comparative advantage because its underlying precepts are unrealistic.

The principle of comparative advantage can be explained by the numerical example in Table 17-2. In this case, Mexico has an absolute disadvantage in the production of both goods, because it now produces only 3 tons of coffee instead of 6, indicating that productivity has fallen by 50% from the previous example. The United States can produce 10W compared with 2W for Mexico and 4C compared with 3C for Mexico. To find the comparative advantage of each nation, the relative productivity of labor must be determined. The United States can produce 10 tons of wheat with one unit of labor, compared with only 2 tons for Mexico, yielding a ratio of 10:2 or 5. The comparable ratio for coffee is 4:3 or 1.33. Although the United States is more productive in the production of both, its relative advantage is greatest in wheat production, because 5 > 1.33. Therefore, the United States has a comparative advantage in wheat production, and Mexico has the least comparative disadvantage in coffee production. Conversely, Mexico's labor productivity yields output to labor input ratios of 0.2 and 0.75 for wheat and coffee, respectively. Based on this example, we may conclude that Mexico has a smaller relative disadvantage in the production of coffee than in the production of wheat. Or conversely, Mexico has a comparative advantage in coffee production.

To examine the gains from this trade relationship, the same example can be expanded. If the United States traded 10W for 10C, it would gain 6C because the domestic rate of exchange in the United States is 10W for 4C. The 6C gained represents labor savings of 1.5 man-hours. The 10W that Mexico receives would require 5 man-hours to produce. These 5 man-hours could be used to produce 15C. By trading 10C for 10W, Mexico gains 5 tons of coffee. Because the United States could exchange 10W for 4C internally, it would gain from trade if it could get more than 4C for 10W. For Mexico to gain from trade, it must receive more than 2W for 3C. By exchanging 10W and 10C, both countries gain from trade.

A further extension of comparative advantage was made in 1936 by G. Haberler in *The Theory of International Trade*. In relaxing the value theory of labor assumption, Haberler was able to demonstrate that the law of comparative advantage was still valid and more universally applicable by employing the concept of **opportunity cost**. According to Haberler, opportunity cost reflects the cost of a good as measured by the amount of a second good that must be given up to release just enough resources to produce one additional unit of the first good. No underlying assumptions are made about the homogeneity of labor or that labor is the only factor of production. Haberler's extension is often called the **opportunity cost theory**, which states that a nation has a comparative advantage in the production of the good with the lowest opportunity cost. For example, the United States must give up 0.4 units of coffee to release just enough resources to produce one additional unit of wheat. In this case, the opportunity cost of wheat is 0.4 ton of coffee, calculated as 10W = 4C or 1W = 0.4C. In

Opportunity cost reflects the cost of producing a good measured by the amount of a second good that must be forgone in order to release enough resources to produce one additional unit of the first good.

Table 17-2
EXAMPLE OF COMPARATIVE ADVANTAGE MEASURED AS UNITS OF OUTPUT PER UNIT OF LABOR

	United States	Mexico
	Tons/Man-Hour	
Wheat	10	2
Coffee	4	3

Mexico, the opportunity cost of wheat in terms of the coffee given up to produce one additional ton is 1.5, because $2W = 3C$; therefore, $1W = 1.5C$. Because the United States has the lowest opportunity cost for wheat $(0.4 < 1.5)$, it has a comparative advantage in the production of wheat.

The same analysis can be conducted to determine the opportunity cost of coffee in Mexico. The amount of wheat that must be given up by Mexico to produce one additional unit of coffee is 0.67 tons. For the United States, the amount of wheat forgone to produce one additional ton of coffee is 2.5 tons, or $4C = 10W$. Therefore, Mexico has the lowest opportunity cost of producing coffee, 0.67 tons compared to 2.5 tons in the United States. We may conclude that Mexico has a comparative advantage in coffee production, and the United States has a comparative disadvantage in coffee production. According to the law of comparative advantage, the United States should specialize in wheat production, exporting the surplus to Mexico in return for coffee. This is the identical result found when the analysis was based on the labor theory of value, but now it is based upon the less restrictive and widely applicable concept of opportunity cost.

Factors Affecting Comparative Advantage

As noted by James Houck (1986), comparative advantage is a real, not a monetary, concept. Because comparative advantage is based on the structure of relative opportunity costs among nations, it is not affected by changes in exchange rates or inflation. While exchange rates may vary among different currencies, they act to translate comparative advantage into absolute advantages, which are compared by individual buyers and sellers worldwide. If all resources could be readily transferred among nations, opportunity costs would be altered and incentives to trade would disappear. However, because productive resources such as land, climate, and labor are not mobile, comparative advantage still may be used to explain much of the world's trade in agricultural products. Due to the increased international mobility of capital and technology, however, opportunity costs and comparative advantage do change over time, leading to different trade patterns.

National differences in the opportunity cost of resources determine comparative advantages and the patterns of trade. Opportunity costs are dependent on several factors. First, the availability of productive resources varies among nations, leading to different costs. The cost of using plentiful resources will be lower than the cost of using scarce resources. Land, labor, capital, technology, and management skills are all affected by the abundance with which they exist in a given nation. Second, the production of different goods and services requires different resources in different proportions. Third, most products can be produced by more than one process, resulting in many different resource combinations. Finally, resource mobility differs among nations. Any of these factors, alone or combined, can affect the opportunity costs within nations and their comparative advantages.

Comparative Advantage and Competitive Advantage

In the long run, comparative advantage, or the relative efficiency with which resources are combined, is reflected in the economic incentives or disincentives for resource use. Agricultural exports represent additional demand for U.S. products and, therefore, depend on the level of economic development and growth in other countries. However, U.S. products must compete with goods from Europe, Australia, and Canada. How successfully those products compete depends on price, quality, timeliness of delivery, and many other factors. In competitive markets, however, international market prices reflect the long-run costs of the human, natural, and manufactured resources required to produce the goods.

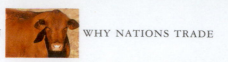

Competitive advantage reflects the absolute cost of a given product in a certain market at a particular time. It accounts for supply as well as demand factors shaping trade patterns and the direction of trade.

In the short run, however, relative absolute costs among nations may not reflect comparative advantage. Michael Porter (1985) was the first to use the term *competitive advantage*. Market distortions brought about by government intervention have been especially prevalent in agricultural trade. **Competitive advantage**, which is defined as the pure economic competitiveness of a nation reflected by the absolute cost of a given good in a given market at a particular point in time, may provide a better understanding of historical trade patterns in agriculture than the more restrictive law of comparative advantage. Further, the existence of historical social ties, common ideology, common defense treaties, and agreement on basic human values may be as important in understanding trade patterns today as reliance on the law of comparative advantage. For example, Cuba is not allowed to ship sugar to the United States, regardless of its comparative advantage or the potential cost savings to U.S. consumers. This section highlights some of the most important factors and forces affecting the competitive advantage in international trade of agricultural products.

Domestic agricultural policies, in combination with the use of import duties and export subsidies, were instrumental in transforming the European Community from a net importer of grains into a net exporter. Subsequently, the United States lost large shares of the world grain market. Many analysts questioned whether the United States had lost its comparative advantage in grain production. The United States responded to these EC policy actions by implementing more competitive farm prices and export subsidies of its own to regain lost markets. In reality, the United States had not lost its comparative advantage but had become less competitive relative to the EC, thereby losing its competitive advantage and the ability to compete in international markets at competitive world prices.

Import quotas can mask the true comparative advantage of a nation. An import quota forces the effective demand for a good to zero once the quota is filled. When the United States implemented the meat import quota in 1991, it made little difference that Australia had a comparative advantage in meat production. Once the United States had accepted all agreed-to tonnage, the market was closed.

Exchange rates have had major impacts on agricultural trade. For example, we may argue that U.S. borrowing in foreign money markets to fund growing budget deficits in the early 1980s drove up interest rates, and increased the demand for dollar-valued assets, and, hence, the value of the U.S. dollar. The resultant higher prices of U.S. goods in international markets perpetuated the loss of competitive advantage for U.S. agriculture. Over time, however, exchange rates declined and the United States regained its competitive advantage. Although the United States maintained its comparative advantage in agriculture during this period, it became much less competitive relative to other countries, losing markets and export sales.

A lack of marketing infrastructure, such as storage and handling facilities, communications systems, and other logistical factors, can alter the competitive advantage of nations, increasing the absolute cost of accessing world markets. High cost and lack of available transportation, along with other transfer costs, can result in short-run distortions in competitive advantage. In summary, competitive advantages that arise from nonmarket factors and forces such as tariffs, quotas, and subsidies are no less real and effective than those reflected by lower opportunity costs. However, competitive advantages based on nonmarket factors are dependent upon politics rather than economics for sustained existence. These basic differences between comparative and competitive advantage should be considered when analyzing trade and competition among nations.

GAINS FROM TRADE

For trade to occur, both nations must benefit from the transaction. Real incomes are raised by trading with other nations. By importing, goods can be purchased at a lower cost and in greater variety and quantity than without trade. Lower prices allow consumers to buy more with their limited resources, thereby increasing disposable incomes for each consumer individually and increasing the overall welfare of the nation. Prices of exported goods may be higher, with additional economic activity and income spread throughout the economy. Finally, consumers with a given amount of income are able to achieve more satisfaction from those incomes with trade than without trade. However, these gains from trade are dependent upon increased specialization in production and exchange of goods.

The gains from trade For trade to occur, all nations involved must benefit. These benefits from trade are referred to as the gains from trade.

The Importance of Exchange and Specialization

The law of comparative advantage can be applied to demonstrate the gains from trade and the subsequent increase in world welfare. Table 17-3 is an extension of the previous hypothetical example. Combinations of wheat and coffee can be produced in the United States and Mexico by fully employing all available resources, the most productive technology, and a maximum of 20 man-hours of labor per year in each country. These production possibilities can be combined to form the complete **production possibility schedule** in Table 17-3. For instance, the United States can produce 200 tons of wheat and no coffee, 160 tons of wheat and 16 tons of coffee, 80 tons of wheat and 48 tons of coffee, or no wheat and 80 tons of coffee. For every 10 tons of wheat produced, the United States releases enough resources to produce 4 tons of coffee. If the United States increases wheat output from 120 tons to 160 tons, it must give up the opportunity to produce 16 tons of coffee, or coffee production must decline from 32 tons to 16 tons. The opportunity cost of wheat for coffee is constant at $10W = 4C$ for every combination of U.S. production possibilities.

The same exercise can be done for Mexico, revealing that the set of production combinations ranges from $40W = 0C$, $32W = 12C$, $16W = 36C$, to $0W = 60C$. The cost of wheat production in Mexico, in terms of the amount of coffee production forgone, is constant at $2W = 3C$ for every combination of production possibilities. For Mexico to increase wheat output by 16 tons, from 8 tons to 24 tons, requires that the opportunity to produce 24 tons of coffee be forgone. The relationship of $2W = 3C$, or $1W = 1.5C$, is constant throughout the production possibility schedule for Mexico.

Table 17-3
PRODUCTION POSSIBILITY SCHEDULE FOR WHEAT AND COFFEE IN THE UNITED STATES AND MEXICO

United States		Mexico	
Wheat (Tons/Year)	Coffee (Tons per Year)	Wheat (Tons per Year)	Coffee (Tons per Year)
200	0	40	0
160	16	32	12
120	32	24	24
100	40	20	30
80	48	16	36
40	64	8	48
0	80	0	60

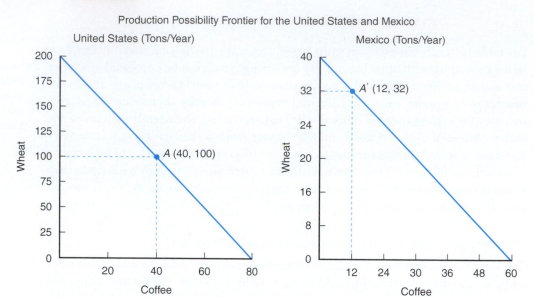

Figure 17-1

These relationships depict the combinations of wheat and coffee that can be produced in Mexico and the United States. At point A, the United States produces 100 tons of wheat and 40 tons of coffee. At point A', Mexico produces 32 tons of wheat and 12 tons of coffee.

The United States and Mexico production possibility schedules are graphed in Figure 17-1, with production possibility frontiers reflecting the combinations of wheat and coffee that can be produced by each nation. For example, at point A, the United States produces 100 tons of wheat and 40 tons of coffee. At A', Mexico produces 32 tons of wheat and 12 tons of coffee. The negative sloping frontiers indicate that for either nation to increase output of one good, some output of the other good must be forgone. For example, if the United States increased wheat production by an additional 10 tons, it would need to give up the opportunity to produce 4 tons of coffee. For Mexico to increase coffee output by 6 tons, it must be willing to give up 4 tons of wheat production.

Distribution of the Gains from Trade

As noted earlier, the basis for trade is the differing opportunity costs among nations. This concept is illustrated in Figure 17-2. In the absence of trade, assume the United States and Mexico produce at A and A', respectively. With trade, the United States would specialize in wheat production (point B), the good for which it has a comparative advantage, and Mexico would specialize in coffee (point B'). The United States could trade 50W for 45C and reach point C, where 150W and 45C are consumed. Under these conditions, the United States is producing 200 tons of wheat, trading 50 tons of wheat to Mexico for 45 tons of coffee, and reaching a higher level of consumption through trade.

Mexico could produce 60 tons of coffee, trade 45 tons of coffee to the United States, and import and consume 50 tons of wheat, thereby reaching point C'. Comparing point A with point C, we may determine that the United States gained 50 tons of wheat and 5 tons of coffee through trade. Likewise, comparing A' and C', Mexico gained 18 tons of wheat and 3 tons of coffee by trading with the United States.

By specializing in the production of the good for which it has a comparative advantage and trading with the other, each nation can increase consumption of both goods. Before trade, the United States produced 100W and Mexico produced 32W, a

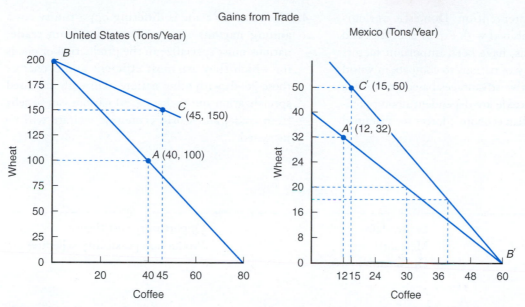

Figure 17-2

Through trade, the United States specializes in wheat production at point B (200 tons), trades 50 tons of wheat to Mexico for 45 tons of coffee, and consumes more of both goods at point C (45 tons of coffee and 150 tons of wheat). Mexico specializes in coffee production (60 tons) and produces no wheat (point B'), trades 45 tons of coffee for 50 tons of wheat, and consumes more of both goods (50 tons of wheat and 15 tons of coffee) than before trade occurred.

total of 132W. With specialization and trade, 200W is produced, all in the United States. Further, in the absence of trade, Mexico produced 12C and the United States produced 40C, a total of 52C. After specialization and trade, 60C is produced, all in Mexico. The net increase in production of 68W and 8C from specialization in the goods for which each had a comparative advantage represents the gains from trade shared by each nation. These gains also reflect the increase in world welfare from increased specialization and trade.

> Through trade, nations specialize in the production of goods for which they have a comparative advantage, trade with each other, and expand consumption and national welfare.

SUMMARY

The purpose of this chapter is to examine the major factors affecting trade patterns and to determine why nations trade. The law of comparative advantage was reviewed and contrasted with absolute advantage and competitive advantage. The gains from trade were examined, with emphasis on how those gains were distributed between nations and the importance of specialization and exchange. The major points made in this chapter are summarized as follows:

1. Trade occurs because individuals, firms, governments, and nations anticipate economic gains from the exchange of goods and services. International commerce is lucrative because goods can be purchased where they are abundant and cheapest and then sold where they are more scarce and highly valued. Differences in national resource endowments

give rise to these differences in costs and prices, and provide the basis for trade.

2. The theory of international trade has evolved over the past four centuries. Mercantilist theory explained trade as a zero-sum game in which all nations strive to maintain a trade surplus at the expense of other nations. Next, the basic concept of absolute cost advantage was used to account for why nations traded with each other. Absolute advantage is not a necessary condition for trade, but the law of comparative advantage actually determines why nations trade and how the gains from trade are distributed.

3. Comparative advantage was contrasted with the concept of competitive advantage to explain how agricultural trade patterns are influenced

by government intervention. Domestic agricultural policies, combined with export subsidies and import restrictions, have been important factors affecting the ability of nations to compete in world markets. Competitive advantages based on the role of government in trade are dependent upon political factors rather than economic forces for sustained existence.

4. The basis for trade is differing opportunity costs among nations. To receive gains from trade, nations must specialize in the production of goods for which they are most efficient and exchange those goods with other nations. Through increased specialization and exchange, all nations can benefit from trade, and world economic welfare will be increased.

KEY TERMS

Absolute advantage

Comparative advantage

Competitive advantage

Laissez-faire

Mercantilism

Opportunity cost

Opportunity cost theory

Production possibility schedule

TESTING YOUR ECONOMIC QUOTIENT

1. The mercantilist philosophy argues that nations can become rich and powerful by exporting more than they import.   T   F
2. Comparative advantage states that a nation will export the foods that it can produce more cheaply than others and import the goods that other nations can produce more cheaply.   T   F
3. Opportunity costs reflect the cost of a good as measured by the amount of a second good that must be given up in order to produce one additional unit of the first good.   T   F
4. Comparative advantage is a monetary concept and is affected by changes in exchange rates or inflation.   T   F
5. Competitive advantages based on nonmarket factors are dependent on politics rather than economics for sustained existence.   T   F
6. The basis for trade is differing _____ among nations.
7. In order to achieve gains from trade, nations must specialize in the _____ of goods for which they are most efficient and _____ those goods with other nations.
8. The _____ shows all combinations of national output that can be produced by fully employing all available _____ and the most productive _____.
9. The laissez-faire philosophy held that
   a. strict government control of all economic activity was necessary.
   b. little or no government control of the economy was needed.

   c. nations need the government to establish export laws to stimulate the economy.
   d. none of the above.
10. According to comparative advantage, a nation should
    a. not import any goods if the nation does not have a relative disadvantage.
    b. specialize in production of the good for which it has the least relative disadvantage.
    c. import only the goods in which the nation has a disadvantage.
    d. not produce any goods if the nation is at a relative disadvantage.
11. Opportunity cost theory states that a nation has
    a. an absolute advantage in the production of the good with the lowest opportunity cost.
    b. no advantage in the production of any good with an opportunity cost.
    c. a comparative advantage in the production of the good with the lowest opportunity cost.
    d. none of the above.
12. Identify the factors affecting comparative advantage.
13. Discuss the difference between comparative advantage and competitive advantage.
14. If the United States produces 20 tons of grain and 10 tons of coffee per man-hour and Mexico produces 15 tons of grain and 3 tons of coffee per man-hour, what commodity should each country produce according to the law of competitive advantage?

REFERENCES

Haberler G: *The theory of international trade*, London, 1936, W. Hodge and Company.

Houck JP: *Elements of agricultural trade policies*, Prospect Heights, Ill, 1986, Waveland Press.

Munn T: *England's treasure by foreign trade*, Oxford, 1928, Basil Blackwell.

Porter M: *Competitive advantage: creating and sustaining superior performance*, New York, 1985, The Free Press.

Ricardo D: *On the principles of political economy and taxation*, Homewood, Ill, 1963, Richard D. Irwin.

Smith A: *An inquiry into the nature and causes of the wealth of nations*, New York, 1937, The Modern Library.

🍎18
Agricultural Trade Policy and Preferential Trading Arrangements

Chapter Outline

Trade liberalization
Trade liberalization reflects the reduction or elimination of import tariffs, export subsidies, and trade-distorting farm and other domestic policies.

The formulation and implementation of effective trade policy poses a crucial yet complex dilemma for policymakers. Large gains in income and employment can be achieved by reducing or eliminating policies that restrict the trade of food and agricultural products. The difficult choices facing international trade policymakers are many. The first problem is how to disengage from trade-distorting domestic policies and still support farm incomes at home. The second problem is how to continue the process to significantly reduce trade barriers, which would foster large economic gains to most nations. The purpose of this chapter is to demonstrate that, although clear economic gains to trade may exist, it is often difficult to pursue those gains because of strong vested interests that pursue the status quo. Although the nation as a whole benefits from trade, there are both gainers and losers within nations. It is the potential losers who effectively protect domestic markets using trade barriers.

TRADE AND WELFARE

In Chapter 17, we learned that in a two-nation, two-good world, each nation could specialize in the production of the good for which it had a comparative advantage and trade with the other nation, and the consumption of both nations was increased. Comparative advantage was used to explain why nations trade. In this chapter, we will use partial equilibrium analysis to determine the specific prices and quantities at which trade occurs. The gains to trade are illustrated using consumer and producer surplus, or welfare analysis, which demonstrates the importance of interdependence between nations resulting from trade.

Autarky or the Closed Economy

First, assume there are two nations: the United States and Japan. One commodity—for example, wheat—is produced by both nations. Figure 18-1 illustrates the hypothetical supply and demand relationships. Initially assume that each nation operates as a closed economy. Also, each nation is self-sufficient, or an **autarky**, and no trade occurs between it and any other nation. Initial market conditions are reflected by the intersection of supply and demand in each nation, in which P_{US} and Q_{US} represent competitive equilibrium in the United States, and P_j and Q_j are market equilibrium in Japan. In the absence of trade, the quantities of wheat produced and consumed in each nation represent market equilibrium.

Next, assume that traders from the United States travel to Japan and discover that the price of wheat, P_j, is much higher than that at home, P_{US}. Traders realize that they can now buy wheat in the United States and resell it in Japan at a profit, ignoring transfer costs. When wheat is exported from the United States to Japan, P_j will begin to decline, and P_{US} will begin to rise. This process, known as **arbitrage**, will continue as long as the price in Japan exceeds the U.S. price, indicating a profit from trade. Finally, exports from the United States to Japan will be sufficient to result in the same wheat price in both nations, hence establishing one market for the trade of wheat.

Autarky occurs when a nation has no trade or is self-sufficient.

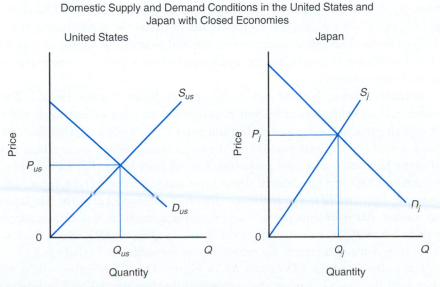

Figure 18-1
Under autarky, or complete self-sufficiency, market equilibrium is determined solely by the interaction of domestic supply and demand conditions. No trade between nations is assumed to occur.

Impacts of International Trade on the Economies of the United States and Japan

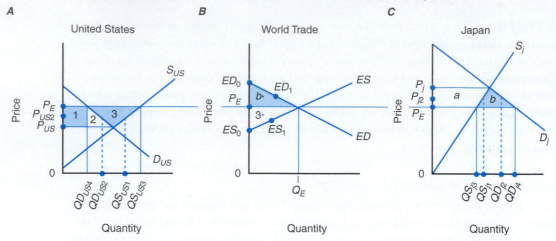

Figure 18-2

In an open economy, trade is allowed to take place. Prices rise in the United States and decline in Japan. Producers in the United States and consumers in Japan benefit from trade. Consumers in the United States and producers in Japan experience losses in economic welfare.

Trade and Partial Equilibrium

To determine the precise price and quantities at which trade will occur, a partial equilibrium model is constructed in Figure 18-2. First, it is necessary to derive the excess supply curve for the United States. **Excess supply** indicates the amount by which quantity supplied exceeds quantity demanded for each price level above P_{US}. Goods will always move from where prices are low to where prices are higher. When S_{US} and D_{US} are equal, competitive equilibrium exists and excess supply is zero. ES_0 can be plotted on the vertical axis of Figure 18-2B, to represent the initial point on the excess supply curve for the United States. When prices rise, more wheat will be shipped, so a second point on the excess supply curve can be determined by plotting the horizontal distance between any two quantities. For example, the difference between QS_{US1} and QS_{US2} can be plotted at P_{US2} in Figure 18-2B, and labeled ES_1. When ES_0 and ES_1 are connected, ES results and indicates the excess supply curve for the United States. This relationship is often referred to as export supply or exportable surplus. The ES function indicates the quantity of wheat supplied to the export market at each price above P_{US}.

Similarly, excess demand can be derived for Japan (Figure 18-2C). **Excess demand** indicates the amount by which quantity demanded exceeds quantity supplied at each price below P_j. At equilibrium price P_j, $S_j = D_j$ and, therefore, excess demand is zero and is plotted as ED_0 in the trade sector. When more wheat is supplied to the Japanese market, prices decline. For any price below P_{j1}, for example, P_{j2}, excess demand, $QD_{j2} - QS_{j1}$, can be determined and plotted as ED_1 in Figure 18-2B. Connecting ED_0 and ED_1 yields ED, or the excess demand curve for the Japanese wheat market. ED represents the quantity demanded in excess of domestic supply at each price below P_j. Japan will import the quantity $QD_{j4} - QD_{j3}$ from the United States. Excess demand is commonly called import demand.

The point at which ED equals ES in Figure 18-2B indicates world wheat market equilibrium. One price, P_E, prevails in both countries and indicates the price at which wheat is traded internationally, or the world market price for wheat. In the United States, the higher price, P_E, causes quantity supplied to increase to QS_{US3} and the quantity demanded to decline to QD_{US4}. The quantity exported

Excess supply is the amount by which a country's quantity supplied exceeds its quantity demanded for each price above autarkic equilibrium. Excess supply is zero at the point where domestic supply and demand are equal.

Excess demand is the amount by which a country's quantity demanded exceeds its quantity demanded for each price below autarkic equilibrium. Excess demand is zero at the point where domestic supply and demand are equal.

from the United States, or export supply, would equal $QS_{US3} - QD_{US4}$. In Japan, at the lower price P_E, the quantity demanded increases to QD_{j4} and the quantity supplied declines to QS_{j3}. The difference equals excess demand and is identically equal to excess supply from the United States and the amount entering world trade in Figure 18-2B, quantity Q_E.

Trade impacts are quite different, yet still have important implications for producers and consumers in both countries. Higher wheat prices cause U.S. producers to expand production by bringing additional land, labor, capital, and technology into use. Consumers, however, reduce wheat consumption because its cost to them has risen. The difference between what is produced and consumed generates the excess supplies shipped to Japan.

Lower prices in Japan mean that wheat consumers will increase their purchases, resulting in greater quantity demanded. Producers will reduce output, however, since prices have fallen. Greater excess demand causes imports to rise. The amount of wheat imported from the United States at price P_E will increase to quantity Q_E. It is important to note that with trade, internal market prices are jointly determined by both domestic supply and demand conditions in both countries and the world market.

Welfare Gains from Trade

The welfare effects of free trade mirror the quantitative effects discussed earlier (Figure 18-2). In the exporting nation, consumer surplus declines by area 1 + 2. Producer surplus, however, increases by area 1 + 2 + 3. Society in the exporting country is better off by the amount of area 3 as a result of trade. In the importing country, consumers gain areas *a* and *b*, and producer surplus declines by the amount of area *a*. There are net societal gains of area *b* for the importing nation. The net gains to trade can be illustrated in the trade sector. Area 3, net gains to the exporting country, is shown by 3* in Figure 18-2B, and represents the amount by which producer gains exceed consumer losses. Net gains for the importing country are shown by *b** and represent the amount by which consumer gains exceed producer losses. The net gains to trade are summarized in Table 18-1. If gainers compensate losers, then economic agents in both nations will be made better off from free trade.

Several key assumptions are necessary in order to achieve consistent results when applying the partial equilibrium analysis to trade. First, a constant exchange rate is assumed between the currencies in the exporting and importing nations. Second, all input and output prices, per capita incomes, population, production technology, and consumer tastes and preferences are held constant. Third, transfer costs between nations are assumed to be zero. If any of these conditions change, it will be necessary to derive new world market equilibrium conditions. Finally, this partial analysis can handle only one product or commodity at a time.

To summarize, producers in the United States and consumers in Japan gain from free trade. Producers in Japan and consumers in the United States will be forced to adjust to the more open market. If those who gain compensate those who lose, however, free trade makes everyone better off. Because trade has potentially very

Table 18-1
GAINS TO TRADE

	United States	Japan
Consumer gains	−(1 + 2)	+(a + b)
Producer gains	+(1 + 2 + 3)	−(a)
Net gains to society	+(3)	+(b)

different impacts on special interest groups, it is often not a popular policy option. In this case, producers in Japan would most certainly resist moves to liberalize trade. Although U.S. consumers will be adversely affected, it is unlikely that any strong resistance to freer trade would occur primarily because no one consumer is forced to bear a large share of the burden of adjustment. The prospects for strong consumer reaction are, therefore, diluted.

WHY RESTRICT TRADE?

It has been demonstrated that there are significant gains to freer trade. The **free trade** argument states that resource use will adjust *across* and *within* national boundaries, so that marginal value products for land, labor, and capital will be equal in all uses. Optimal world welfare will result. Why is trade restricted? Very simply, the economic hardship generated by adjustments to more open borders often leads some sectors or industries to effectively lobby for protection. **Protectionism** occurs when government policy is implemented to remedy domestic economic problems associated with excessive imports. Protectionist policies reduce competitive pressures on producers and allow them to forego the adjustments brought about by freer trade.

Protectionism reduces competitive pressure but also reduces the efficient use of resources.

Protectionism in Agriculture

Although the URA was important in reducing agricultural trade barriers, protectionism in agriculture still abounds. In 2008 to 2010, for example, the level of support to producer incomes from government averaged 22% for OECD countries.[1] During this same period, it was estimated that U.S. farmers received nearly 10% of their income from government programs. Producers in Japan depended upon government for 50% of their income, and farmers in the EU received 22% of income from government sources. The perceived need for a country to be self-sufficient in food production or to have a high degree of food security often transcends the economic benefits of trade, leading to high protective tariffs or the implementation of other restrictive import policies.

Although tariffs and quotas are the traditional methods used to restrict trade, many nations employ less overt measures. The use of support prices above the world price, for example, was common during the 1960s and led to overproduction and lower import demand. Surplus disposal policy may employ a trade-distorting subsidy on exports. Taxes on exports may be used to keep products from leaving a country, resulting in greater supplies and lower prices for domestic consumers, but higher prices for foreign consumers. More subtle measures, such as exchange rate controls, may be used to stimulate a country's exports and improve its competitive position. Although the intent of these measures may not be to disrupt trade, the effects are trade distorting nonetheless.

Arguments against Trade

Arguments against trade can be classified into the following general categories:

1. To protect a new or an infant industry.
2. To counter unfair foreign competition.
3. To improve the balance of payments.
4. To protect national health, the environment, and food safety.

---

[1] The Organization for Economic Cooperation and Development (OECD) is a 34-member organization established in 1961 to promote the economic well-being of member countries, while contributing to the overall development of the world economy.

The argument to protect a new or an infant industry is often made because when production of a product first begins, firms are usually less efficient than they will be after economies of scale are achieved. Costs of production tend to be higher and labor may be less skilled and therefore less productive than it will become over the long term. These infant industries call for protection for a period of time so that they do not fail before they have the opportunity to become efficient. Over time, economies of scale will result, allowing the protected firms to become globally competitive. This argument has appeal for nations in the early stages of industrial development. In fact, the United States used the infant industry argument to support tariffs in the 1800s. Most often, however, the argument is used by developing countries that want to protect smaller, less-efficient businesses from large industrial firms in the United States, Japan, and Europe. Although this argument is justified in many cases, it has been used to perpetuate trade barriers and to maintain inefficient industries well beyond their infancy.

Unfair competition by foreign governments and businesses is often cited by special interest groups who seek protection. Much of the recent U.S. trade law has been in response to calls for protection against unfair competition. **Section 301** of the Trade Act of 1974 provides presidential authority to impose duties on products from nations whose trade practices are deemed unfair or found to restrict U.S. commerce.

Another component of this argument is that U.S. wage rates are higher than those in developing countries, and therefore, U.S. businesses face a competitive disadvantage. The U.S. Congress was urged by organized labor interests to vote against the establishment of a North American Free Trade Agreement because Mexico has relatively low-cost labor, which might lure some U.S. manufacturing plants to Mexico, taking jobs with them.

In Chapter 16, we learned that a nation's balance of payments position can serve as an indicator of its overall international well-being. When payments to foreigners chronically exceed earnings, confidence in the nation's currency and economic strength may be undermined. Foreign investment may level off, followed by a decline in the value of currency. To stem the outflow of currency, a nation may attempt to reduce its payments to foreigners by limiting imported products. If export earnings remain constant, reduced imports will bring the international payments account more into balance. In reality, however, these actions often invite retaliation by foreign governments and firms, resulting in reduced export earnings. Protectionist policies rarely result in higher, sustained exports over the long run.

Trade in food products is also subject to restrictions related to the protection of human, animal, and plant health; the environment; and food safety. For example, the United States prohibits the import of fresh/chilled meat from nations that have endemic aftosa, or foot-and-mouth disease. U.S. meat imports from Mexico are limited to shipments from those facilities that have been inspected and approved by the USDA. Likewise, U.S. meat exports to the EU must be from facilities approved by the EU. These health provisions are called **sanitary and phytosanitary regulations (SPS)**. For U.S. exports of plant products, a phytosanitary certificate is issued by the Animal and Plant Health Inspection Service, which certifies that a particular shipment is free of harmful diseases or pests. Although many of the SPS currently in effect around the world are based upon valid health concerns and scientific evidence, it has become increasingly popular for special interest groups to attempt to convince governments that health or food safety concerns exist when none actually do. The EU has been criticized for implementing a ban on imported meat containing growth hormones. This action effectively eliminated the EU market for U.S. beef.

Important trade legislation Section 301 of the Trade Act of 1974 allows the imposition of import duties on products from countries that engage in unfair trade practices.

SPS Regulations Sanitary regulations are used to ensure the safety of imported meats. Phytosanitary regulations protect plant products from harmful insects and diseases.

Success in the Uruguay Round to reduce tariff and nontariff barriers to trade has resulted in greater use of technical barriers to trade by special interest groups to influence trade in their favor. Technical barriers to trade (TBT) are a special form of nontariff barrier that includes not only SPS but also restrictive food safety and labeling regulations or other peculiar laws that inhibit trade. As duties and import quotas are reduced and eliminated through trade negotiations, it appears likely that TBT will be used by many countries to restrict imports.

TRADE RESTRICTIONS

Trade restrictions may take many different forms. For discussion purposes, we will classify them into three groups: (1) import policies, (2) domestic food and agriculture policies, and (3) export policies. Import policies relate to those measures designed to restrict imports to protect domestic industries or to raise government revenue. Domestic food and agriculture policies may attempt to influence production and/or consumption and may take the form of support prices, consumption taxes, stock policies, input use restrictions, or input subsidies. Export policies attempt to stimulate the export of products, often below the world market price.

Import Policies

Border policies can be classified as tariff or nontariff barriers to trade. Tariffs, or duties, were once the dominant restrictions affecting trade. Recent efforts to liberalize trade through the General Agreement on Tariffs and Trade (GATT) and the World Trade Organization (WTO), however, have resulted in a lower overall level of tariff protection in many developed nations. Nontariff barriers, such as quantitative restrictions limiting imports, increased in importance from 1950 until 1990 because they were often more politically acceptable, and nations have sought ways to restrict trade and still comply with international obligations to reduce tariff protection. Consequently, nontariff barriers have become the most widely used means of restricting agricultural trade.

Tariff Barriers A tariff is a tax levied on goods when they enter a country. Tariffs may be imposed for protection or to generate revenue. Protective tariffs are implemented to insulate domestic producers from import competition. Protective tariffs are not intended to prohibit imports. In negotiations to liberalize trade, tariffs often go through a process of tariff binding, in which their initial level is agreed upon and can be reduced over time, but not raised without additional consultation. Revenue tariffs represent a tax on imports and are implemented to raise revenue for the government. The general effect of tariffs is to raise the domestic price and reduce the quantity demanded. The U.S. tariff on imported melons, for example, results in higher consumer prices for melons and substitute goods, regardless of whether the melons are from foreign or domestic sources.

Tariffs may be classified into three broad categories. An ad valorem tariff is assessed as a percentage of the value of imported goods. Cantaloupes imported from Mexico, for example, were assessed a tariff of 35% ad valorem. A specific tariff is assessed as a fixed amount of money per unit of imported product. Imports of fresh grapefruit were assessed a tariff of $0.029 per kilogram, for example. A compound tariff is calculated on both an ad valorem and a specific basis. Goods may be assessed at the rate of 5% ad valorem and $20 per ton. Ad valorem and specific tariffs are the most common tariffs applied in the trade of food and fiber products.

A nation's prices, production, and consumption can all be affected by a tariff. Figure 18-3 illustrates these effects. It is assumed that this is a small nation whose

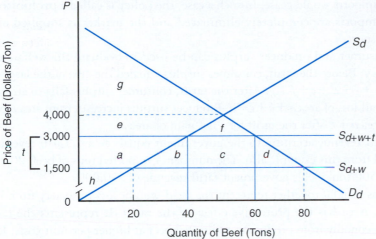

Figure 18-3

Free trade equilibrium is denoted by the intersection of S_{d+w} and D_d. Imports equal 60 tons at a price of $1,500/ton. The imposition of a tariff, t, on imports would lower imports to 20 tons (60 tons – 40 tons) and increase domestic production to 40 tons. A tariff has the overall effect of protecting domestic producers from import competition but also causes higher prices for consumers. In this small-nation case, the full burden of the $1,500/ton tariff is passed on entirely to consumers. Consumption of domestic output declines to 60 tons.

imports represent only a small share of world market supply. This nation is therefore a price taker, facing a constant world price for imported beef. Under these assumptions, the nation is not large enough to influence the world price, a fairly realistic case for many countries. If D_d represents domestic demand for beef and S_d is domestic supply, then in the absence of trade, market price is $4,000 per ton, the quantity supplied is 50 tons, and the quantity demanded is 50 tons. With international trade, S_{d+w} is the free trade supply function, indicating the tons of beef available to the small nation from domestic and foreign sources combined. The world price, $1,500 per ton, prevails and quantity supplied declines to 20 tons. Quantity demanded rises to 80 tons. Excess demand (80 tons – 20 tons) of 60 tons of beef is supplied by imports. Domestic beef prices have fallen as a result of free trade, making consumers better off because they can buy more at lower prices. Producers, however, are worse off because they sell less at a lower price than before trade.

Under free trade, the domestic beef industry is being damaged by imported products. Sales and profits are declining, producers are going out of business, and workers are losing their jobs. Assume labor and management unite and successfully lobby Congress to impose a specific tariff of $1,500 per ton on beef imports. Because this is a small nation and, therefore, a price taker on the world market, world beef supplies remain constant. This new policy raises the domestic beef price by the full amount of the tariff, placing the burden of adjustment entirely on beef consumers. Available supply increases by the amount of the tariff to S_{d+w+t}.

With the protective tariff, a new world market equilibrium price, $3,000 per ton, is established. Due to higher prices, domestic consumption declines by 20 tons, from 80 tons before the tariff to 60 tons after the tariff. Production increases by 20 tons, from 20 tons to 40 tons. Because domestic production increases and domestic consumption declines, imports fall to 20 tons. The overall effects of the tariff are to restrict imports, raise domestic prices, and increase production. Domestic beef producers are protected from import competition because prices are raised and imports

Types of import tariffs
Import tariffs may be classified as ad valorem, specific, or compound. The terms *tariff* and *duty* are often used to mean the same thing: a tax on imports.

are reduced. Note that if the tariff were high enough to raise the price to $4,000 per ton, beef imports would cease. In such a case, the policy is called a **prohibitive tariff**, because imports are completely eliminated and the market is supplied only from domestic sources.

Consumer and producer surplus can be used to evaluate the welfare effects of tariff policy. Before the tariff, consumer surplus equaled the area of the large triangle, area $a + b + c + d + e + f + g$. After the tariff, consumer surplus falls to area $e + f + g$, or an overall loss of area $a + b + c + d$. Producer surplus increases from area h before the tariff to area $a + h$ after the tariff for a net gain of area a.

Tariff revenue accruing to the government equals area c. Tariff revenue can be calculated from this example also. The number of imported tons of beef, 20, times the tariff, $1,500/ton, yields government tariff revenue of $30,000.

Areas b and d together represent the dead-weight loss to society from imposing the tariff. Area b is the protective effect of the tariff. It represents the loss to the domestic economy from producing additional beef at higher per-unit costs. Resources less suited to beef production are employed when tariff-induced output expands. The protective effect occurs because less efficient domestic beef production is substituted for more efficient imported beef. Area d is called the consumption effect and results from the tariff artificially increasing the price of beef. Domestic consumers are denied the opportunity to purchase beef at the lower world price of $1,500 per ton and must therefore pay the higher cost of $3,000 per ton. This loss in welfare represents a real net cost to society, which is not captured by any other economic agent.

The net welfare effects of a tariff imposed by a small nation are to

1. reduce consumer welfare,
2. increase producer welfare,
3. raise government revenue, and
4. cause a dead-weight loss to society.

A tariff imposed by a small nation reduces net social welfare.

Nontariff barriers (NTBs) are classified as any government policy, other than a tariff, that limits imports but does not reduce domestic production.

Nontariff Barriers   In recent years, barriers to trade other than tariffs have become much more important. **Nontariff barriers (NTBs)** include any government policy, other than tariffs, that reduces imports but does not limit the domestic production of goods that substitute for imports. Governments have sought ways to limit imports without violating their international obligations to reduce tariffs. Politicians are often reluctant to levy a tariff because it acts as a tax not only on imports but also on domestic consumers. NTBs may, therefore, offer a more politically acceptable alternative. NTBs include restrictions at national borders, laws and regulations that discriminate against imports, export subsidies, price supports, favorable exchange rates and credit terms for exporters, and other programs or production assistance that substitutes domestic output for imports or stimulates exports.

An import quota   limits the amount of a product that may be imported during a specified period of time.

Import quotas restrict the physical amount of a product that may be imported during a specified period of time. Presently, the single most important NTB is the import quota. Quotas may be global, meaning that they do not discriminate among nations that supply imported goods. Quotas may also be allocated, meaning that each exporting nation is allocated a maximum amount of product that can be shipped.

The U.S. Meat Import Act of 1964 required the President to consider the imposition of import quotas on frozen, chilled, or fresh veal, mutton, beef, and goat meat when it was estimated that imports would reach the level of an adjusted base quota, near 7% of domestic production. The law was amended in 1979 to allow more imports in years when U.S. domestic meat production is low and less imports in years when U.S. meat production is high. This feature is referred to as a countercyclical measure, which attempts to account for the biological nature of the cattle production cycle and weather variability. The law was replaced by a tariff-rate quota under the URA in 1995.

The effects of a quota are similar to those of a tariff. In Figure 18-3, for example, if a quota of 20 tons were imposed, the domestic price would rise to $3,000 per ton. By restricting available supplies on the market, domestic prices for beef could not fall below $3,000 per ton, assuming supply and demand conditions remain unchanged. When domestic prices rise due to limited supplies of beef, consumer surplus declines and producer surplus increases as before. An import quota has been referred to as one of the most insidious trade restrictions employed because it satisfies special interest groups by affording higher prices, but limits the physical quantity available on the market, often causing shortages of critical goods and exorbitantly high prices.

Rents from a quota are quite different from the tariff case, however. Area c represented tariff revenue to the government imposing the tariff. With a quota, however, this rent may accrue to importers, exporters, or government. Beef consumers must now pay an additional $1,500 for each of the 20 tons imported under the quota. If U.S. importers organize collectively and bargain with foreign exporters to buy beef on the world market at $1,500 per ton and resell at $3,000 per ton, quota revenue would accrue to the import firms in the form of monopoly, or windfall profits.

Alternatively, foreign exporters could organize as sellers and limit market supplies to 20 tons. This action would drive up the price to $3,000 and allow the exporters to capture and share the profits. However, as with any cartel scheme, the incentive for each individual exporter to cheat and attempt to undersell the market at a lower price would be great and might lead the collective action to disintegrate.

Finally, the government of the importing nation could auction off import licenses to domestic importers according to the highest bidder. The government could then capture the quota revenue that would have accrued to the importers.

Under a quota, the distribution of rents among the economic agents—domestic importer, foreign exporter, government—is indeterminate. Windfall profits from the quota are shared among the agents depending upon the degree of bargaining power each possesses. These results lead to the often surprising conclusion that in some instances, exporters may actually prefer to limit their sales to a particular country.

Voluntary export restraints (VERs) are agreements between the government of an importing country and exporters that limit exports to a specified amount. Authority to negotiate VERs is provided to the President in Section 203 of the Trade Act of 1974. The U.S. Trade Representative conducts the actual negotiation. VERs have been used most recently to limit U.S. imports of soft wood lumber from Canada, steel, automobiles, and textiles. Under a VER, the importing nation does not establish a limit on imports, but exporters agree to voluntary limits on the amount shipped during a specified time period. Exporting nations often agree to participate in VER schemes out of fear that the importing country may adopt more severe import restrictions. In practice, VERs are often negotiated between an exporting nation and the U.S. government as Congress is considering the imposition of an import quota. VERs can also benefit exporters.

By limiting the available supplies on the market, a VER leads to higher prices in the importing nation. Higher prices cause a decline in consumer surplus within the importing country. Because the exporting nation limits the amount shipped, it can allocate the rights and determine who receives the windfall profits. Exporters receive those rights in order to not retaliate against the importing country. It is not so surprising to see that some nations would agree to limit their own exports to another country. Together, these exporters behave in much the same way as a monopoly.

Voluntary export restraints limit the amount of exports during a given time period.

The variable levy was a key nontariff barrier that was used by the EC to protect domestic agriculture while transforming the European Community from one of the largest net grain importers to one of the largest net grain exporters. A **variable levy** is a variable import tariff equal to the difference between a designated domestic price and the lowest landed import price. The variable levy in the EC was adjusted daily for grains and sugar; weekly for dairy, beef and live cattle, and rice; monthly for olive oil; and quarterly for pork, poultry, and eggs. The common agricultural policy (CAP) of the EC set the minimum price at which any import could enter the market. If the import price fell below this level, a levy was imposed that equaled the difference between the minimum price and the import price. Using this mechanism, the EC has effectively regulated the import of food products into the European Community and, therefore, protected agricultural producers from lower-cost, more-efficient foreign competition. The EC has adopted major reform of the CAP by agreeing to reduce support prices, institute direct payments to producers to compensate them for lower prices, and implement supply controls. CAP reform resulted in a fixed duty, adjusted biweekly.

Another NTB used by the United States and many other countries to protect agriculture is the **tariff-rate quota (TRQ)**. This measure is used extensively by WTO members in order to comply with the URA provisions requiring the conversion of import quotas to their tariff equivalent. The TRQ allows a specified amount or number of products to enter an importing country at one tariff rate, often zero (the within-quota rate), whereas imports above this level are assessed a higher rate of duty (the over-quota rate). Although not often used to restrict imports of manufactured goods, the TRQ has been used to limit the import of milk, cattle, fish, brooms, tobacco products, coconut oil, and, more recently, sugar to the United States. In the early 1970s, a TRQ on fluid milk was implemented that set a nonbinding import quota of 3.0 million gallons annually. The within-quota tariff was set at $0.02 per gallon, and the over-quota tariff was $0.065 per gallon.

> **A tariff-rate quota (TRQ)** allows a specified amount of imports at low or zero duty, whereas imports above that level are charged a higher duty.

The TRQ is viewed as a compromise between consumers, who desire low-cost imports, and producers, who want higher prices and protection from foreign competition. Compromise between these two sets of diverse interests results because the extremely adverse impacts and higher prices from quota are averted, and at the same time producers are offered some insulation from the impacts of severe import competition by the higher over-quota tariff.

To liberalize trade in the future, it is likely that the WTO will rely upon the use of tariff-rate quotas for the tariffication of nontariff barriers. **Tariffication** is the process whereby quotas, licenses, variable levies, and other nontariff barriers to trade are converted to their tariff equivalents. These tariffs are then bound at negotiated tariff rates and reduced over a specified period of time.

Figure 18-4 illustrates the effects of a tariff-rate quota on trade and economic welfare. Domestic supply and demand for a small-nation importer are denoted by S_d and D_d, with an autarkic equilibrium price of $300. With free trade, this nation produces 10 tons of corn, consumes 50 tons, and imports the balance, 40 tons, at a price of $100 per ton.

Now assume that a tariff-rate quota of 5 tons is imposed. The within-quota tariff is reflected by S_{d+w+t_1}, and the over-quota tariff is higher and denoted by S_{d+w+t_2}. Because imports initially exceed the quota amount, both the within-quota and over-quota rates apply. The TRQ results in an increase in corn prices from $100 per ton to $200 per ton. Domestic production increases to 20 tons, domestic consumption falls to 40 tons, and imports decline to 20 tons. Higher domestic prices and increased production result in an increase in producer surplus equal to area e, and dead-weight losses equal to area $f + g$.

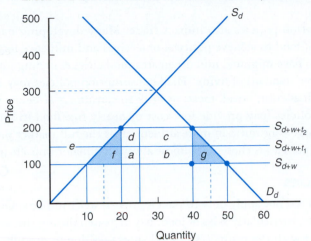

Figure 18-4
The tariff-rate quota combines both a tariff and a quota. Imports are allowed to enter the market at a low duty rate up to the quota, and imports over the quota are assessed a higher duty rate. The TRQ represents a compromise between consumers who desire low prices and producers who want high prices.

Revenue generated by the TRQ, however, is apportioned between the government as tariff revenue and businesses as windfall profits. In this case, 20 tons of corn is imported after the TRQ is imposed. The government collects area a when tariff revenue is equal to the within-quota tariff, $t_1 = \$50$ per ton, times 5 tons. Area $b + c$ also accrues to the government and is found by multiplying the over-quota tariff, $t_2 = \$100$ per ton, times 15 tons.

Area d represents a windfall profit to domestic or foreign businesses. Following the imposition of the TRQ, the domestic price for the first 5 tons of corn increased to $150, reflecting the world price of $100 per ton plus the tariff of $50 per ton. If importers can obtain foreign corn at $150 per ton and resell on the domestic market for $200 per ton, or the over-quota price, area a would be captured as windfall profit by the importers. However, if foreign exporters can restrict corn shipments, they can force up the price of corn and extract profits from importers, and capture area a by raising their supply price to $200 per ton. Any portion of area a captured by foreign exporters represents a welfare loss to the importing nation.

Domestic Agriculture and Food Policies

Trade is heavily influenced by the type and degree of domestic agriculture and food policies adopted by countries. To achieve domestic policy objectives related to food security, market stability, or economic and social development, many nations, either explicitly or implicitly, subsidize domestic agricultural production or food consumption. Policies to achieve these often-conflicting objectives usually are

1. those that alter market price, thereby influencing production and consumption,
2. those that only affect the production and/or processing of foods, and
3. those that affect consumption or utilization of raw commodities, semiprocessed products, or consumer-ready products.

Direct payments to producers, deficiency payments, payments in kind, marketing loans, and cash bonuses all directly affect the planting and production decisions of producers. Policies that subsidize production inputs, such as fertilizer, credit, transportation, irrigation, and the development of infrastructure, such as roads, bridges, and storage facilities, affect agricultural production either directly or indirectly. Consequently, these policies often have the same trade-distorting impacts as tariffs and nontariff barriers. Many countries decouple farm program payments to

comply with URA provisions. These are less distorting but can still distort resource use in agriculture.

Consumption policies also impact trade. Many developing nations subsidize consumption of food to achieve political objectives and influence large shares of the population who have migrated into urban areas and cities searching for higher-paying jobs and a better standard of living. Food consumption policies may include providing staples such as corn, bread, meat, or milk to consumers at below the free market price. Other policies may provide low-cost or nearly free food to some parts of the nation, but charge consumers in more affluent areas the full market price. Whichever form or mix of policies is chosen, it almost always has major impacts on trade.

Export Policies

Trade barriers need not apply only to policies that restrict imports. Government control over exports, particularly sales of food, has increased during the past two decades. Fears associated with the prospects of food shortages due to drought and high consumer prices resulted in the widespread use of measures to limit the export of food in Ukraine and India. Because of large crop surpluses during the late 1970s and early 1980s, policies designed to expand or promote exports became popular tools for increasing food shipments.

Policies designed to either restrict or promote agricultural exports are undertaken to

1. dispose of surplus production or stored commodities,
2. reduce or halt the export of domestic commodities, thereby limiting the potential for consumer price increases,
3. develop or maintain domestic processing industries and increase employment rather than export raw materials,
4. limit the economic or military capability of another nation, and
5. encourage another country to undertake policy reform by denying critical food or fiber supplies.

Because the EC sets minimum internal producer prices well above the world price, surplus production results. This surplus is disposed of on the world market using another nontariff barrier, the export subsidy, or restitution. An **export subsidy** is a direct cash or in-kind payment provided by a government to encourage export sales.

The Market Access Program (MAP) was authorized by the 1996 FAIR Act to stimulate the export of high-value food and agricultural products. The MAP provides assistance in cash or commodities to trade promotion associations to partially fund foreign market development, advertising, and promotional activities. Although MAP was designed to create commercial markets for U.S. products, it has been criticized because a few large multinational firms have received substantial government assistance, perceived as a subsidy under this program. Funding for MAP was subsequently reduced by nearly one-half. MAP was referred to as the Market Promotion Program in the 1990 Farm Bill and as the Targeted Export Assistance program in the 1985 Farm Bill.

Export policies can have major impacts on the trade of food and agricultural products. Policies designed to restrict exports obviously affect the perception of the exporter as a reliable supplier, leading to greater uncertainty, reduced purchases by other nations, and possible retaliation by importing nations. An export tax is one example. Due to rising prices, numerous countries implemented taxes on grain exports in 2003–2008 in order to retain domestic supplies and reduce prices. Market disruption occurs, which leads to greater variability in trade volumes, increased

An export subsidy is a direct or in-kind payment provided to exporters in order to stimulate exports of a product.

uncertainty by private and public decision-makers, and resource misallocation and inefficiencies in production and marketing. Domestic consumers and foreign producers may experience short-term gains from these policies because domestic prices fall and world prices rise; however, domestic producers and foreign consumers generally lose. Export embargoes can be especially disruptive and painful for foreign consumers if imposed during times of tight world food supplies. These measures may also damage the reputation of the implementing nation as a serious advocate of freer trade.

Policies that subsidize exports can be equally disruptive to world markets. Foreign consumers and domestic producers gain from export subsidies, and domestic consumers and foreign producers lose. Although export subsidies may be effective in achieving short-term policy goals, they are almost always costly in terms of both money and image.

AGRICULTURAL TRADE POLICY MAKING

The United States faces critical policy choices as it adjusts to changes emanating from the international marketplace. Agricultural trade policy has taken on added importance as trade has become crucial to the future prosperity of U.S. agriculture. Policy formulation related to trade, domestic programs, and economic growth and development occurs in a **multilateral** setting. Major international institutions and policy-making bodies now provide an important linkage between efforts to reduce distortions to trade and accomplish social goals to support sustainable domestic agricultural production, and the need to ensure consumers of a low-cost, safe food supply. The industrial nations, particularly the United States, Japan, and the EU countries, have received harsh criticism from trading partners for failing to eliminate agricultural subsidies that distort trade.

The General Agreement on Tariffs and Trade and the World Trade Organization

The **General Agreement on Tariffs and Trade (GATT)** resulted from the Bretton Woods Agreement negotiated at Bretton Woods, New Hampshire, in 1947. The GATT was one of the three global institutions created after World War II to assist in rebuilding nations devastated by war and fostering economic recovery worldwide. Although GATT was replaced by the **World Trade Organization (WTO)** in 1995, its principles were incorporated into the WTO. The International Monetary Fund (IMF) and the International Bank for Reconstruction and Development, or the World Bank, were the other two. The GATT emerged as the multilateral body governing trade after the failure of many of the 22 member nations and the United States to ratify the International Trade Organization. GATT's primary objective was to encourage economic growth and development by liberalizing world trade. Although GATT was a multilateral agreement establishing the rules governing international trade, it also served as an institutional forum for negotiating reductions to trade barriers and for settling trade disputes. GATT was replaced by the World Trade Organization and has 159 member nations, which account for more than 90% of world merchandise trade.

Several key principles governed the operation of GATT, and subsequently the WTO. First, and most important, is the **most-favored-nation (MFN)** clause, which states that trade must be nondiscriminatory. Concessions granted to one nation automatically apply to all other WTO members. Second, tariffs, rather than nontariff barriers, should be used to protect domestic industries. Third, agreed-upon tariff levels are binding. If duties are raised by one country, compensation may be required by its

The World Trade Organization, a specialized agency of the United Nations, was formed in 1995 and has 159 members. The WTO is headquartered in Geneva, Switzerland, and has a staff of 600 people. The goal of the WTO is to help producers of goods and services conduct business.

trading partners. Finally, imported goods should be treated no less favorably than domestically produced goods.

In seven previous rounds of negotiations, GATT was effective at reducing tariff protection in the manufacturing sector from an average of 50% in 1947 to 5% in the mid-1980s. However, protection for agriculture in industrial countries rose from 21% in 1965 to more than 40% during the same period (OECD).

Despite the overall increase in protection of agriculture, some important concessions were granted in agricultural trade. Most notable was during the Dillon Round when the EEC granted the United States zero tariff binding on imported soybeans and corn gluten feed. Next was the Tokyo Round, which led to Japanese concessions to liberalize its market for imported beef and citrus. Major attempts to bring agriculture under GATT discipline were unsuccessful prior to 1986.

Important exceptions to GATT rules existed. Many of these exceptions relate to agricultural trade. The United States requested and received a waiver from GATT to employ quotas and other import restrictions on imported agricultural products that interfere with the operation of domestic farm programs. Other nations have also used this exemption, most notably the EC, to impose a variable levy that blocks many agricultural imports. Although GATT prohibited the use of export subsidies, agricultural products are exempt if such subsidies do not distort "historical" market shares. These exemptions led to a proliferation of nontariff barriers to trade and the current level of distortion in international trade of food and agricultural products.

The Punta del Este Declaration of September 1986 launched the eighth and final GATT round, referred to as the Uruguay Round. All GATT contracting parties affirmed their commitment to liberalize agricultural trade. It was anticipated that the Uruguay Round would allow a multilateral reformulation of agricultural policies, leading to lower taxpayer costs and less market distortion. The Uruguay Round represented the first serious attempt to liberalize agricultural trade. Previous piecemeal approaches to reform domestic and agricultural trade policies had failed, and by the mid-1980s, the United States and other world powers were concerned that the cost of protecting agriculture had become excessive.

After 3 years of negotiation, the parties issued a midterm review reflecting the status of progress in the Uruguay Round. It was significant that all parties could agree to "substantial, progressive reduction" in those policies that distort trade. Differences among the United States, the EC, and Japan again resulted in little real progress to reform trade. A special session was held at Blair House, Virginia, in November 1992 to settle the differences between the United States and the EC. During this session, the United States and the EC agreed to adopt the following measures to liberalize agricultural trade:

1. Reduce domestic support by 20%—U.S. and EC deficiency payments would be exempt.
2. Reduce subsidized export tonnage by 21% and expenditures by 36%.
3. Implement tariffication of nontariff barriers and bindings under GATT.

This dispute became so heated that French farmers organized blockades of major roadways to stall traffic, attract public attention, and develop public support for their position. France threatened to exercise its right to veto the agricultural trade reforms proposed by the EC. The Uruguay Round was 3 years beyond its scheduled completion date when finalized. Although the final outcome fell short of expectations, it represents an important first step to agricultural **trade liberalization**. The stage was set to further reduce barriers to agricultural trade in further future negotiations.

The Doha Development Agenda (Chapter 16) is now being negotiated within the framework of the WTO. While the final outcome is unclear, consensus was

The role of the GATT   The GATT effectively lowered tariffs on industrial goods, but did little to liberalize trade in agriculture until the Uruguay Round Agreement on Agriculture in 1995.

Trade liberalization   is the reduction and eventual elimination of tariff and nontariff barriers to trade.

reached among all 159 members on several major issues affecting agriculture. First, countries with the highest tariffs may make bigger reductions than countries with lower tariffs, affording increased and more equitable access to many markets now virtually closed to imports. Second, a date will be agreed upon by which export subsidies must be eliminated. This will have the largest impact on the EU. Finally, trade-distorting domestic support to agriculture will be substantially reduced. This could have major implications for U.S. farm programs since countries with the highest levels of farm support may be required to make the largest reductions. For more information, see the World Trade Organization website, www.wto.org. Negotiations faltered in 2008 as countries failed to agree on market access and domestic support reductions. The outcome of the Doha Development is now uncertain.

The Doha Round   The most recent WTO round, the Doha Development Agenda, will operate on three pillars of agricultural reform: market access, export competition, and trade-distorting domestic support.

The United Nations Conference on Trade and Development

The United Nations Conference on Trade and Development (UNCTAD) was formed in 1964 as an agency of the General Assembly of the United Nations to stimulate the economies of developing countries through trade. These nations, concerned about domination of GATT and IMF by industrial countries, believed that their needs were not being met (Tweeten, 1992). Many developing countries were also convinced that a conspiracy existed because prices for agricultural products had fallen relative to prices for manufactured products. With a majority of votes in the United Nations, developing countries successfully established the UNCTAD in 1964.

UNCTAD has the goal to eliminate or reduce tariff and nontariff barriers to trade levied by industrial countries. In recent years, UNCTAD has served as a forum for industrial and developing countries to discuss trade issues. Any member of the United Nations may join UNCTAD.

The most significant policy development occurring under the auspices of UNCTAD was the granting of nonreciprocal trade concessions to developing countries by industrial countries. The **generalized system of preferences (GSP)** permits many manufactured and agricultural products from developing countries to enter developed country markets duty-free. The objective of the GSP is to expand exports by developing countries to bring about economic development. The United States grants duty-free entry to developing countries for about one-half of the items classified in the tariff schedule of the United States. The basic U.S. legislation granting these tariff concessions was the Trade Act of 1974. Agricultural products entering the U.S. market under the GSP include okra, citrus, melons, nuts, cucumbers, and sugar. More than 100 developing nations now receive preferential tariffs under the GSP for products shipped to the United States.

The GSP was legal under GATT because a waiver of most-favored-nation status was granted in 1971. Under this waiver, industrial countries may grant special low tariffs to developing countries without extending those same tariff preferences to other industrial countries.

U.S. Agricultural Trade Policy Formulation

Although the President and Congress are both responsible for U.S. agricultural trade policy, Congress has delegated much of its authority to the President. The executive branch has the overall responsibility to formulate and administer the everyday operational details of trade policy. The Office of the U.S. Trade Representative (USTR), the U.S. International Trade Commission (USITC), and the U.S. Department of Agriculture have specific roles in implementing various aspects of U.S. trade policy related to agriculture.

The United States Trade Representative (USTR) is responsible for U.S. trade policy and leads the trade-negotiating team in all trade negotiations.

The U.S. Trade Representative The USTR is the primary agency responsible for all U.S. trade policy matters. The USTR is a cabinet-level position created by executive order in 1980 when the President moved the office from the Department of State to the cabinet. The USTR holds the rank of ambassador. The USTR is responsible for all trade activities related to the WTO, bilateral and multilateral trade negotiations, direct investment incentives and disincentives, and negotiations before the United Nations that relate to trade and before any other organization where trade is the primary issue. Public input from business, individuals, and members of Congress is received by the USTR on matters related to trade and investment.

The Economic Policy Council

Interagency coordination is critical to the effective formulation and implementation of U.S. trade policy. The Economic Policy Council (EPC) is the primary unit of the Executive Branch for coordinating trade policy (Tweeten, 1992). The EPC is composed of the secretaries of agriculture, commerce, defense, labor, transportation, treasury, and state. The secretary of treasury serves as the president pro tempore. The director of the Office of Management and Budget and the chairman of the Council of Economic Advisers also serve on this interagency committee.

The President also has more than 40 other committees that serve in an advisory role in evaluating and formulating trade policy. The Advisory Committee for Trade Policy (ACTP) Negotiations is the lead committee charged with advising the President on trade policy negotiations. The ACTP is composed of private sector business interests and is usually administered by a departmental staff assistant. The Agricultural Policy Advisory Committee represents the interests of agriculture. It is composed of private sector representatives from producer, processor, cooperative, trade, and other organizations. The Agricultural Technical Advisory Committee is composed of 10 sector committees related to cotton, dairy, grain and feed, livestock and products, oilseeds and products, poultry, processed foods, sweeteners, and tobacco.

The U.S. International Trade Commission The U.S. International Trade Commission (USITC) is an independent government agency charged with investigating U.S. trade law violations. A chairman, vice-chairman, and four commissioners appointed by the President make up the USITC. Commissioners serve 9-year terms and may not be reappointed. The USITC conducts inquiries into the laws of the United States and other countries, examines complaints related to foreign competition, and compares imports with U.S. consumption levels. Injuries to domestic industries are investigated, and when appropriate, the President may impose import protection.

Investigations of possible agricultural trade law violations usually involve complaints related to Sections 201 and 301 of the Trade Act of 1974 and the Tariff Act of 1930. Section 201 allows the President to impose duties or other import restrictions on imports that threaten U.S. industries producing like goods. This provision is intended to provide temporary relief from injurious competition. The Tariff Act of 1930 authorizes investigation of cases in which imports may be subsidized and dumped at less than fair value on the U.S. market. This act, commonly referred to as the Smoot–Hawley Act, established the highest tariff rates ever imposed by the United States, largely in response to high domestic unemployment during the Great Depression. It remains the cornerstone of U.S. import legislation (Miller, 1985).

Before the 1980s, the United States depended upon GATT for increased market access and resolution of trade disputes. During the second half of the 1980s, however, the United States began to move toward preferential bilateral or regional trading arrangements known as **preferential trading arrangements (PTAs)**. PTAs provide for the elimination of tariff and nontariff barriers (import quotas and licenses) to trade. Investment laws, transportation regulations, and mechanisms for resolving trade disputes are often included.

THE IMPORTANCE OF PREFERENTIAL TRADING ARRANGEMENTS

In many cases, PTAs are used to strengthen existing trade ties among nations, creating opportunities for additional employment and stimulating investment. More recently, concerns about air and water pollution, worker safety, and unfair labor regulations have led to the inclusion of environmental and labor issues in the negotiation of PTAs.

The **Reciprocal Trade Agreements Act of 1934 (RTA)** authorizes the President to fix tariff rates, and has resulted in a liberal tariff stance for the United States. From 1934 to 1947, in fact, the United States negotiated bilateral tariff reduction agreements with 29 nations. However, when the GATT emerged as the primary trade forum, the RTA declined in importance as a mechanism for trade liberalization.

In September 1985, the United States and Israel entered into a bilateral agreement to reduce tariff and nontariff trade barriers in merchandise trade. Services trade was liberalized, and provisions were made to protect intellectual property rights.

The Canada–United States Trade Agreement went into effect on January 1, 1989. It provided for the elimination of both tariff and nontariff barriers to trade between the two countries, specified rules governing trade in services, and reduced restrictions on investments by both nations over a 10-year transition period.

The North American Free Trade Agreement (NAFTA), submitted to the U.S. Congress on November 3, 1993, was implemented on January 1, 1994. NAFTA eliminated tariff and nontariff barriers to trade among the United States, Mexico, and Canada over 5-, 10-, or 15-year periods. NAFTA substantially reduces or eliminates most agricultural trade barriers between the United States and Mexico, but did not fully liberalize United States–Canada agricultural trade. Other agreements have been implemented with Australia, Bahrain, Central America–Dominican Republic, and Morocco, for example.

FORMS OF ECONOMIC INTEGRATION

For the most part, PTAs entered into by the United States have taken the form of free trade areas. A **free trade area** consists of provisions to remove both tariff and nontariff barriers to trade with members, while retaining individual trade barriers with nonmember nations. Trade barriers with the rest of the world differ among member nations and are determined by each member's policymakers. The **North American Free Trade Agreement (NAFTA)** is an example of a free trade area. However, free trade areas represent only one form of economic integration.

A **customs union** requires that members eliminate trade barriers with each other and takes economic integration a step further by establishing identical barriers to trade against nonmembers, most often in the form of a common external tariff. The European Community was probably the world's most well-known customs union. It includes a CAP for all 15 member states. CAP is characterized by a variable levy system that limits imports from nonmembers by applying a common variable duty.

A **common market** includes all the aspects of a free trade area and customs union, but takes integration even further by permitting the free movement of goods and services, labor, and capital among member nations. In January 1993, the EC moved closer to full common market status by implementing the Single European Act (SEA). Further unification of the EC by creating a single, barrier-free market for most goods and services will be attempted. The SEA calls for elimination of barriers to trade in goods and services, labor, and capital.

Finally, an **economic union** represents the most complete form of economic integration. Member nation social, taxation, fiscal, and monetary policies are harmonized or unified. Belgium and Luxembourg formed an economic union in the 1920s. Customs documents and procedures, value-added taxation, and nontariff barriers to trade have been harmonized to facilitate the movement of goods among EC member nations. However, more contentious issues, such as implementing a common currency, delayed the ratification of the Treaty of Maastricht, the document that completes the economic integration of the EU. Finally in late 1993, plans to move the EC to an economic union were approved, calling for common banking laws, coordinated macroeconomic policy, and a common EU currency.

Various forms of economic integration have occurred within the framework of the GATT, now the WTO. Although Article I of GATT prohibits the use of preferential tariff rates, an important exception is allowed by Article XXIV, which specifies the conditions under which signatories may form and legally operate PTAs. These conditions are as follows: (1) the elimination of trade barriers on substantially all trade among members, (2) remaining trade barriers against nonmembers are not more severe or restrictive than those previously in effect, and (3) interim measures leading to the formation of the agreement are employed for only a reasonable period of time. When these three conditions are met, international trade agreements do not violate any of the WTO articles.

Vague phrases such as "substantially all trade" and "reasonable time period" have allowed much latitude for interpretation of the articles. Since its inception, GATT has been notified of more than 70 PTAs, some establishing interim agreements with no final date for completing the PTA being specified. None have been formally disapproved (Jackson, 1989). PTAs were allowed to proliferate under GATT, and it appears that the WTO will not take a less conciliatory stance toward PTA formation.

REASONS FOR PREFERENTIAL TRADING ARRANGEMENTS

The emergence of the United States as a major player in the formation of regional trading blocs has raised questions about why PTAs have become such a popular policy tool and why the United States is seeking to participate. Reasons might include the following:

1. To provide for special trading arrangements with countries that are economically or politically important to the strategic interests of the United States
2. To achieve timely, substantial reduction in barriers to trade, particularly agriculture, intellectual property rights, services trade, nontariff barriers, and dispute settlement procedures
3. To counter the economic and political power created in Europe by further integration of the EU and the prospects for trade and economic cooperation with former Eastern bloc nations, Central and South America, and Asia
4. To reduce the flow of illegal immigration and other side effects of trade barriers by stimulating economic growth and development
5. To foster political stability and economic prosperity, thereby supporting the continuation of the democratic process and reducing the likelihood of political and social disruption (Rosson, Runge, & Hathaway, 1994)

Counter Economic and Political Power in Other Parts of the World

NAFTA created one of the world's largest free trade areas, consisting of 450 million people, $13 trillion in economic output, and more than $2.0 trillion in trade. The EU has 370 million people, $18.4 trillion in economic output, and trade of $5.8 trillion. However, the EU already has extended preferential market access to Poland, Hungary, the Czech Republic, Estonia, Slovenia, and Croatia, which expanded to 28 members in 2013. As economic and political integration in Europe continues, it is likely that the United States will experience more pressure to form larger and economically stronger PTAs in the Western Hemisphere.

In 1991, Argentina, Brazil, Paraguay, and Uruguay initiated the Southern Cone Common Market, MERCOSUR, to foster economic growth and mutual interests among the member nations. Although smaller in economic strength than NAFTA, MERCOSUR has stimulated interest and concern among many publics.

Economic and political interests of the United States may dictate that, as economic blocs emerge around the world, steps must be taken to ensure that existing and future trading relationships are maintained, and even strengthened, by participation in one or several forms of economic integration. As Salvatore (1993) points out, any free trade area or customs union acting as a single unit in international trade negotiations will likely possess much more bargaining power than the total of its members acting separately. Further, reaching a compromise with a unified group of nations may sometimes be easier than dealing with each nation individually.

Reduce Side Effects

PTAs may be used to encourage member nations to undertake economic, social, or political reform. For instance, over the long run, NAFTA will create an economic incentive for Mexican workers to seek employment in Mexico, rather than in the United States. It is doubtful that any other policy action taken by Mexico could or would have been as instrumental in instituting policy reform as NAFTA. This could have other consequences for the United States, such as reducing the poverty in border colonies and eliminating the strain on border health care, water, sewage, and other facilities required to accommodate a near doubling of population over the last decade. It will also reduce the flow of low-wage labor to the United States and increase costs in some industries.

Other reforms include an increased effort on the part of the Mexican government to control air and water pollution in its major cities and along its border with the United States. There is evidence that the government shut down the state-owned-and-operated petroleum refinery in Mexico City to indicate to the United States that Mexico was serious about environmental degradation and its consequences. Sewage treatment facilities have been constructed in border cities to stop the dumping of untreated waste into the Rio Grande and its tributaries. Meat-packing facilities have been relocated outside of cities to improve air quality and reduce problems with disposal of animal waste.

Foster Political Stability and Economic Prosperity

Probably an overriding goal of the United States in liberalizing trade with Mexico was to foster the continuation of democratic government and political stability of the past few decades. More open markets and the ensuing economic growth may be one of the most effective means of improving the standard of living in Mexico and ensuring that economic prosperity leads to a stable political environment.

Although various reasons may cause the United States to participate in PTAs, it is likely that strong economic and political interests will be the driving force behind such efforts. In the future, leverage may be needed to negotiate continued or additional access to major markets. Negotiating from the strength of an economic bloc may prove to be one effective alternative.

DO PREFERENTIAL TRADING ARRANGEMENTS CREATE OR DIVERT TRADE?

Economic integration that leads to the formation of regional trading blocs can have both trade-creating and trade-diverting effects. One argument against PTAs is that they may lead to increased trade within the bloc but less trade outside the bloc. Arguments in support of PTAs state that they may lead to the merger of regional blocs and, therefore, multilateral trade liberalization more quickly than other forms, such as the WTO. Although both arguments have merit, theory indicates that the effects of PTAs are largely empirical and depend upon several important conditions, which will be discussed in the next section.

With the prevalence of PTAs, a question of critical importance is whether the United States will become better off economically as the economies of North America become more closely integrated. Each case of PTA creation must be examined empirically and on its own merits. It is impossible to assert unequivocally that PTAs may be more efficient than free trade among all nations. Given that trade is less than free, the creation of a PTA may or may not result in welfare gains to society.

Recent advances in international trade theory, however, allow some conclusions to be made about the likely outcomes of PTA creation. Viner (1950) and Johnson (1965) were among the first to explain the costs and benefits of PTAs through the example of a customs union. These writers focused primarily on trade creation and trade diversion, along with total welfare gains. Later efforts by Salvatore (1993) focused on the dynamic effects of PTA creation.

Static Effects

The static effects of creating a PTA are measured by trade creation and trade diversion. Trade creation occurs when some domestic production of a member nation is replaced by lower-cost imports from another member nation. Assuming full employment of domestic resources, trade creation increases the economic welfare of member nations, because it leads to greater specialization in production and trade, based on comparative advantage. PTA members will import from one another certain goods not previously imported at all due to high tariffs. The trade-creation effect results in efficiency gains for member nations because some members shift from a higher-cost domestic source of supply to a lower-cost foreign source. A trade-creating PTA may also increase the welfare gains of nonmembers, because some of the increase in its economic growth will produce real increases in income that will in turn translate into increased imports from the rest of the world.

The impacts of PTAs can be determined using the partial equilibrium analysis (Tweeten, 1992). Assume that the case to be examined is that of Mexico and the United States, in which Mexico is a small-nation importer. In Figure 18-5, it is assumed that the United States is the low-cost supplier of a product imported by Mexico. Domestic supply and demand are S_M and D_M, respectively. Before the free trade area is created, the price in Mexico is P_{M+t} and the supply of U.S. imports with the tariff is M_{US+t}. Domestic production and consumption are Q_{S1} and Q_{D1}, respectively. Quantity imported with the tariff is $Q_{D1} - Q_{S1}$. Creating a free trade area

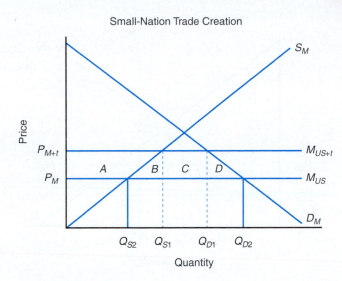

Small-Nation Trade Creation

Figure 18-5
When tariffs imposed by a small-nation importer are removed, consumers experience welfare gains, and producers and government experience welfare and revenue losses, respectively. The nation as a whole experiences welfare gains.

removes the tariff and increases imports from the United States to M_{US} and lowers the price to P_M. When the price in Mexico declines, domestic production falls and domestic consumption increases. Imports from the United States rise to $(Q_{D2} - Q_{S2})$, which exceeds the quantity imported with the tariff.

Trade Creation Welfare impacts of forming a trade-creating free trade area can be determined by examining changes in producer and consumer surplus. Mexican consumers gain area $A + B + C + D$ after the free trade area is implemented. Mexican producers, on the other hand, lose area A, and the government loses tariff revenue equal to area C. Total welfare gains to Mexico are the sum of areas B and D. The free trade area allows the importing nation to regain the dead-weight loss from a tariff.

A summary of the welfare effects of a small-nation, trade-creating free trade area is given as follows:

Consumer gains	$A + B + C + D$
Producer gains	$-A$
Government revenue	$-C$
Net gains to society	$= B + D$

Gains from a free trade area are expected to be large if the tariff that is to be removed is large and as domestic supply and demand become more elastic over the long run. Finally, consumer income gains within a free trade area can be expected to create trade with nonmember countries.

Trade Diversion Trade diversion occurs when lower-cost imports from a nonmember nation are replaced, or diverted, by higher-cost imports from a member nation of a free trade area. This occurs naturally with a PTA, due to the preferential trade treatment provided by member nations. Trade diversion reduces global welfare because it shifts production from more-efficient producers outside the PTA to less-efficient producers within the PTA. The international allocation of resources becomes less efficient, and production shifts away from comparative advantage. Member nations may gain or lose from a PTA, as shown in this section.

The trade and welfare effects of trade diversion are demonstrated in Figure 18-6. Assume that both the United States and the EU compete for the Mexican market. The EU is initially the largest supplier to this market, noted by M_{EU}, which exceeds U.S. imports denoted by M_{US}. With an equal tariff of amount t applied to imports from both supplying countries, the EU is the sole source of import supply because P_{EU+t} is

Figure 18-6

Trade diversion reduces global welfare because production is shifted from more efficient nonmembers to less efficient member producers. The small-nation importer gains from trade diversion only when the recovery of dead-weight losses due to the tariff exceeds losses of government revenue from the tariff.

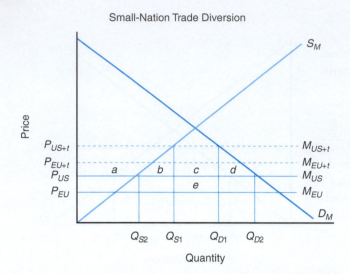

Small-Nation Trade Diversion

below P_{US+t}. In the absence of a free trade area, the price in Mexico is P_{EU+t}, and domestic production and consumption are Q_{S1} and Q_{D1}, respectively. Imports from the EU are equal to $Q_{D1} - Q_{S1}$.

If the United States and Mexico enter a free trade area, the tariff, t, is removed from imports supplied by the United States, but the tariff is not removed from imports supplied by the EU. P_{US} now prevails in the Mexican market, giving the United States a competitive advantage relative to the EU price, P_{EU+t}. Mexican consumption increases to Q_{D2}, and domestic production falls to Q_{S2}. Imports from the United States increase to $Q_{D2} - Q_{S2}$. Imports from the EU are no longer competitively priced, and it no longer supplies the Mexican market. The result of the free trade area was to replace EU imports with those from the United States.

Welfare impacts of the trade-diverting free trade area reveal that consumers gain at the expense of producers and government. Because market price declines from P_{EU+t} to P_{US}, consumers gain areas a, b, c, and d. Producers lose area a. Government tariff revenue declines by the sum of areas c and e. Therefore, net gains to Mexico from entering a free trade area with the United States occur only if the sum of areas b and d exceeds area e.

A summary of the welfare effects of a small-nation, trade-diverting free trade area graphed in Figure 18-6 can be summarized as follows:

Consumer gains	$a + b + c + d$
Producer gains	$- a$
Government revenue	$- (c + e)$
Net gains to society	$= b + d - e$

Most attempts to create PTAs contain both trade-creation and trade-diversion effects and can increase or decrease member welfare depending on the relative strength of the two opposing forces. PTAs will most likely lead to trade creation and increased welfare of member nations under the following conditions:

1. High pre-PTA trade barriers increase the probability that trade will be created among members, rather than diverted from nonmembers to members.
2. The more countries included in the PTA and the larger their size, the more likely that low-cost producers will be found among its members.
3. A PTA formed by competitive, rather than complementary, economies is more likely to produce opportunities for specialization in production and trade creation.

4. When member nations are in proximity to one another, transportation costs become less of an obstacle to trade creation.

5. If the free trade area contains countries with the lowest-cost source of goods and services consumed by member nations, gains can be expected.

Dynamic Effects

The formation of a PTA can be expected to have major dynamic benefits that should be considered important to the participating nations. In fact, it has been recently estimated that the dynamic gains from forming a PTA often exceed the static or welfare gains by a factor of five or six (Salvatore, 1993). The more important dynamic gains include increased competition, economies of scale, stimulus to investment, and more efficient use of economic resources.

Increased Competition   Possibly the most important single gain from a PTA is the potential for increased competition. Producers, especially those in monopolistic and oligopolistic markets, may become sluggish and complacent behind barriers to trade. With the formation of a PTA, trade barriers among members are eliminated, and producers must become more efficient to effectively compete. Some may merge with other firms; others will go out of business. The higher level of competition is likely to stimulate the development and adoption of new technology. These forces combined will likely reduce costs of production and, in turn, consumer prices for goods and services. Importantly though, the PTA must ensure that collusion and market-sharing arrangements are minimized if competitive forces are to operate efficiently.

Economies of Scale   Another major benefit of PTAs is that substantial economies of scale may become possible with the expanded market. If firms were serving only the domestic market, the expanded market with the PTA would likely create substantial export opportunities, resulting in more output, lower costs per unit, and greater economies of scale. For instance, it has been determined that before joining the EEC, many firms in small nations such as Belgium and the Netherlands were comparable in size to U.S. plants, and thus enjoyed economies of scale by producing for the domestic market and for export. However, after becoming members of the EEC, significant economies of scale were gained by reducing the range of differentiated products manufactured in each plant, thereby gaining from increased specialization and greater reliance on comparative advantage.

Stimulus to Investment   The formation of a PTA is likely to stimulate outside investment in production and marketing facilities to avoid the discriminatory barriers imposed on nonmember products. Further, in order for firms to meet the increased competition and take advantage of the enlarged market, investment is likely to increase. In most cases, investment is an alternative to the export of goods—a benefit provided to PTA members. The large investments made by U.S. firms in Europe after the mid-1950s and in the years and months leading up to the Single European Act were fostered by their desire not to be excluded from this large potential market; investing ensured that their products would not be restricted by tariff and nontariff barriers.

Efficient Resource Use   Finally, if the PTA is a common market, the free movement of labor and capital is likely to stimulate more efficient use of the economic resources of the entire community. Efficiency of industries and individual firms will likely increase with increased access to lower-cost capital and additional labor. Lower consumer costs and higher real incomes should follow.

SUMMARY

The purpose of this chapter is to compare free trade with more restrictive trade policies such as tariffs and quotas. Social welfare resulting from these policies was examined, emphasizing the net gains to both exporting and importing nations. Protection of agriculture was discussed, focusing on reasons why government intervention in agricultural trade is so prevalent. The role and importance of the General Agreement on Tariffs and Trade (GATT) and the World Trade Organization (WTO) was examined, along with a discussion of waivers granted for agriculture, which have led to the proliferation of nontariff barriers to trade. Significant reductions in agricultural trade barriers in each of the seven previous GATT rounds were noted. The major points made in this chapter are summarized as follows:

1. Free trade affects exporting and importing nations differently. With free trade, prices in the exporting country rise, production increases, consumption declines, and exports rise. Prices in the importing nation decline, resulting in lower production, higher consumption, and increased imports. In summary, producers in the exporting nation and consumers in the importing country gain from free trade, and consumers in the exporting country and producers in the importing country lose. There is a net gain in social welfare associated with free trade. If gainers are willing to compensate losers, both sets of economic agents will become better off from free trade.

2. Although there are gains to trade, protectionism in agriculture abounds. Domestic policy goals related to food security and self-sufficiency often transcend the economic benefits of free trade. Protective nontariff barriers now dominate much of the trade in food and agricultural products. Arguments against free trade are often based on the need to protect a new industry, counter unfair foreign competition, improve the balance of payments, and protect national health, the environment, and food safety.

3. Trade restrictions can take the form of tariff or nontariff barriers. A tariff is a tax levied on goods when they enter a country. When a tariff is imposed by a small nation, consumer welfare declines, producer welfare and government revenue both increase, but net social welfare is reduced when compared with free trade. Nontariff barriers are government policies, other than tariffs, that reduce imports but do not limit domestic production of goods that substitute for imports. Import quotas, variable levies, voluntary export restraints, tariff-rate quotas, and other domestic policies are included. Nontariff barriers are now the most important set of barriers limiting agricultural trade. Revenue generated by a quota may accrue to government, importers, or exporters as windfall profits depending upon the bargaining power of each economic agent.

4. Agricultural trade policymaking has taken on added importance as trade has become more crucial to the future prosperity of U.S. agriculture. The General Agreement on Tariffs and Trade (GATT), now the World Trade Organization (WTO), is a specialized agency of the United Nations designed to encourage economic growth and development by liberalizing international trade. WTO's 159 member nations account for more than 90% of total world trade. Seven rounds of GATT negotiations failed to make substantial progress to liberalize trade in agriculture, because many trade barriers are linked to domestic policies supporting farm income or prices. Previous rounds have also focused on the reduction of tariffs on industrial products. The Uruguay Round Agreements (URA) represented the first serious attempt to liberalize agricultural trade. Progress to substantially reduce trade-distorting domestic agricultural policies led to a major confrontation between the United States—a proponent of trade liberalization—and some of its major trading partners and allies, such as Japan and France, which oppose reductions in farm support measures. Divisiveness characterized the Uruguay Round and limited its ability to completely eliminate trade barriers affecting agriculture. It is unclear if the Doha Round will succeed in further liberalizing agricultural trade.

5. U.S. agricultural trade policy formulation and implementation is the joint responsibility of the executive branch and Congress. Day-to-day operations are delegated to the President and operationally handled by the Office of the U.S. Trade Representative, the U.S. International Trade Commission, and selected secretaries and interagency advisory committees. Agriculture has specially appointed committees that interact with the USTR and Congressional committees to provide input on agricultural trade policy issues

and technical matters. Coordination of trade policy is necessary to ensure effective and timely implementation of U.S. laws and regulations designed to influence trade in food and agricultural products.

6. Preferential trading arrangements (PTAs) function legally within the framework of existing GATT and WTO rules. Because of the vague nature of these rules, nations have been able to discriminate rather freely against nonmember nations with little fear of retaliation.

7. PTAs may take the form of a free trade area, the least complete type of economic integration, a customs union, a common market, or an economic union, the most complete form of economic integration.

8. A trade-creating, free trade area results in welfare gains for consumers in a small-nation importing country. Losses in economic welfare accrue to producers and government. Net gains to society occur because economic gains offset losses. In a trade-diverting trade area, net gains to society are realized only if consumer gains offset losses in government revenue from tariff elimination.

9. Trade theory provides only ambiguous conclusions regarding the consequences of PTA formation. Free trade is certainly more efficient than discriminatory trade, but in a world of less-than-free trade, PTAs may permit major economic gains under certain conditions. It is likely that dynamic gains, such as increased competition, economies of scale, and greater investment will far exceed static gains from trade creation or trade diversion.

10. PTAs will most likely lead to trade creation and increased welfare of member nations under the following conditions:

- High pre-PTA trade barriers increase the probability that trade will be created among members, rather than diverted from nonmembers to members.
- The more countries included in the PTA and the larger their size, the more likely that low-cost producers will be found among its members.
- A PTA formed by competitive, rather than complementary, economies is more likely to produce opportunities for specialization in production and trade creation.
- When member nations are in proximity to one another, transportation costs become less of an obstacle to trade creation.
- If the free trade area contains countries with the lowest-cost source of goods and services consumed by member nations, gains can be expected.

11. It appears likely that PTAs will take on added importance as policy tools for negotiating fewer barriers to trade among groups of nations. Although the United States may participate in the continued formation of PTAs for many reasons, it is likely that strong economic and political interests will be the driving force behind these actions. In the future, it will likely become more important to negotiate continued or additional access to major markets using the additional leverage provided by nations acting as a single unit.

KEY TERMS

Arbitrage
Autarky
Common market
Customs union
Economic union
Excess demand
Excess supply
Export subsidy
Free trade
Free trade area
General Agreement on Tariffs and Trade (GATT)

Generalized system of preferences (GSP)
Import quotas
Most-favored-nation (MFN)
Multilateral
Nontariff barriers (NTB)
North American Free Trade Agreement (NAFTA)
Preferential trading arrangements (PTAs)
Prohibitive tariff
Protectionism

Reciprocal Trade Agreements Act of 1934 (RTA)
Sanitary and phytosanitary regulations (SPS)
Section 301
Tariff
Tariffication
Tariff-rate quota (TRQ)
Trade liberalization
Variable levy
Voluntary export restraints (VERs)
World Trade Organization (WTO)

TESTING YOUR ECONOMIC QUOTIENT

1. Under autarky, each nation operates as a closed economy and market equilibrium is determined solely by the interaction of domestic supply and demand conditions.   T   F

2. Excess supply is the amount by which quantity demanded exceeds quantity supplied for each price level above equilibrium.   T   F

3. Protectionism occurs when government policy is implemented to remedy domestic economic problems associated with excessive exports.   T   F

4. The infant industry argument says that new firms need to be protected until their products become widely known in the market.   T   F

5. Nontariff barriers reduce imports and limit the domestic production of goods that substitute for imports.   T   F

6. Tarrification is the process whereby quotas, licenses, variable levies, and other nontariff barriers to trade are converted to their tariff equivalents.   T   F

7. Some of the net welfare effects of a tariff imposed by a small nation are
   a. to reduce consumer welfare, increase producer welfare, and raise government revenue.
   b. to increase consumer welfare, decrease producer welfare, and raise government revenue.
   c. to increase consumer and producer welfare, and raise government revenue.
   d. none of the above.

8. A quota represents
   a. a quantitative restriction on the amount of a good imported.
   b. a combination of ad valorem and specific duties.
   c. a compromise between taxpayers and consumers.
   d. none of the above.

9. Nontariff barriers (NTB) include which of the following?
   a. Export subsidies, specific tariffs, restrictions at national borders
   b. Export subsidies, ad valorem tariffs, price supports
   c. Price supports, restrictions at national borders, favorable credit terms for exporters
   d. Price supports, favorable credit terms for exporters, specific and ad valorem tariffs

10. The three global institutions created after World War II to assist in rebuilding nations devastated by war and fostering economic recovery worldwide are
    a. GATT, World Bank, International Bank for Reconstruction and Development.
    b. United Nations Conference on Trade, GATT, International Monetary Fund.
    c. GATT, International Monetary Fund, International Bank for Reconstruction and Development.
    d. International Monetary Fund, International Bank for Reconstruction and Development, United Nations Conference on Trade.

11. The World Trade Organization
    a. is a multilateral agreement establishing the rules for governing international trade.
    b. serves as an institutional forum for negotiating reductions to trade barriers and trade disputes.
    c. has the main objective of encouraging economic growth and development by liberalizing world trade.
    d. all of the above.

12. Identify and discuss the four general categories of arguments against trade. Discuss at least four reasons for undertaking policies designed to either restrict or promote agricultural exports.

13. Preferential trading arrangements (PTAs) provide for the elimination of tariff and nontariff barriers to trade.   T   F

14. Trade creation occurs as some domestic production of a member nation is replaced by lower-cost imports from another member nation.   T   F

15. A trade-creating preferential trading agreement can increase the welfare gains of nonmember nations as well as member nations.   T   F

16. Under Article I of GATT, now the WTO, preferential tariff rates are prohibited, but an exception is allowed by Article XXVI if which of the following conditions are met?
    a. Trade barriers are eliminated on substantially all trade among members.
    b. Trade barriers remaining against nonmembers are not more severe or restrictive than those previously in effect.
    c. Interim measures leading to the formation of the agreement are employed for only a reasonable period of time.
    d. All of the above.

17. The Southern Cone Common Market, MERCOSUR, was created to foster economic growth and mutual interests among the member nations of
    a. Brazil, Guatemala, Paraguay, Argentina.
    b. Paraguay, Uruguay, Argentina, Brazil.
    c. Argentina, Guatemala, Panama, Brazil.
    d. Brazil, Paraguay, Argentina, Panama.

18. Important dynamic gains from the formation of a preferential trading arrangement include

a. economies of scale.

b. increased competition.

c. stimulus to investment.

d. more efficient use of economic resources.

19. List the five reasons discussed for preferential trading arrangements.

20. Discuss the five conditions under which preferential trading arrangements will most likely lead to trade creation and increased welfare of member nations.

21. Draw two graphs, one to illustrate small-nation trade diversion and one to illustrate small-nation trade creation. Label and explain each one.

22. The current round of multilateral trade negotiations that began in 2001 is referred to as

a. Qatar Charter.

b. WTO Round.

c. Doha Development Agenda.

d. none of the above.

REFERENCES

Jackson JH: *The world trading system: Law and policy of international economic relations*, Cambridge, Mass, 1989, MIT Press.

Johnson HG: An economic theory of protectionism, tariff bargaining, and the formation of customs unions, *Journal of Political Economy* (September 1965).

Miller WJ: *Encyclopedia of international commerce*, Centerville, Md, 1985, Cornell Maritime Press.

Rosson CP, CF Runge, and D Hathaway: *Food and agricultural policy issues and choices for 1995*, Boulder, Colo, 1994, Westview Press.

Salvatore D: *International economics*, 4th ed., New York, 1993, Endowment of International Peace.

Secretariat, Organization for Economic Cooperation and Development: *Agricultural policies, markets, and trade: Monitoring and outlook 1993*, Paris, 1993.

Tweeten L: *Agricultural trade: Principles and policies*, Boulder, Colo, 1992, Westview Press.

Viner J: *The customs union issue*, New York, 1950, Carnegie Endowment for International Peace.

World Trade Organization, www.wto.org, Geneva, Switzerland.

absolute advantage   exists when one nation can produce goods more cheaply than another nation.

absolute unit cost advantage   barrier exists if the unit cost of production for an established firm is lower at all levels of output than would be the case for a new entrant, which means there is no level of output at which a new firm could operate competitively.

Acreage Reduction Program (ARP) percentages   spelled out the set-aside requirements facing farmers desiring to participate in the 1990 FACT Act. An ARP percentage of 25%, for example, required the farmer to idle 25% of the base acreage established by the farmer over time.

agricultural bargaining associations   (e.g., cooperatives) formed by producers to improve the degree of competition in markets for agricultural commodities.

agricultural economics   an applied social science that deals with how producers, consumers, and societies use scarce resources in the production, processing, marketing, and consumption of food and fiber products.

Agricultural Act of 2014   represents the current farm program legislation in force until 2018. This legislation provided the new safety net for crop producers and included new programs for specific livestock producers.

Agricultural Marketing Agreement Act of 1937   a legislative act favorable to agricultural organizations passed by Congress in 1937. This law established agricultural marketing orders. Marketing orders and agreements refer to arrangements among producers and processors of agricultural commodities. The chief goal of a marketing agreement or order is to improve producer income through the orderly marketing of a commodity by a group of producers and to avoid price fluctuations faced by an individual operating alone in the marketplace. Commodity prices may be controlled through negotiations with different groups involved in the marketing process or limitations on the supply placed on the market. Marketing orders have historically been used by the federal government to control the quantity coming into specific markets and, hence, support the market prices and incomes received by farmers.

Agricultural Risk Coverage (ARC)   the producer receives a payment under the 2014 farm bill when either revenue for all crops or revenue for a single crop falls below a specified percentage of a revenue benchmark. This payment supplemented what the producer received from a crop insurance policy.

arbitrage   the process of purchasing commodities in one market at a low price and rapidly selling them in another market at a higher price.

arc elasticity   a measure of elasticity between two distinct points on a demand curve.

area of producer surplus   producer surplus is found graphically by calculating the area lying above the market supply curve and below the market equilibrium price.

asset   something of value owned by a farm or ranch. Assets are generally divided into either physical and financial assets or current (short-term) and fixed (intermediate and long-term) assets.

asset fixity   refers to the difficulty that farmers have in disposing of tractors, plows, and silos when downsizing or shutting down their operations.

autarky   each nation is self-sufficient, and no trade occurs between it and any other nation.

automatic fiscal policy instruments   those taxation and spending policies that take effect without an explicit action by the President and Congress. Examples are the progressive income tax system and unemployment compensation program.

average cost   average cost is defined as cost incurred by the business in the current period divided by output or simply cost per unit of output.

Average Crop Revenue Election (ACRE)   an ACRE payment was based on the state-level difference between actual revenue and an ACRE guarantee times a percentage of the farm's planted acres.

average fixed costs (AFC)   the fixed costs incurred by the business in the current period per unit of output. Average fixed costs are calculated as AFC = TFC ÷ output or AFC = ATC − AVC. The average fixed cost curve, or AFC, associated with specific levels of output declines as output is expanded.

average physical product (APP), or simply average product   the level of output or total product produced by a business per unit of input used. Average physical product is calculated as follows: APP_{labor} = output ÷ labor, $APP_{capital}$ = output ÷ capital, and so on.

Examples include yield per acre, gain per pound of feed fed, and so on.

average revenue (AR)   the level of revenue earned per unit of output. Average revenue is calculated as AR = revenue ÷ output. Average revenue is also equal to the market price under the conditions of perfect competition. This suggests that the revenue the business receives per unit is identical no matter how much the business produces.

average total costs (ATC) or average costs   the total costs incurred by the business in the current period per unit of output. Average total costs are calculated as ATC = TC ÷ output or ATC = AFC + AVC. The average total cost curve, or ATC, associated with specific levels of output plays an important role in determining total profit. Figure 6–4 showed that the difference between average revenue (which is the same as marginal revenue) or AR and ATC at O_{MAX} represents the average profit or profit per unit at this level of output. This difference (AR – ATC) multiplied by the level of output O_{MAX} represents the level of total profit. The minimum point on the ATC curve, where the MC curve intersects this curve from below, represents the break-even level of output.

average variable costs (AVC)   the variable costs incurred by the business in the current period *per unit of output.* Average variable costs are calculated as AVC = TVC ÷ output or AVC = ATC – AFC. The average variable cost curve, or AVC, associated with specific levels of output also plays an important role in assessing the economic performance of a business. Figure 6–3B shows that the minimum point on the AVC, where the MC curve intersects this curve from below, occurs at O_{SD}. This level of cost per unit of output corresponds to the lowest the business can afford to see the market price (and AR) fall and continue to operate in the short run.

balance of payments (BOP)   a record of all foreign transactions by private and public entities over a specific period of time, usually 1 year.

balance of trade   the monetary value of a nation's merchandise exports minus the value of merchandise imports for a given period of time.

balance sheet   a financial statement reporting the value of real estate (land and buildings), non–real estate (machinery, breeding livestock, and inventories), and financial (cash, checking account balance, and common stock) assets owned by farms and ranches and also outstanding debt. The difference between total farm assets and total farm debt outstanding represents the net worth of the farm.

barriers to entry   those forces that make it difficult for firms to enter any industry. They may exist due to government policy in the form of extensive regulation, high start-up costs or competition that impedes firms from succeeding in a particular market.

barter   the direct exchange of one good or service for another without the use of money.

barter economy   an economy in which money is not used as the medium of exchange. Instead, households and businesses swap goods and services in satisfying their needs.

bilateral trade-weighted exchange rate index   determined by using U.S. agricultural trade volumes with major trading partners to calculate weights, and then multiplying those weights by the real exchange rate of each country.

biofuels   are fuels that are derived from biological sources, from corn kernels to trees. They include, but are not limited to, ethanol from corn, sugarcane, or other sources, and biodiesel, which is refined from vegetable oil.

biological resources   include livestock, wildlife, and different genetic varieties of crops.

breakeven   level of output at which average total costs equal average revenue.

budget constraint   defined by the income available for consumption and the prices that a consumer faces. This constraint defines the feasible set of consumption choices facing a consumer.

business cycle   reflects the pattern of movements in the economy's real output, interest rates, or unemployment rate (also referred to as business fluctuations).

cap and trade   a policy in which an aggregate limit (a cap) on pollution is established and then polluters can transfer rights and responsibilities between themselves in order to achieve a pollution goal at the lowest possible cost.

capital   can take on different meanings. For example, the value of owner equity or net worth represents the amount of capital the owner has invested in the business. This is a financial concept. Capital can also refer to physical assets. For example, the term *capital stock* appearing in the production function refers to fixed and variable inputs used to produce a product.

capital access and cost   a potential barrier that deals with the requirement of substantial capital investment in order to enter any industry.

capitalism   a free market economic system in which individuals own resources and have the right to employ their time and resources however they choose, with minimal legal constraints from government.

capitalized value the present value of a future stream of income or costs.

Capper–Volstead Act of 1922 the growth of farmers' cooperatives in the United States dates from the passage of the Capper–Volstead Act of 1922, which exempted cooperatives from certain restraints imposed by the Sherman Antitrust Act of 1890 and the Clayton Act of 1914. The Capper–Volstead Act of 1922 was the principal legislation exempting cooperatives from antitrust laws.

cardinal measure in demand theory, a measure that attempts to quantify the amount of satisfaction obtained from consumption. The measure is known as utils.

cartel cooperative pool formed by oligopolists to set an artificially high price.

ceteris paribus the assumption that all other factors that might affect demand are held constant during the time period. This term is the Latin phrase most often used by economists.

Catastrophic Coverage (CAT) minimum coverage option in the federal crop insurance. The subsidized coverage only protected producers against losses exceeding a high magnitude.

change in demand a shift in the demand curve generally caused by changes in the prices of complements or substitutes, income, and tastes and preferences.

change in quantity demanded refers to the change in quantity observed on the horizontal axis associated with a movement up or down the demand curve as opposed to a shift in the demand curve.

characteristics of oligopoly the economic conditions that define oligopoly are the same as those of monopolistic competition except that there are only a few sellers, each of which is large enough to have an influence on market volume and price. Oligopolists are interdependent in their decision-making. Nonprice competition is the main competitive strategy. The fact that oligopolists match all decreases but not all increases in price leads to a kinked demand curve and discontinuous marginal revenue curve.

civilian labor force reflects the size of the working-age population minus those in the military and institutions.

Clayton Act of 1914 a legislative act passed by Congress in 1914 attempting to minimize the potential social costs of imperfect competition. This Act created the Federal Trade Commission, charged with the responsibility of investigating business organizations and practices in order to insure competitive practices. Although the Sherman Antitrust Act was general in identifying what actions were illegal, the Clayton Act was quite specific. The Clayton Act thus plugged loopholes and deficiencies in the Sherman Act.

Clean Water Act the federal law that regulates the quality of the water bodies in the United States, including oceans, lakes, streams, and rivers; groundwater; and wetlands.

cobweb adjustment one way to think of how markets adjust to a new equilibrium is the cobweb theorem, which leaves a pattern much like a spider web.

coincident indicators indicators of changes in economic activity that reflect current activity.

collusion a situation designed to increase or maintain profits through price-fixing and/or to restrict entry of new firms in an industry.

Commodity Credit Corporation (CCC) an agency within the U.S. Department of Agriculture that makes nonrecourse loans to farmers for the purpose of supporting prices at a specific level.

commodity loan rate price per unit (pound, bushel, bale, or cwt) at which the CCC provides nonrecourse loans to farmers to enable them to hold program crops for later sale. Loans can be recourse for dairy farmers and sugar processors.

commodity shortage the amount by which the quantity demanded at a given price exceeds the quantity supplied.

commodity surplus the amount by which the quantity supplied at a given price exceeds the quantity demanded (not to be confused with producer or consumer surplus).

common market includes all the aspects of a free trade area and customs union but takes integration even further by permitting the free movement of goods and services, labor, and capital among member nations. In January 1993, the EC moved closer to full common market status by implementing the Single European Act (SEA).

communism a type of economic system and social order structured upon common ownership and distribution of the production of goods and services based on need. This economic system is in contrast to capitalism and socialism.

comparative advantage ability of a nation to specialize in the production of the good for which it has the lowest opportunity cost.

competitive advantage economic competitiveness of a nation reflected by the absolute cost of a given good in a given market at a particular point in time.

complements goods typically consumed together, such as hamburgers and hamburger buns; goods for which cross-price elasticities are negative.

concentration refers to the number and market power of firms marketing their products in a particular market. A market characterized by a small number of firms accounting for the majority of total sales is said to have a high degree of concentration.

concept of elasticity originated by Alfred Marshall, it is a key input in making sound business decisions. Determinants of the elasticity of demand of a commodity include availability of substitutes for the commodity, the type of market, the level of the marketing channel, the percentage of the budget spent on the commodity, and time.

conditions of monopolistic competition the conditions of monopolistic competition are identical to those of perfect competition except that the products sold are no longer homogeneous. Product differentiation is the key difference between monopolistic competition and perfect competition.

Conservation Reserve Program (CRP) a major provision of the Food Security Act of 1985 designed to reduce erosion and protect water quality on up to 45 million acres of farmland. Landowners who sign contracts agree to convert environmentally sensitive land to approved permanent conserving uses for 10–15 years. In exchange, the land owner receives an annual rental payment up to 50% of the cost of establishing permanent vegetative cover.

consumer equilibrium the consumption bundle that maximizes total utility and is feasible as defined by the budget constraint. The marginal utilities per dollar spent on a good or service must be equal.

consumer price index (CPI) weighted average of the prices consumers pay for goods and services.

consumer surplus a measure of the savings achieved by consumers at the current market price from the price they would have been willing to pay for a specific quantity of a good or service. Consumer surplus is equal to the area below the market demand curve and above the market equilibrium price.

consumption expenditures by consumers for food, nonfood, nondurable goods, durable goods, and services.

consumption bundles quantities of various goods or services that a consumer might consume.

consumption function a mathematical expression of the relationship between the level of consumption and variables such as the level of disposable personal income.

contractionary fiscal policy actions taken by Congress and the administration to contract aggregate demand; examples include raising income tax rates or lowering government spending.

contractionary monetary policy actions taken by the Federal Reserve to contract aggregate demand; examples are selling government bonds, raising the discount rate, and raising reserve requirements.

Cooperative Marketing Act of 1926 a legislative act favorable to agricultural organizations passed by Congress in 1926. This law permitted farmers or their associations to acquire, exchange, and disseminate a variety of price and market information.

cost-push inflation rise in general price level resulting from businesses and unions raising their prices and wage requests (also referred to as market power inflation).

countervailing actions measures to counteract the possible adverse effects of imperfect competition.

cross-price elasticity a measure of the response of consumption of a good or service to changes in the price of another good or service. It is defined as the percentage change in the quantity of good A demanded divided by the percentage change in the price of good B.

cross rate the exchange rate between two currencies as calculated from the value of a third currency.

currency appreciation an increase in the foreign price of the domestic currency.

currency depreciation a decrease in the foreign price of the domestic currency.

currency devaluation action taken by monetary authority to deliberately decrease the value of currency from a fixed level relative to other currencies.

currency drain currency (paper bills and coins) held by the public, whether it is in their wallets or stored in a nonbank location (e.g., hidden in a mattress).

customs union requires that the members eliminate trade barriers with each other and takes economic integration a step further by establishing identical barriers to trade against nonmembers, most often in the form of a common external tariff. The European Economic Community was probably the world's most well-known customs union.

cyclical unemployment unemployment of labor associated with adverse trends in the business cycle.

dead-weight loss   social costs of imperfect competition.

deficiency payment   value of payment per unit equal to the difference between the market price and the target price. This payment is made to participating farmers.

demand curve   a schedule that shows how many units of a good the consumer will purchase at different prices for that good during some specified time in a specified market, all other factors constant.

demand-oriented macroeconomic policies   active policy actions that are designed to shift the economy's aggregate demand curve.

demand-pull inflation   rise in the general price level that occurs when aggregate demand for goods and services is rising and the economy is approaching full employment output.

differentiated product   a product that is made different from others through advertising or quality variation.

direct crowding out   when government activity supplants private business activity on a dollar-for-dollar basis.

direct payments   payments in the form of cash or commodity certificates made directly to producers for such purposes as production flexibility contract payments, deficiency payments, annual land diversion, or Conservation Reserve payments.

discounting   the process of converting future income or costs to their present value.

discount rate   the rate the Federal Reserve charges when it lends to member commercial banks (all national banks plus those state banks that choose to be members of the Federal Reserve System).

discretionary fiscal policy instruments   those taxation and spending policies that require an explicit action by Congress and the President before they are effective. Examples include the Tax Reform Act of 1986 and the Balanced Budget and Emergency Deficit Control Act.

disposable income   personal income after the payment of tax obligations.

economic union   represents the most complete form of economic integration. Member nation social, taxation, fiscal, and monetary policies are harmonized or unified. Belgium and Luxembourg formed an economic union in the 1920s. Customs documents and procedures, value-added taxation, and nontariff barriers to trade have been harmonized to facilitate the movement of goods among EEC member nations.

economics   a social science that studies how consumers, producers, and societies choose among the alternative uses of scarce resources in the process of producing, exchanging, and consuming goods and services.

economies of scale   in production are said to exist if the cost per unit of output *declines* as the level of production *increases*. The costs per unit of output are decreasing with increasing scale of the firm as fixed costs are spread out over more units of output as a result of expansion.

elastic   a demand or supply curve is said to be elastic if the percentage change in quantity is greater than a given percentage change in price. The flatter the curve, for example, the more elastic it is said to be.

elasticity of supply   percentage change in quantity supplied with respect to a percentage change in the price of the product.

Endangered Species Act (ESA)   the federal law that seeks to protect animal and plant species that are in danger of extinction or threatened. The law, passed in 1973, restricts all activities that could place in danger a species identified as threatened or endangered by the U.S. Fish and Wildlife Service.

Engel curve   the schedule that shows how many units of a good the consumer will purchase at different income levels, all other factors constant.

Engel's Law   as disposable income of a consumer increases, the percentage of income spent for food decreases if all other factors remain constant.

equilibrium output   the level of aggregate output at which aggregate demand equals aggregate supply.

equity   also commonly referred to as net worth, equity represents the owner's share of the business. This value is found by subtracting total liabilities from total assets.

euro   common currency of 11 EU member countries that participate in the European Monetary System.

eurodollars   U.S. dollars on deposit outside the United States.

excess demand   the amount by which quantity demanded exceeds quantity supplied at each price below equilibrium.

excess reserves   difference between legal reserves and required reserves.

excess supply   the amount by which quantity supplied exceeds quantity demanded at each price level above equilibrium.

exchange rate   the number of units of foreign currency that can be exchanged for one unit of domestic currency; also referred to as the rate of foreign exchange.

expansionary fiscal policy actions taken by Congress and the administration to expand aggregate demand; examples include lowering income tax rates or raising government spending.

expansionary period phase of business cycle during which the nation's output is expanding.

expansionary monetary policy actions taken by the Federal Reserve to expand aggregate demand; examples include buying government bonds, lowering the discount rate, or lowering reserve requirements.

export dependent when a relatively large share of a country's agricultural production is exported.

Export Enhancement Program (EEP) a program that targets export assistance to recover lost export markets. The program provides cash subsidies to exporters to help them compete for sales in specific countries. The United States in recent years has agreed to reduce the level of these subsidies in return for concessions made by the European Union.

export subsidy a direct cash or in-kind payment provided by a government to encourage export sales.

exports the quantity and value of goods or services sold and shipped to other countries.

externality a cost or benefit that flows from the use or ownership of a resource and is felt by someone other than the owner of the resource or participant in a market transaction.

FAIR Act see Federal Agriculture Improvement and Reform Act.

fallacy of composition economic reasoning that is true for one individual but not for society as a whole.

farm sector sector of the economy comprising farms and ranches producing raw agricultural products. This sector is often referred to as the farm business sector.

Farm Service Agency (formerly Farmers Home Administration or FmHA) federal agency in the U.S. Department of Agriculture that makes subsidized or guaranteed loans to farmers and loans or grants to rural communities.

Farmer-Owned Reserve (FOR) Program mechanism used in past legislation that fostered storage of surplus commodities owned by the producer as opposed to the government. This program was eliminated in 1996 with passage of the FAIR Act.

Federal Agriculture Improvement and Reform (FAIR) Act commodity legislation that decoupled planting decisions from government payments, giving farmers greater planting flexibility.

federal budget statement of the federal government's receipts and expenditures for the year.

federal budget deficit reflects the extent to which annual federal government expenditures exceed federal government revenue.

Federal Crop Insurance Corporation (FCIC) an entity within the U.S. Department of Agriculture which administers federal crop insurance programs.

Federal Crop Insurance Reform Act of 1994 legislation that made it made it mandatory for producers to participate in crop insurance programs in order to be eligible to receive other farm program benefits. Minimum participation was CAT coverage.

federal expenditures sum of social security and income security payments, national defense expenditures, medicare expenditures, interest on the national debt, and the costs of other federal government programs during the year.

federal funds market an interbank market from which banks borrow the excess reserves of other banks to meet their own short-term reserve requirements.

Federal Open Market Committee (FOMC) consists of members of the Board of Governors of the Federal Reserve System plus selected district Federal Reserve Bank presidents who meet periodically to assess the appropriate action the Fed should take (buy or sell) in the open or private secondary bond market for government securities.

federal receipts sum of personal and corporate income taxes, social security taxes and contributions, and other sources of revenue received by the federal government.

Federal Trade Commission (FTC) an agency of the federal government charged with the responsibility of prohibiting companies from acting in concert to increase their market control and prohibiting false and deceptive trade practices.

fiduciary monetary system value of currency issued by government based on the public's faith that currency can be exchanged for goods and services.

final demand expenditures by consumers, investments by businesses, government spending, and net exports. Final demand is equal to total output intermediate demand.

financial structure refers to the right-hand side of the balance sheet or the amount of debt the business has relative to the amount of equity the owner(s) has (have) invested in the business.

firm supply curve represented by the segment of the firm's marginal cost curve that lies above the average variable cost curve or shutdown level of output.

fixed costs (FC) specific form of current production costs that do *not* vary with the level of output or input use. This is a short-run cost concept; all costs are considered variable in the long run. Fixed costs are calculated by outside entities. Examples include the individual value of the business's current property tax bill, the insurance premium due this year, or the interest portion of the business's current mortgage payment.

floating exchange rate system whereby currency values are determined by market supply and demand conditions, with minimal government intervention.

Food, Agriculture, Conservation and Trade (FACT) Act of 1990 the omnibus food and agriculture legislation signed into law on November 28, 1990, that provided a 5-year framework for the secretary of agriculture to administer various agricultural and food programs.

Food and Drug Administration (FDA) an agency within the U.S. Department of Health and Human Services charged with the responsibility of monitoring the safety of U.S. food supply; has the broadest authority of ensuring the safety and wholesomeness of food and beverage products.

food and fiber industry consists of business entities that are involved in one way or another with the supply of food and fiber products to consumers.

food and fiber system an economic system consisting of business entities that are involved in one way or another with the supply of food and fiber products to consumers.

Food Security Act (FSA) of 1985 the omnibus food and agriculture legislation signed into law on December 23, 1985, that provided a 5-year framework for the secretary of agriculture to administer various agricultural and food programs.

Food Stamp Program (FSP) federal program administered by the Food and Nutrition Service of the U.S. Department of Agriculture that provides food assistance to needy households. This agency also administers the National School Lunch Program, the School Breakfast Program, and the Special Supplemental Food Program for Women, Infants, and Children. Since 2008, the Food Stamp Program has been known as the Supplemental Nutrition Assistance Program (SNAP).

free trade the absence of direct or indirect government intervention to alter market prices and quantities.

free trade area consists of provisions to remove both tariff and nontariff barriers to trade with members, while retaining individual trade barriers with nonmember nations. Trade barriers with the rest of the world differ among member nations and are determined by each member's policymakers. The North American Free Trade Agreement is an example of a free trade area.

frictional unemployment occurs when persons change jobs and are currently unemployed.

full employment high degree of employment of nation's resources (an unemployment rate of 5% to 6% for labor and a capacity utilization rate for capital of 86% to 88% are generally thought to constitute full employment).

full employment GDP the level of aggregate output at which labor and capital in the economy are employed at their natural or noninflationary rate.

full employment output see full employment GDP.

future profit effect business expectations rising (declining), future profits will shift the investment function to the right (left).

General Agreement on Tariffs and Trade (GATT) an agreement originally negotiated in Geneva, Switzerland, in 1947 to increase international trade by reducing tariffs and other trade barriers. The agreement provides a code of conduct for international commerce and a framework for periodic multilateral negotiations on trade liberalization and expansion. The Uruguay Round Agreement established the World Trade Organization (WTO) to replace the GATT. The WTO officially replaced the GATT on January 1, 1996.

general equilibrium regards all sectors of the economy as being interdependent. The events taking place in both the money market and the aggregate product market are considered when discussing general equilibrium.

generalized system of preferences (GSP) permits duty-free entry of many manufactured and agricultural products into industrial countries.

global climate change the process through which increasing atmospheric concentrations of gases lead to a warming of the planet. This effect is frequently called the *greenhouse effect.*

greenhouse gases are those gases that reflect radiation back to Earth. The most abundant of these is carbon dioxide (CO_2), which increases when fossil fuels such as oil or coal are burned.

gross domestic product (GDP) referred to as the nation's output, GDP is equal to consumer expenditures, business investment, government spending, and net exports (exports minus imports).

gross farm income   annual level of income received from farming activities before farm expenses, taxes, and withdrawals have been deducted.

Human Nutrition Information Service   an agency within the U.S. Department of Agriculture that distributes dietary guidelines designed to improve nutrition and health of consumers.

human resources   the services provided by laborers and management to the production of goods and services.

imperfect competition   market structure when one or more of the characteristics of perfect competition are not present.

imperfectly competitive market structures   imperfectly competitive forms of market structure can be classified according to the number of firms and size distribution in the market, the degree of product differentiation, the extent of barriers to entry, and the economic environment within which the industry operates. The profit-maximizing level of output for any imperfectly competitive firm is determined where marginal revenue equals marginal cost.

import dependence   occurs when a relatively large share of a country's domestic food or fiber consumption is accounted for by imported products.

import quotas   a policy that restricts the physical amount of a product that may be imported during a specified period of time.

imports   the quantity and value of goods or services purchased from other countries and shipped to the United States.

incentive-based policies   are those that use economic incentives to achieve their objectives. Taxes, subsidies, or transferable permits can be used to achieve environmental goals, and all of these have the potential to be less costly than command-and-control policies in which the government dictates what should be done.

income effect   decrease (increase) in the price of a product means the consumer can afford to buy more (less) of the product.

income elasticity   a measure of the relative response of demand to income changes. It is defined as the percentage change in the quantity demanded divided by the percentage change in income.

independent goods   goods whose consumption is independent from the consumption of other goods; examples include toothpaste and milk or any meat product and any detergent; goods for which cross-price elasticities are zero.

indifference curve   a graph of the locus of consumption bundles that provide a consumer a given level of satisfaction.

indirect crowding out   occurs when the U.S. Treasury sells new government bonds to finance federal budget deficits, which leads to higher interest rates in money markets and reduces scheduled private investment expenditures.

inelastic   a demand or supply curve is said to be inelastic if the percentage change in quantity is less than a given percentage change in price. The steeper the curve, for example, the more inelastic it is said to be.

inelastic demand   is the inelastic nature of the demand curve for raw agricultural products, which, when combined with a bumper crop and rising imports, drives down the price of these commodities. Revenue falls sharply because of steepness of the demand curve.

inferior goods   goods for which consumption falls (rises) when income increases (decreases).

inflation   sustained rise in the general price level.

inflation-adjusted   refers to a nominal dollar value that has been converted to a real dollar value using a price deflator like the consumer price index or CPI.

inflationary gap   amount by which planned aggregate spending in the economy exceeds full employment GDP.

input categories   consist of land, labor, capital, and management.

input costs   all inputs to production (land, labor, capital, and management) have a cost. The cost of two inputs can be captured in something called an iso-cost line.

interest rate   refers typically to the unit cost of borrowing money. Also see nominal and real interest rate.

interest rate effect   the slope of the investment function reflects the change in investment expenditures if the rate of interest changes.

intervention currency   currency used by other nations' monetary authorities to keep currencies within their predefined ranges.

iso-cost line   much like the budget line for consumers, this line reflects the particular level of expenditure for two inputs. The slope of the iso-cost line is the ratio of the prices of the two inputs.

isoquant   a curve that reflects the combinations of two inputs that will produce a specific level of output.

iso-revenue line the rate at which the market is willing to exchange one product for another.

isoutility curve same as indifference curve.

labor one of four general classes of inputs captured in a production function. This category includes hired and unpaid family labor, but specifically excludes management.

lagging indicators indicators of changes in economic activity about one or two quarters after they occur.

laissez-faire economic philosophy of little or no government control of a nation's economy.

laissez-faire macroeconomic policy a decision to let the economy work its way out of its current problems; avoid taking active policy actions.

land one of four general classes of inputs captured in a production function. This category refers to the soil and other attributes of land used to produce a raw agricultural product.

law of diminishing marginal returns as successive units of a variable input are added to a production process with the other inputs held constant, the marginal physical product eventually decreases.

law of diminishing marginal utility marginal utility declines as more of a good or service is consumed during a specified period of time.

leading indicators indicators of changes in economic activity about one or two quarters before they occur.

least-cost criteria the point of tangency of an iso-quant and an iso-cost line, and not where they might cross, represents the least-cost combination of two inputs to produce a particular level of output.

leftward shift in the consumer's demand curve refers to movement of a demand curve to the left from its initial position as incomes or other factors increase (decrease) demand for a good or service.

legal reserves the minimum amount of money that banks and other financial institutions are required to keep in specific form. For example, a bank's reserves consist of currency in its vault plus the balance it maintains on account at a Federal Reserve District Bank. Also see required reserves and excess reserves.

legislative acts passed by Congress such as the Clayton Act of 1914, the Packers and Stockyards Act of 1921, and the Capper–Volstead Act of 1922 are vehicles designed to minimize the social costs of imperfect competition. Countervailing measures including price regulation and taxation, as well as marketing orders and marketing agreements, also exist for offsetting potential adverse effects of imperfect competition.

long-run average cost curve also known as long-run planning curve; is comprised of points on a series of short-run average cost curves, illustrating profitability for different sizes of operations.

lump-sum tax a fixed tax (e.g., a license fee), irrespective of the level of output, used by regulatory agencies to eliminate or reduce the profit of a monopoly.

luxuries normal goods whose income elasticity exceeds one.

macroeconomics branch of economics that focuses on the broad aggregates such as the growth of gross domestic product, the money supply, the stability of prices, and the level of employment.

macro policy objective the theoretical goal of macroeconomy policy is to eliminate inflationary or expansionary gaps when they occur.

managed float the government intervenes periodically to change currency values.

management another form of human resources which provides entrepreneurial services.

manufactured resources resources such as plows, tractors, tools, buildings, and other improvements to land that are manufactured by human beings; often referred to as capital.

manufacturing capacity utilization rate actual output of a nation's manufacturing firms divided by their potential output.

marginal cost (MC) the change in total cost of production as the output or total product of the business is expanded. Marginal cost is calculated as $MC = \Delta$ cost $\div \Delta$ output. Marginal cost represents the total cost of producing another unit of output. Marginal cost to a wheat producer is the change in total costs of producing another acre of wheat. This is an important statistic because the profit-maximizing level of output for a business under conditions of perfect competition occurs at the point where the marginal cost of production is identical to the price of the product or marginal revenue. The portion of the business's marginal cost curve lying above AVC represents the firm's supply curve.

marginal input cost (MIC) the change in the cost of a resource used in production as more of this resource is employed. Marginal input cost is set by the market for the resource. The marginal input cost for labor, for example, is equal to the wage rate the business faces in the hired labor market. The marginal input cost for fertilizer is the price of fertilizer in the marketplace.

marginal propensity to consume   the slope of the consumption function, tells us how much consumption will change if consumer disposable income changes.

marginal physical product (MPP), or simply marginal product   the change in output or total product the business would achieve in the current period by expanding the use of an input by another unit. Marginal physical product is calculated as $MPP_{labor} = \Delta$ output ÷ Δ labor, $MPP_{capital} = \Delta$ output ÷ Δ capital, and so on. Examples include an increase in alfalfa production from the application of additional lime and the additional number of cases of fruit canned in the current period from the addition of another canning line.

marginal rate of product transformation   the rate at which a unit of one product must contract to produce a unit of a second product.

marginal rate of substitution   the rate of exchange of pairs of consumption goods or services to leave utility or satisfaction unchanged, or the absolute value of the slope of an indifference curve.

marginal rate of technical substitution   the rate of substitution or trade-off between two inputs in the production of a specific product; also represents the slope of an isoquant curve.

marginal revenue (MR)   the change in the revenue earned (from the production if the business is expanded). Marginal revenue is calculated as $MR = \Delta$ revenue ÷ Δ output. Marginal revenue under conditions of perfect competition is identical to the price the business takes in the marketplace. This means that the additional revenue received from the marketplace will be unaltered by changes in the quantity the business produces; it is a price taker. For example, suppose the price of corn is $2.50 per bushel. This price will remain unchanged by the production decision of an individual producer.

marginal utility   the change to utility or satisfaction as consumption of a good is increased by one unit.

marginal value product (MVP) or value of marginal product   the change in the revenue earned by the business as it employs an additional unit of a resource, holding other resource use constant. Marginal value product is calculated as $MVP = MPP \times$ market price of product. The marginal value product of labor to a wheat producer, for example, is equal to the MPP_{labor}, or change in wheat output resulting from the employment of an additional farm laborer, multiplied by the price of wheat. Stated another way, this represents the change in revenue the wheat producer would receive by hiring another laborer.

market demand curve   this curve, along with the supply curve, determines the equilibrium price for a given commodity or service. This captures the individual demand schedules of consumers participating in this particular market.

marketing bill   the value of food expenditures contributed by firms beyond the farm gate.

marketing orders   arrangements among producers and processors of agricultural commodities in which the chief goal is to improve producer income through the orderly marketing of a commodity. Historically, marketing orders have been used by the federal government to control the quantity of a commodity coming onto specific markets.

market supply curve   represents the summation of the quantities desired by all consumers at specific prices.

materials   this term has different meanings in different settings. In the context of "time and materials," this refers to the various inputs used to produce a good or provide a service.

measuring market activity   is done by measuring aggregate economic activity in product markets (national product) or resource markets (national income). Both values are identical in the national income and product accounts maintained by the federal government.

medium of exchange   benefit where money allows businesses and individuals to specialize in their endeavors and to purchase the goods and services they need with money they receive for their efforts.

mercantilism   the economic and political philosophy that national wealth and power were dependent upon a nation being able to export more than it imported.

merger   the combination of two or more firms into one.

microeconomics   branch of economics that focuses on the economic actions of individuals or specific groups of individuals.

mixed economic system   an economic system in which markets are not entirely free to determine price in some markets but are in others. Government controls in selected markets and welfare programs are indicative of a mixed economic system.

monetary economy   parties trade goods and services in a barter economy when paying for other goods and services. Money typically changes hands in a monetary economy when paying for goods and services.

money multiplier   the reciprocal of the fractional required reserve ratio if there are no currency drains and no excess reserves.

monopolistic competition a market structure in which a large number of firms produce a differentiated product. It is relatively easy to enter such a sector.

monopoly a market structure that has only one firm selling to buyers.

monopsonistic competition a market structure in which there are a large number of buyers of resources, but there exists the capacity of the buyers to differentiate services. Differentiation of services includes convenience of distribution and location of processing facilities as well as willingness to provide credit and/or technical assistance.

monopsonistic exploitation the difference between the prices paid for an input under perfect competition and monopsony.

monopsony a market structure that has only one firm buying from sellers.

most-favored-nation (MFN) GATT principle that states that trade among member nations must be non-discriminatory. Concessions granted to one nation automatically apply to all other WTO members.

multilateral refers to negotiations among many nations, as opposed to two nations, which is bilateral; working together for a common purpose such as reducing trade barriers. The WTO is an example of a multilateral institution.

Multiple Peril Crop Insurance (MPCI) the general name given to crop insurance policies that cover a number of perils including rain, hail, frost, and drought.

national bank bank that is chartered by the federal government.

national debt sum of the outstanding federal government securities and other claims on the federal government.

national income national income is equal to gross national product less depreciation. It is also equal to personal income less transfer payments from government plus business retained earnings.

national product see gross domestic product.

National School Lunch Program (NSLP) provides financial and community assistance for meal service in public and nonprofit private schools and other specific institutions.

natural resources resources such as land and mineral deposits, which are available without additional effort on the part of the owners.

near monies assets that are almost money or can be converted to money quickly with little or no loss in value.

necessity normal goods whose income elasticity is less than one.

net farm income gross farm income minus farm expenses and taxes.

nominal dollar values refers to the current price or value of a good, service, or asset.

nominal GDP value of aggregate output in the economy measured in that period's prices or current prices.

nominal interest rate market rate of interest unadjusted for the current rate of inflation.

nominal values refer to economic measures for which no adjustments to inflation have been made.

nonprice competition attempts to increase the demand for a product or service via product differentiation and to make the demand for the product less price-elastic.

nonrecourse loan an amount of money equal to support price times the quantity offered as collateral lent by the Commodity Credit Corporation (CCC). The loan is considered paid in full, when turned over to the CCC, when the market price falls below the support price.

nontariff barriers (NTBs) government policies, other than tariffs, that reduce imports but do not limit domestic production of goods imported as substitutes.

normal goods goods for which consumption rises (falls) when income increases (decreases).

normative economics a branch of economics that focuses on determining what-should-be issues and questions. Unlike positive economics, it assigns specific values with specific goals or objectives.

North American Free Trade Agreement (NAFTA) free trade area among the United States, Mexico, and Canada.

oligopoly (oligopsony) a market structure in which there are a small number of sellers (buyers). Each seller or oligopolist (buyer or oligopsonist) knows how the other sellers (buyers) will respond to any changes in quantity marketed or prices he or she might initiate.

open market operations the buying and selling of government securities by the FOMC's fiscal agent at the New York District Federal Reserve Bank.

opportunity cost the cost of producing a good as measured by the amount of a second good that must be forgone to release just enough resources to produce one additional unit of the first good.

opportunity cost theory a nation has a comparative advantage in the production of the good with the lowest opportunity cost.

ordinal measure a measure that implies only a ranking of the amount of satisfaction in some sort of ordered fashion.

output index an index that shows the level of current production or output of a particular commodity or group of commodities relative to a particular year, or base period.

own-price elasticity of demand a measure of the relative response of consumption of a good or service to changes in price. It is defined as the percentage change in the quantity demanded divided by the percentage change in price.

Ozone is a gaseous molecule that contains three oxygen atoms (O_3). High in the atmosphere it plays a positive role by shielding the Earth from harmful ultraviolet rays from the sun, but close to the ground it is a harmful component of smog.

Packers and Stockyards Act of 1921 a legislative act favorable to agricultural organizations passed by Congress in 1921. This law reinforced anti-trust laws regarding livestock marketing.

partial equilibrium analysis assumes the events taking place outside the market under analysis remain constant.

payment rate the amount paid per unit of production to each participating farmer for eligible payment production under the FAIR Act.

perfect competition market structure characterized by a large number of producers selling a homogeneous product, each with perfect information, and no barriers to entry or exit.

perfectly elastic/inelastic demand curve a perfectly elastic (inelastic) demand curve is parallel (perpendicular) to the horizontal or quantity axis.

point elasticity a measure of elasticity at a given point on a demand curve.

positive economics a branch of economics that focuses on what-is and what-would-happen-if questions and issues; does not involve value judgments or policy prescriptions to reach a particular objective.

potential GDP gross domestic product at full employment of labor and capital.

precautionary demand for money demand to hold money to pay for unexpected expenditures that may arise.

preferential government policies preferential government policies may take the form of tax incentives or other measures that provide firms with a competitive advantage.

preferential trading arrangements provide for the elimination or phased reduction of tariff and nontariff barriers among nations. PTAs may be bilateral or multilateral and often include the liberalization of investment laws, transportation regulations, protection of intellectual property, and provisions for resolving trade disputes. PTAs may be a free trade area, a customs union, a common market, or an economic union.

preferential government policies Preferential government policies may take the form of tax incentives or other measures that provide firms with a competitive advantage.

premature inflation when demand-pull inflation occurs before full employment.

present value the value today that is equivalent to a future income or cost, taking into account the interest that could accrue between the present and the future. For example, if the interest rate is 5%, then the present value of $105 in 1 year is $100.

price–consumption curve the connection or locus of all tangency points between budget lines and indifference curves. Each tangency identifies a point on the demand curve.

price flexibility the reciprocal of own-price elasticity. This entity measures the percentage change in price due to a 1% change in quantity demanded.

price index an index that shows the level of current price of a particular commodity or group of commodities relative to a particular year, or base period.

Price Loss Coverage (PLC) the producer under the 2014 farm bill receives a payment when the market price for a covered crop falls below an established fixed price. This payment supplemented what the producer received from a crop insurance policy.

price taker a business is said to be a price taker when its actions have absolutely no effect on the price of the input it is buying or the price of the product it is selling.

prior appropriation water rights a system of water rights in which rights to particular quantities are established and can typically be transferred to other parties.

producer price index (PPI) weighted average of the prices producers pay for goods and services.

producer surplus a measure of the economic rent or returns above total costs accruing to businesses participating in a market during the current period.

production function relationship between output and the factors of production (labor, capital, land, and management).

production possibilities frontier reflects the physical trade-off in production between two products. The firm is said to be technically efficient at all combinations of two products along this curve.

production possibility schedule all combinations of national output that can be produced by fully employing all available resources and the most productive technology.

productivity level of output per unit of input. Crop yield per acre is a measure of productivity.

profit total revenue minus total expenses.

profitability returns on capital invested in farm assets represent a measure of profitability. Profitability may be expressed in dollar terms (e.g., net farm income) or in percentage terms (e.g., rate of return on farm capital).

prohibitive tariff a tariff so high that imports are completely eliminated and the market is supplied only from domestic sources.

property rights the rights, privileges, and limitations on use that are enjoyed by the owner of a resource. Property rights often include elements such as the right to use and sell but may be subject to limitations such as zoning restrictions or where you can fire a gun.

protectionism any government policy implemented to remedy domestic economic problems associated with trade.

purchasing power reflects what $1 today would have purchased in goods and services in a particular base period.

quota a ceiling or limit placed on the amount of a particular good that may be imported during the period.

rate of inflation inflation, or sustained rise in the general price level, expressed as a percentage change from one period to the next.

real dollar values refers to the current price or value of a good, service, or asset adjusted for inflation.

real exchange rate the value of one currency relative to other countries, accounting for different inflation rates between countries.

real GDP value of aggregate output in the economy measured in the prices of another (i.e., base) period or constant prices.

real economy refers to activity in nonfinancial markets such as consumption, investment, government spending, and net exports as well as production and labor markets.

real interest rate nominal interest rate minus the rate of inflation.

recessionary gap amount by which full employment GDP exceeds equilibrium GDP.

recessionary period phase of business cycle during which the nation's output is declining.

real values inflation-adjusted values.

Reciprocal Trade Agreements Act of 1934 (RTA) authorizes the President to fix tariff rates and has resulted in a liberal tariff stance for the United States. From 1934 to 1947, in fact, the United States negotiated bilateral trade agreements with 29 nations. However, when the GATT emerged as the primary trade forum, the RTA declined in importance as a mechanism for trade liberalization.

rental rate of capital the cost of capital broadly defined; the price you would have to pay to rent all the inputs used to produce the business's product.

required reserves minimum amount of deposits that banks and other depository institutions must hold in reserve.

Revenue Protection (RP) plan a form of crop insurance that provides protection against the loss of farm revenue due to adverse fluctuations in prices and yields.

rightward shift in the consumer's demand curve refers to movement of a demand curve to the right from its initial position as incomes or other factors increase (decrease) demand for a good or service.

Risk Management Agency (RMA) an agency within the U.S. Department of Agriculture created to administer federal crop insurance programs and promote the use of risk management products and decision tools.

Robinson–Patman Act of 1936 a legislative act favorable to agricultural organizations passed by Congress in 1936. This law primarily covered price discrimination practices.

sanitary and phytosanitary regulations (SPS) domestic regulations that require that food, animal, and plant products are free of harmful diseases or pests.

savings that part of disposable personal income not spent on current consumption.

scarce resources can be decomposed into natural and biological resources, human resources, and manufactured resources.

scarcity refers to the fixed quantity of resources that are available to meet society's needs.

seasonal unemployment unemployment of labor associated with changes in business conditions that are seasonal in nature.

Section 301 provision of the Trade Act of 1974 and its counterpart in the Trade and Competitiveness Act of 1988 that was designed to provide presidential authority to impose duties on products from nations whose trade practices were deemed unfair or that restrict U.S. commerce.

sequester the setting aside of funds previously authorized to fund federal government expenditures required under the Gramm–Rudman Act, if projected deficits exceed targets.

set-aside the amount of land that had to be set aside—either left idle or planted to another crop.

Sherman Antitrust Act of 1890 a legislative act passed by Congress in 1890 in an attempt to minimize the potential social costs of imperfect competition. This act explicitly prohibited monopoly and other restrictive business practices. Section 1 of this historic act made it illegal to act in restraint of trade by conspiring with other individuals or firms. The act did not allow restraining trade through price-fixing arrangements or controlling and sharing industry output by collusive agreement. Section 2 of the act forbade the use of economic power to exclude competitors from any market.

shifts in the demand curve refers to movement of a demand curve to the right (left) from its initial position as incomes or other factors increase (decrease) demand for a good or service.

shutdown level of output at which average variable costs equal marginal cost.

slippage extent to which farmers participating in government programs "overproduce" by retiring marginal acres rather than highly productive acres, or decide not to participate at all.

slope of an iso-cost line is represented by the ratio of two inputs. This line allows us to use economics to determine the least-cost combination of two inputs.

socialism an economic system in which resources are generally collectively owned and government decides through central planning how human and nonhuman resources are to be utilized in different sectors of the economy; prices are largely set by the government and administered to consumers and producers.

Soil Bank Program was a program operated by the U.S. Department of Agriculture in the mid-1950s that removed land from production in exchange for government subsidies. This was justified largely on conservation grounds. This program was very costly to operate and was discontinued in favor of other programs that attempted to control production by paying farmers not to produce on certain land.

special drawing rights (SDR) an artificial currency unit in the form of accounting entries to supplement reserve assets of the International Monetary Fund and its member countries. The SDR has no commercial uses.

Special Supplemental Food Program for Women, Infants, and Children (WIC) integrates health care, nutrition education, food distribution, and food stamps into a single comprehensive health and nutrition program. It is targeted toward mothers with children who already participate in other welfare programs.

specialization the separation of productive activities between persons or geographical areas in such a manner that none of these persons or regions is completely self-sufficient.

speculative demand for money demand to hold money as an asset in expectation that other asset prices will fall.

Stacked Income Protection Plan (STAX) a supplemental coverage option designed specifically for cotton producers by the 2014 farm bill.

stage I of production begins at zero input use and continues to the level of input use where the marginal physical product is equal to the average physical product. The average physical product will have reached its peak at the end of stage I.

stage II of production represents the range of interest to economists. Why stop in stage I where the APP is rising, and why produce in stage III where the MPP is negative?

stage III of production begins at the point where the marginal physical product reaches zero. Increasing the use of the input beyond this point is no longer rational since the marginal physical product is negative.

stagflation existence of increasing inflation during a period when the economy is stagnant or experiencing little or no economic growth.

state bank bank that is chartered by a state government.

store of value value of holding money as an asset.

structural unemployment unemployment of labor associated with structural changes in the economy.

substitutes goods *A* and *B* are substitutes if the cross-price elasticity of demand is positive.

substitution effect substitution of a product for another because the price of the former has declined or increased.

Supplemental Coverage Option (SCO) a feature of the 2014 farm bill that enabled crop producers to expand their coverage against production losses and price declines.

supply curve ranges the aggregate supply curve for the economy has three ranges: the Keynesian or depression range, the normal range, and the classical range.

supply-side economics promotion of macroeconomic policies that increase productivity and thereby shift the current aggregate supply curve to the right, which is believed to promote higher output at lower price levels.

target price mechanism the target price establishes the per-unit ceiling on program payments to growers of program crops such as corn and wheat.

tariff a tax levied on goods when they enter a country. Tariffs may be imposed for protection or to generate revenue (also referred to as a duty).

tariffication the process whereby quotas, licenses, variable levies, and other nontariff barriers to trade are converted to their tariff equivalents.

tariff-rate quota (TRQ) allows a specified amount or number of products to enter an importing country at one tariff rate (the within-quota rate), whereas imports above this level are assessed at a higher rate of duty (the over-quota rate).

tastes and preferences these factors, along with prices, incomes, and wealth, influence the demand for a particular good or service. Consumer attitudes toward caloric intake and cholesterol can affect the demand for one product over another.

the Keynesian cross illustrates the equilibrium level of aggregate expenditures in the economy, or where they equal aggregate output.

the Phillips curve was developed to describe the trade-off between inflation and unemployment in the short run. Studies have shown that it is less well suited to this task in the longer run.

tonnage metric tons equal to 2204.62 pounds per ton.

total costs (TC) sum of all individual categories of production costs during the current period. Total costs are calculated as TC = TVC + TFC. See examples for total fixed costs and total variable costs. Total costs are an important statistic used to calculate accounting profit.

total economic surplus the sum of producer surplus and consumer surplus.

total fixed costs (TFC) sum of all current production costs that do *not* vary with the level of output or input use. Total fixed costs are calculated by adding up all individual fixed costs. Total fixed costs can also be measured residually by subtracting total variable production costs from total costs. Examples include the total value of all fixed costs, or the business's current property tax bill, plus its current insurance premium due, plus the current mortgage interest payment due, and so on.

Total Maximum Daily Load (TMDL) TMDL refers to both a limit and a process. The limit sets allowable loadings for a water body to meet water quality standards. The governmental process that follows includes allocating the responsibilities for reducing the loads to the various emitting groups in the watershed.

total physical product (TPP), or simply total product the total output of goods or services produced by the firm during the current period. The total product of a wheat farmer is the yield per acre multiplied by the number of acres harvested. Examples include total wheat produced by a wheat producer, total pounds of milk produced by a dairy farmer, and total number of cases canned by a canning factory.

total revenue (TR) sum of all money received by the business from the sale of the products it markets during the current period. Total revenue is calculated as $TR = (P_{corn} \times Q_{corn}) + (P_{wheat} \times Q_{wheat}) + \ldots$ or the sum of the cash receipts from marketings of the business's individual products. Examples include cash receipts from wheat marketed by a wheat farmer, cash receipts from flour marketed by a miller, cash receipts from bread marketed by a baker, and so on. Total revenue is another important statistic used to calculate accounting profit.

total utility the total satisfaction derived from consuming a given bundle of goods and services.

total variable costs (TVC) sum of all individual categories of production costs that do vary with the level of output or input use. Total variable costs are calculated by adding up all individual variable costs. Total variable costs can also be measured residually by subtracting total fixed costs from total costs. Examples include the total value of all variable costs, or the current fuel bill, plus the fertilizer bill, plus wages paid to hired labor, plus the rental payment on farmland, and so on.

trade deficit the monetary amount by which a nation's merchandise imports exceed exports during a given time period, usually 1 year.

trade liberalization   the removal of government policies that restrict a nation's imports or exports.

trade surplus   the monetary amount by which a nation's merchandise exports exceed imports during a given time period, usually 1 year.

transaction demand for money   demand to hold money to pay for expected expenditures.

transferable discharge permits (TDP)   rights to discharge pollution that can be bought and sold in order to achieve a pollution goal at the lowest possible cost.

Treaty of Maastricht   created the European Monetary System, with the euro as its common currency.

types of consumer expenditures   consumer expenditures are classified as durable goods, nondurable goods, and services. A new television is an example of a durable good, while food is an example of a nondurable good.

unemployment   that portion of the total labor force that is actively seeking employment but currently without a job is *unemployed*. *Unemployment* is an aggregate term that refers to a group of individuals that are unemployed.

unemployment rate   number of unemployed persons divided by the size of the total civilian labor force.

unemployment rate for labor   total labor force minus the number of employed persons in the economy.

unit of accounting   money provides a basis with which the relative value of goods and services can be assessed and with which the profitability and financial position of businesses can be assessed.

unitary elastic   the demand curve is unitary elastic where the percentage change in quantity is identical to the percentage change in price.

U.S. Department of Agriculture (USDA)   a cabinet-level department of the executive branch of the federal government. Its vast programs address the production, safety, and marketing of agricultural products as well as the administration of various welfare programs designed to provide food and nutrition services in domestic and foreign markets.

U.S. Environmental Protection Agency (USEPA)   the federal agency that enforces environmental laws including the Clean Water Act and the Clean Air Act.

U.S. Meat Import Act of 1964   a law authorizing the imposition of quantitative poundage quotas on imports of fresh, chilled, and frozen beef, veal, mutton, and goat if imports are expected to exceed certain prespecified minimum quantities.

utility function   a mathematical or functional representation of the satisfaction a consumer derives from a consumption bundle.

utils   imaginary units of satisfaction derived from consumption of goods or services.

value   million or billion U.S. dollars.

variable costs (VC)   level of specific current production costs that *do* vary with the level of output or input use. It is a short-run cost concept; all costs are considered variable in the long run. Variable costs are calculated by taking the price of the input multiplied by the quantity of the good or service used (e.g., hourly wage rate multiplied by the hours of hired labor employed by the business). Examples include current fuel bill, fertilizer bill, wages paid to hired labor, rental payment on farmland, and repair costs for machinery and motor vehicles.

variable levy   a variable import tariff (duty) imposed by the European Union equal to the difference between a predesignated domestic price and the lowest landed import price.

voluntary export restraint (VER)   a voluntary agreement, in lieu of more restrictive measures such as higher duties or lower quotas, between two or more countries to limit the export of a certain product or commodity to a prespecified national market.

weaker dollar   a decline in the value of the U.S. dollar relative to other currencies so that the dollar buys less foreign currency than before.

wealth effect   the consumption function will shift to the right (left) if consumer wealth increases (decreases).

Wetlands Reserve Program (WRP)   designed to restore and protect up to 1 million wetland acres.

World Trade Organization (WTO)   created in 1995 to provide a single institutional framework encompassing all articles of the GATT and all agreements and arrangements concluded under the Uruguay Round of the GATT. Timely dispute settlement and establishment of a trade policy review mechanism are two important functions of the WTO.

Index